Workplace Communications

FOR ENGINEERING TECHNICIANS AND TECHNOLOGISTS

Workplace Communications
FOR ENGINEERING TECHNICIANS AND TECHNOLOGISTS

DAVID W. RIGBY
North Seattle Community College

The Wordworks™ Series

Prentice Hall

Upper Saddle River, New Jersey
Columbus, Ohio

Library of Congress Cataloging-in-Publication Data

Rigby, David W.
   Workplace communications for engineering technicians and technologists / David W. Rigby.
      p. cm. – (Wordworks)
   ISBN 0-13-490129-0
    1. Business communication. 2. Business writing. 3. Business presentations. I. Title.

HF5718 .R543 2001
651.7–dc21

00-034706

Vice President and Publisher: Dave Garza
Editor in Chief: Stephen Helba
Executive Editor: Debbie Yarnell
Associate Editor: Michelle Churma
Production Editor: Louise N. Sette
Production Supervision: Clarinda Publication Services
Design Coordinator: Robin G. Chukes
Text Designer: Ceri Fitzgerald
Cover Designer: Ceri Fitzgerald
Production Manager: Brian Fox
Marketing Manager: Jimmy Stephens

This book was set in Optima by The Clarinda Company. It was printed and bound by Victor Graphics, Inc. The cover was printed by Victor Graphics, Inc.

The Wordworks trademark is the registered property of David W. Rigby. © 2000 by David W. Rigby.

10 9 8 7 6 5 4 3 2 1
ISBN 0-13-490129-0

Welcome to Wordworks™

Wordworks™ is a series of four communication skills manuals. The manuals consist of three writers' guides for engineering and technical applications and an additional guide to in-service spoken communication. The manuals are designed to provide in-demand information in a readable fashion. They are matter of fact and use oriented.

Each manual focuses on specific exit skills that are necessary for job performance. In this respect the texts are unique. They were inspired by the carefully tailored goal orientation of corporate seminar manuals. This strategy is at the heart of the streamlined manuals of corporations where specific skill outcomes are the narrow focus of company class time. For skills manuals this strategy can accelerate and focus the learning process, and the approach is particularly useful in college programs where English components are never more than a course or two of the total learning experience.

The *Wordworks*™ manuals are conversational, visual, and practical, so that the learning experience is accessible. Chapter-by-chapter discussions encourage a learning curve of understanding. Important concepts and practical applications are identified and explored. Models and other illustrations are also features of the texts.

The texts rely on an extensive use of graphics to conceptualize ideas, and models are provided to draw attention to desirable skills. This approach is intended to help students build a strong understanding of logical design features that they can use to construct any writing project.

Three of the texts in the series deal with the basics of the craft of writing, and the series uses a writer-to-writer strategy to explore and explain this craft. The manuals are intended to be learning tools, but because they focus on a craft, they are not intended to be overly academic. To the extent to which streamlining can be achieved in a college environment, the *Wordworks*™ series provides a practical approach to skills training that is compatible with the limited time available for communications offerings in technical programs. The manuals encourage curiosity and provide a learning-oriented climate, but they also simplify the path to practical knowledge and skills.

Each title in the *Wordworks™* series is intended to complement the other titles:

Basic Composition Skills for Engineering Technicians and Technologists. This first text of the series is intended to help upgrade fundamental skills in writing. It is a thorough discussion of the problems that are encountered by writers and the solutions to those problems. This book is uniquely designed to build upon existing skills that are part of every workday.

Writer's Handbook for Engineering Technicians and Technologists. The second title in the series is a writer's handbook of the rules and practices of writing. Part style guide, part grammar book, part technical writing reference, the *Writer's Handbook* is designed to be a bridge that can support both *Basic Composition Skills* and *Technical Document Basics*. The *Writer's Handbook* is specifically intended for engineering and technical students.

Technical Document Basics for Engineering Technicians and Technologists. The third title in the *Wordworks™* series develops a concentrated focus on basic technical writing know-how. The text is designed to identify and explore the documentation standards that are used to develop and produce technical projects. The basic skills are condensed into a thoroughly illustrated and readable text.

Workplace Communications for Engineering Technicians and Technologists. An additional text supports the *Wordworks™* concept with an exploration of spoken communication. Studies reveal that 80% or more of work-related communcation in trade and industrial settings is handled in conversation. The absence of training tools in this area is an invitation for communication problems if only because spoken messages outnumber our memos by four to one! *Workplace Communications* helps identify and improve in-service communication skills.

About Workplace Communications

New employees face a major challenge in adaptation when they accept a job with a new company. Success within a company will depend on a great deal more than applying technical skills to the circumstances of the business. Companies view an employee's social adaptation and behavior as serious matters on a par with education. An employee must be keenly aware of the "people skills" involved in workplace dynamics.

There is profit and loss in behavior. The course for which you are using this text is very likely required by your college because corporations want to see that potential employees have given some attention to very basic job skills: a positive attitude, a team-player spirit, and solid performance technique. The presence of such a course on your transcript will show that you have given some training time to nontechnical skills that matter to the success of a business—and to *your* success. The idea is that trainees should be required to read about and discuss corporate practices, work behavior, task-group skills, and social interaction as part of workforce training.

Companies want employees who carry themselves well. They want employees who know how to handle the social situations they will confront in a company environment. Company life is not just a matter of being on time. An on-call repairman, for example, is an important figure for a company. He has a social role and a social responsibility. He may also wear a company uniform. He is likely to drive a company truck with the company name. He is responsible for a valuable company contract every time he is on a call. The truck may say "How am I driving?' but the supervisor is asking herself, "How is he performing?"

Workplace Communications is divided into a number of discussions that should help build an awareness of the complexities of communication in the workplace. The first two chapters are designed to examine the basis of power structures and how they operate in the business world. Power is control, and communication is the vehicle of this force that oversees any organization.

In addition to the formal channels of power and communication, there are also informal channels that are part of every workplace. Chapter 2 explains the unique character of the well-worn paths of language politics and explains the delicate balance maintained in the daily routines of company life.

Once the groundwork for the workplace environment is identified, the text shifts attention to the employees and examines the complex matter of interpersonal communication.

Chapter 3 looks into the nature of visual perception and the way we see the world around us. The following chapter examines communication as an activity and looks at the many elements that control it, including such features as voice, body language, and the many subtle dimensions of one-to-one dialogue.

Once the mechanics of interpersonal communication patterns are understood, the discussion shifts to suggestions that can improve the mechanics and the effectiveness of conversation. The fundamental truths of Chapter 5 are obvious enough: we must listen with care; we must speak with care. The chapter details the less obvious practices that encourage better listening and better speaking.

There are predictable barriers to communication in the workplace. Conversation is easily blocked by noise, for example, but the workplace is more than an office setting, a shop, or an assembly plant. Because the workplace is a community, communication barriers include interaction as well as the environment of interaction. From the plant to the business offices to the boardroom, office politics can result in troublesome barriers. Chapter 6 identifies typical environmental and social difficulties that affect workplace communication.

The text also explores specific skills activities whenever possible. Chapter 7 is devoted to group decision-making and the committee meetings. Group action is a fundamental tool in many company environments, and effective participation can be learned. Chapter 8 continues the discussion with a close look at group roles.

The following two chapters focus on the committee presentation as a specific activity. It is highly likely that any professional technical employee will deliver an occasional presentation in a small-group situation. Chapters 9 and 10 discuss the proper handling of brief technical presentations. Two subsequent chapters then examine the importance of sales techniques and the craft of sales management. Chapter 11 identifies the unique situation in which technical services are marketed, and Chapter 12 provides additional suggestions that will help both employees and small-business owners handle sales with skill.

An additional chapter in *Workplace Communications* deals with the employment interview. There is little to discuss in workplace communications until employment has been secured, and a new job is usually the result of a successful interview. Chapter 13 discusses the employment interview process and explores strategies for successful interviews.

The workplace sometimes involves communication problems that are more social or psychological than mechanical in nature. Chapters 14 and 15 examine the social matrix in which all communication is suspended. These chapters do not pretend to solve the enormous variety of communication barriers we encounter; however, with awareness and sensitivity, we can bridge many of the social and psychological difficulties so that our conversations are shared and understood and so that *we* share and understand the problems of others. The chapters are intended to encourage that awareness. Chapter 14 examines practical solutions to conflict at both the employee-to-employee level and in group-to-group confrontations. Chapter 15 is a brief discussion of cross-cultural values.

Workplace Communications is designed to be an approachable and practical examination of workplace realities and the often subtle difficulties that may not be apparent. The text will help build awareness, which is the first step to improved communication.

Dealing with people is a challenge. We adopt a practical approach to meeting the challenge:

1. Together we look at in-service applications of client and employee relationships in general.
2. We then identify a variety of service-oriented practices that you can use for routine work-day activities.
3. Then, *you* have to construct the attitude and style that fits your business needs.

Contents

Chapter 7
Group Decision-Making   121

Chapter 8
Group Roles and Leadership   147

Chapter 9
Committee Presentations   177

Chapter 14
Conflict Resolution   355

Chapter 15
Cross-Cultural Values   381

Introduction

At the college where I teach, the engineering technology faculty members meet with corporate representatives through special committees to learn the companys' needs, their short-term goals, their long-range objectives, and their requirements in the employment marketplace. Your college will have similar committees. Your technical program wasn't designed randomly as a survey of basics in your field. The program was carefully constructed to meet the marketplace for jobs. Educators try to design college programs to suit *your* need to find productive and interesting employment, but they achieve this goal by meeting market *demands* for certain skills in today's technologies: computer information systems, construction engineering, biomedical equipment technology, digital electronics, AutoCAD, environmental control technology, microcomputer management, biotech, avionics, and so on. If the market is strong, your opportunity is strong.

You then take a program that was designed for a specific job setting where there is a demand for specific theoretical and practical knowledge. Representatives from industry help us design the programs, and they demand, among other types of training, coursework in writing and speaking skills. The on-site service technician must be able to handle herself in front of an unpleasant customer. She needs "service skills" as well as technical skills. At the very least she must be cordial and helpful—all day, every day. That is not easy. Perhaps such service skills as patience or cordiality and enthusiasm can be taught. Perhaps they cannot. Yet if you think about it, you can see what the industries are demanding. Suppose you take an electronics course rather than a writing course. Companies might argue that they can teach you the electronics if they have to, but they certainly are not in a position to teach you English. They have high hopes that you will arrive with both technical and service skills.

In terms of practical in-service skills, the only material any company is likely to offer is a sales program, and sales strategies are intended for only a small group of employees. Besides, in-house training is costly for companies. They would rather hire the complete product.

The problem with trying to teach service skills is not apparent in the more popular terms "speaking skills" and "communication skills." What companies really are asking for are "social skills," but no one wants to say it. They do *not* mean public speaking; they have only an occasional use for the public speaker behind a podium, and that person is likely to be an upper-level executive. Suppose a repair tech is a computer service technician. He arrives at a corporation for the third time in a week. The boss he encounters is hopping mad. The secretary is in tears. The office is in an uproar. Obviously our tech does not need "speaking skills." He needs "people savvy," and people skills are not easy to teach. Perhaps such traits as patience cannot be taught. Heart attacks have been associated with the well-known Type A personality. If we cannot survive our own traits, educators are probably not in a position to teach us to overcome them. However, if nothing else, we need to learn to cope; there is a lot of stress out there. There is a lot of misunderstanding. Certainly, many pitfalls *can* be avoided with a little help from a classroom.

In sum, we need some way to approach company policies about behaviors—which are never written down except for such matters as policy concerning absenteeism—and we need some way to look at specific service skills, the ones that matter in the marketplace; they include your communication skills and the *attitude* and *style* of your job skills. Such skills certainly will not be required in your job description, but they will be expected. What should we call this attitude and style companies are looking for? Call it "a job well done," through the communication of the right image.

One approach to image is obvious among professionals. Attitude is part of their job. Professional people often refer to their business as a "practice." Notice that many professionals *practice* confidence, understanding, interest, politeness, and enthusiasm while they ply their trades. Professionalism is company policy most everywhere: professional knowhow *and* a professional manner. These elements of a positive attitude are obvious to anyone, but less because we possess them than because we admire them in other people. We respond to confidence and understanding and enthusiasm with our trust. Our trust is more important than profit. Profit may be adequate for the short haul, but trust keeps a company in business for the long haul. Attitude is not a secret to success; it is well known as the *key* to success.

Which professionals am I talking about? Pick any pros you admire. They don't have to be doctors or lawyers. How about baseball players? The attitude of professional athletes is particularly interesting. Off the field they usually show all the professional style that we see on the field: Patience, humility, teamwork, respect for competition. And they *share* the trust they build, both among themselves and with their fans.

Can a classroom situation be of help? A positive image is a matter of self-awareness. You cannot very easily act it out. Confidence is something you have; interest is something you show; enthusiasm is something you share. However, your image has a lot to do with "style," which is going to be reflected in your service skills. One way to build trust, then, is for you to work on style so that you look like a pro. The *image* you create goes hand in hand with *in-service skills* in communication. You will have to take charge of the former, but a classroom environment can help you focus on the latter.

Understand, then, that this text does not focus extensively on image or attitude. These matters are largely in your hands. You might say that this business of enthusiasm and confidence and understanding is obvious enough. Maybe so—when the going is easy. And if the going gets tough? Well, you have heard that the tough get going, but that comment is not much of a strategy. Let's say the tough get tactical. And the tough stay tactful. You have to stay in control. A positive attitude is a little more difficult to sustain in a crisis. Some days the job is easy; other days the job is difficult. The technical job at hand *and* the job of communicating can be difficult. The job of being professional can be difficult. Self-management is the key. You have to take control. You will have plenty to do.

Let's look at two strategies. A college textbook cannot package something called "Company X service skills," but it can categorize the ways you will deal with people at work so that we can discuss these daily encounters. We don't have to look at the issues as "social activities" and "correct behavior" as such. **We can isolate relationships and identify tactical approaches: we can look for *people skills*.** We should think of the issue as a kind of management skill—self-management. First, there are activities that are fairly predictable and that involve very real tools you can learn and put to use. Interviewing is an example. Being a member of a committee is another. Good technique in sales or service, or other outreach areas that affect consumers, are examples. These activities depend on people skills.

Interviewing skills will help you get the job you want and the promotions you want. Learning *how to function well on committees* or in any work-related group will help you keep your job and advance in your purposes at work. *Sales techniques* can obviously be taught and practiced also. In areas of communications such as these, skills can be learned. That is why secretarial programs often teach proper technique for managing phone conversations. However, most of your other activities at work may have no such predictable patterns. A talk with the boss is a talk with the boss. Lunch with your coworkers is just lunch. Chat with the secretaries is just chat. So it seems.

There is a second strategy we can take to look at the broader spectrum of the endless work-related speaking or communication activities, which can appear impossible to analyze at first. We will examine how communication specialists see corporate settings in terms of communication. There have been many studies by sociologists, psychologists, speech communication theorists, and others, and there are countless self-help materials that offer advice on handling yourself in business. It is difficult, however, to tell you what to do in all the vast areas where speaking skills and people skills must respond to situations that vary. The interview, for example, is a fairly fixed event. With practice an interview can be controlled with skill. But the bulk of an employee's activities are less defined and depend on the situation of the moment.

Consider service technicians. Suppose you find an attractive position as a service tech. You may be on a call in a walnut-paneled office on the thirtieth floor of a skyscraper in Philadelphia. You may be in a suburban office park in Des Moines. You may be rummaging through the back of your truck in a Kansas City amusement park or puzzling over a network server in the basement of a restaurant in Sacramento. The conditions are endlessly varied. "Rules" that fit unique situations probably defy the general laws of business physics. However, service skill suggestions have as much to do with your general behavior as with communication, and your behavior will be with you all of the time. Behavior can be managed: in this respect, we can look at practices that will be helpful most of the time.

Essentially, our communication efforts among employees and clients are either regular events with predictable characteristics, or they are unique and improvised. Monthly meetings are regular. If your supervisor is angry, well, you will have to improvise. So we *can* train as one plan of action, if we can find a few good rules to guide us in routine situations. **We need a second plan of action to control the endless variables. The proper management of behavior and a keen awareness of people skills will shape a useful alternative that can be used in less predictable situations.**

For the second strategy you can put a lot of useful people skills into action almost anywhere, anytime, if you are perceptive about the "business" around you. We will look at the second strategy in the next few chapters because we have to do a little background work to set the stage for a workday of people skills. The background situations are the handiest place to start to see the use of good service-oriented communication.

We must first look at corporations in terms that relate to you and in terms that you can relate to: power structure, company practices, business communication. Your awareness of these systems—the system of power or the system of communication—should give you a handle on *what* to do. You are not going to look up "communication situation 431" and review the rules. As you work you gain experience, and we can add *awareness* to the experience.

With an awareness of what is happening between people, you can help control a situation, or at least understand it. In discussing communication controls, we will cover all the basics: communication in theory, and in practice; how to do it, and how not to do it. And we want to learn how to *look* good because image really counts. We are looking at a management skill: the craft of self-management.

First, let's look at corporations to analyze the basic working environments where the rules are not always clear, but where you should be able to respond with awareness and take charge of what you do. Secondly, we will look at communication as a skill. Later, starting in Chapter 7, we will look at regular, everyday matters such as committee work, interviewing, and similar vital skill areas in the world of business and industry. These routine matters can be approached in practical, "how-to" terms.

For corporations, the emphasis is service. For you the emphasis is skill. Service skills and self-management build trust.

Work in Progress

1. A Case Study

At the beginning of each of the following chapters you will have the opportunity to read a page from the working notes of a network systems designer. Like yourself, the author is an engineering technician. She has completed serveral specialized technical certifications that are offered on the campus of a two-year college, and she is a self-employed contractor, specializing in small LAN systems. She will take you through the paces of her negotiations as she attempts to secure a contract for an installation. You will see the challenges of employee relations and the importance of client-contractor negotiations.

I am the president of NetWorks NorthWest, a small start-up company entering the booming business of setting up local area networks (LANs) for a variety of different clients and in a variety of different settings. My employees just call me by my first name—Jane—but there are only three of us. I have several university degrees, but I'm one of the many people who have subsequently returned to community colleges to pick up a technical specialization. I finished a basic computer hardware certification and also a CCNA (Cisco Certified Network Associate) certification. Cisco Systems provides a large percentage of networking equipment used in LANs and on the Internet.

There are a number of employment options available to networking specialists, and they range from help-desk positions in large companies to installers to self-employed contractors like myself. For personal reasons—mainly scheduling flexibility that will fit my family—I'm more interested in bidding for contract awards where I can install the systems myself with the help of one or two employees. Besides, I have run a contracting business for ten years and I prefer to be self-employed.

I recently completed a small installation downtown; it was the perfect setting for the newcomer who enters the business with schooling as a primary background. I'm going to explain the process step-by-step, from my first encounter with the prospective client to the bank deposit after my final billing. Although technical and computer networking skills are at the heart of the story, you will be surprised at the role of people skills. Success really depends on those skills. Writing proved to be critical also, as you will see. The actual events transpired over a period of several months, and I will touch upon the highlights at the beginning of each of the chapters to follow.

J.Q.C.

The Corporate Cave: the Company as Society

A company is more than a series of products. A company is a society. For this reason, count on a company to be interested in you as a person. Like the company, you are more than a product. You are more than the sum of your newly acquired skills. After a few years in college you might assume that you are going to market your technical skills, but you probably should count on marketing yourself as well. You matter as a person as much as you matter as an engineering technican. From your interview on you become a member of the society of your company when you become an employee.

It may be hard to overcome the manufacturing perspective or production viewpoint, particularly if you have worked for companies that are cold, indifferent operations, but it helps to understand the company-as-society idea because you often will encounter a genuine company culture or company spirit. According to anthropologists and sociologists, a society takes shape anytime a group gets together and agrees to certain conditions. From the dawn of human communities thousands and thousands of years ago, the basic list of practices that build a society hasn't changed.

Imagine that we are nomads who decide to gather at an oasis and build a town, perhaps because we have mastered animal husbandry and seed gathering and now we are willing to try to settle down so we can put this knowledge to work. What do we need? The usual list includes

1. **A DIVISION OF LABOR**   **(TASKS)**

2. **DIVISION OF PROPERTY**   **(SPOILS)**

3. **A HIERARCHY OF SOCIAL ROLES**   **(THE CONTROLLED AND THE CONTROLLERS)**

4. **RULES AND NORMS**   **(LAW)**

5. **JUSTICE**   **(ENFORCEMENT)**

6. **EDUCATION**   **(HISTORY)**

7. **A COMMON LANGUAGE**   **(WE MUST SPEAK OF SHARED GOALS AND DIRECTIONS)**

Nineteenth-century thinkers such as Karl Marx tried to juggle the first three items on the list, and the communist world tried to follow his ideas, but the communist era never really managed to break the historical pattern.

The question is, do corporations follow the same historical social pattern that we see in societies? Is IBM no different from ancient Rome? Is GMC similar to a nation? (Do not forget that the gross earnings of Boeing or General Motors can easily exceed the gross national product of some nation states.) Let's look at these social conditions a little more closely.

The division of labor where we work is a reality that we don't have to discuss. It is all too obvious. The division of property is equally apparent in our wages, and certainly varies. At work, we notice who gets the company car, the window offices, carpeting, the top floor, the basement. The division of property is very real in the business world. The social roles rise up through a building so obviously that the company hierarchy is easier to see than the structure of your city or county government. And it is a government. Companies do have rules and norms. Some typical examples are any number of union contracts that spell out expectations. These are written rules of agreement. Companies also often have strict regulations on tardiness and absenteeism that are written in policy manuals. In addition, "spoken rules" abound concerning the personal use of telephones and such matters.

Unspoken rules also exist and they are probably the toughest to handle for the employee, at least for a new employee.

Justice is another force at work. An employee may be reprimanded. Wages can be docked. Unions use complex appeals processes, and executive files are organized to "show cause" so that someone can be fired. Education is also a key dimension of company growth and development. The company learns. The employees learn: retreats, in-service training, company seminars, paid tuition for night school at local colleges. And finally, there is the "language."

IBM has a language. The US Air force has a language. You have just learned a new language in the last two or three years of college. Our specializations have a tendency to construct language systems—or at least vocabulary systems. Add the jargon of a specific company and you find that there is a common language unique to a group of employees. Sometimes the language is large enough to make it difficult for the parent language, English, to bridge the gap between us when we talk shop. We may barely understand what technical people are saying. They may as well speak French or Urdu. The language of, say, molecular biology can seem as foreign.

You can see, then, that companies contain all the conditions of a society. In fact, they all are little societies, but the extent to which a company actually develops the spirit of community or the feeling of a culture will vary a great deal. In any discussion about this issue that I have with a group of men and women, their "war stories" identify a broad spectrum of businesses and a huge variance in corporate awareness of a company society, but few employees fail to see the community inside the gates. Your place of business is likely to be a lot more than a production site. A great deal of what you do and what you say has a social context or social meaning that goes beyond your role as an employee. Look beyond your hands. You are probably being paid for what your hands do. You do a great deal more. You certainly speak about a great deal more. If you are aware of your company as a society, you will have much greater control of what you do and how you present yourself.

The Top-Down Management Model

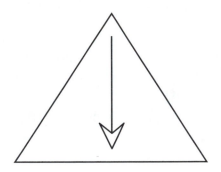

Businesses large and small are designed to allocate tasks. The more tasks a person allocates, the more control that person has over the destiny of the business. A small privately held business of perhaps twenty employees will have a "boss," who is the owner. He is likely to have created the business himself and he will have developed it over a period of years. He is in complete control even though he may have managers. He may have two foremen working for him and perhaps a daughter is in charge of office operations and a clerical staff. She will be the boss in ten years. These are the three midlevel members of his organization. Obviously the middle structure is designed to distribute control. The boss no longer needs to try to handle everything himself.

At some point in the development of a company of this size, the business can run into communication failures. If there is only one effective control of tasks—one boss—then the owner reaches a point at which he or she is overworked with the effort to control, which is simply the effort to allocate tasks and manage the quality of effort. Since the exclusive vehicle for these controls is communication, the boss reaches a point at which the communication is not effective, if it is happening at all. Suppose a line of manufacturing is thousands of units ahead of orders while another unit is weeks behind in production for back orders. Communication is the likely problem if the owner is overextended and has "too much to do." Too much to do means too little time to communicate. The operation is out of control.

As a necessary course, then, the middle levels—and many others—evolve to allocate tasks. Each level of a company has powers. The powers are asserted by all manner of communication that controls or regulates the efforts of workers in the work environment. Out of basic needs for efficiency in good communication, power structure and government emerge.

The little company of twenty is very typical. Its structure reflects a long tradition in business practice. This classical management model maintains power largely at the top levels of a business; the owners are the bosses. **This organizational model is sometimes called a *top-down* system, which means that all the power of the company is found at upper levels, and all the power over employees is delegated downward through the employee structure.** Managers in the middle may not have much control over the corporation or business, but they have control over the employees "below."

What is the basic concept of a worker in top-down systems? The concentration of power isn't simply the privilege of ownership or the right to control. An owner *could* put the tasks of regulation and production in the hands of workers. But, alas, this doesn't often happen because there is a specific concept of the worker built into the top-down system. Since the workers have no vested interest in the quality of production and the quantity of production (because they own no part of the business) they must have someone oversee and control their activities. In fact, if we think back to the Industrial Revolution, when the classical management system began to take shape in northern Europe, we can imagine the origin of the structure as peasants poured into the cities to work in factories. Unable to read or write, having no familiarity with factories and machines, unmotivated except by desperation, seeing no gain to their efforts beyond meager wages, the laborers helped shape the top-down structure we have today. Managers assumed responsibility for everything, perhaps because they wanted to, but also because they thought they had to. Labor was thought to be passive and poorly motivated.

The relationship between you and your coworkers and you and your supervisor is very much controlled by the concept of top-down regulation. Power over people usually stops before it gets to you. You are not likely to have power over employees "below" you. Mid-managers, such as shop foremen, are the bottom rung of the power ladder. Your power is of a different sort. You have the power of horizontal *organization* among your coworkers. You also have power over the *quality* of your skills, and you have power over the *quantity* of your skills. Your supervisor has a type of control over your skills and your production, but supervisors set maximal standards—often beyond any likelihood of being achieved. You are, in reality, the only one who is really in control. You see this fact all around you at work. Some employees labor to minimum standards while others far exceed them. Demands and reality often differ.

The issue of power structure isn't all theory. Management models have a direct control over employees, which includes you. Obviously, managers will always have an effect on the activities of your workday. Can this conventional power structure have any other effects on you? Yes, in many ways.

Employees are constantly influenced by the decisions of managers because of the top-down power structure. It is in your best interest to understand these influences. For example, supervisors can deal with output through very predictable strategies. They can raise the standards; that is, they can hike the production quota. They also control employee promotions and wage increases. These common perks may create a very aggressive workplace where employees are competitive. Some supervisors award output generously. If you are a self-starter or a go-getter, this is the place for you. Other companies bump along. If you are laid-back, you may want to seek a less aggressive style of business setting that basically works as a well-oiled company without a lot of pressure.

Companies also shift from one style to the other. When small local companies are involved in a takeover because of a buyout, the original employees usually do not stay with the new company for long. When management changes, management style changes also. The society of the company changes. Employees quit, because their society has been altered. We can all recall examples of this problem. Debra was involved in a buyout. Here is what happened:

> *I worked for Chadwick Lumber Company. It was privately owned. There were eight of us in the office. All the others worked in the yards or in delivery. We were sold overnight. At least none of us knew it was coming. A month later our new manager arrived from Alabama from ABC, a huge national lumber corporation. Out of the eight I was the only one left after one year—maybe because I was the youngest. I mean I had only worked there for three years, but the old-timers were very unhappy. They all quit. I lasted another year.*

Since a company is a society, takeovers, buyouts, downsizing, and other dramatic events can seriously alter the community that the employees establish.

It goes without saying that you are not invited to share in the power of a company. You may be offered some of the company holdings (stock) but no one will hand you the reins. Employees usually gain power in only one way: by becoming midmanagers through promotions. If you want to see yourself move upward through a company, be sure the option is available. Many professions—police and fire services for example—are rigidly paramilitary. They are very much oriented toward promotions. The rookie someday will be captain—and who knows, perhaps he will be the mayor as well. Many other corporations "hire in" their managers. As a potential employee you want to be aware of your options, and to understand these societies you must analyze the management of power.

Bottom-Driven Companies

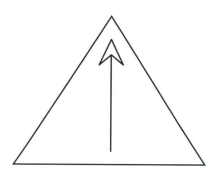

In due time, a reaction to placing control in the hands of a few was bound to occur. The last two hundred years saw the development of one revolt after another. History focuses on the results of the huge revolutions against states and governments, but the factory floor was a key location of much of the discontent. The seeds of these revolutions were based on a simple question: *Why not* put control in the hands of the workers?

The reverse of the top-down management system thus evolved; the model is based on worker control and is called a *bottom-driven* system. How broadly to define the bottom driven model is not clear. If a restaurant is owned and operated by a family, I suppose we could say it is bottom-driven. After all, *control* and *profit* are the pleasures of the entrepreneur. On a much larger scale if the employees of an airline buy the airline from the previous owners, or if a union of steelworkers buys a mill that is threatened with closure, we see that these much larger versions of bottom-driven systems probably lose the voice of any one laborer-owner.

There are practical concepts that drive the bottom-driven model. **The basic idea is that the worker can contribute to the development of production quality and output quantity *if* the worker is given a share in outputs, investment, or ownership.** In other words, bottom-driven models do not fear giving power to the employees.

In practice, bottom-driven companies quickly set up a hierarchy of managers who, on the surface, create a structure that looks like a traditional top-down operation. Every business needs leadership, and it is usually easier to have a few managers than to call for a vote of all members for every decision that has to be made. Nonetheless, there is a profound difference in the philosophies of top-down systems and bottom-driven systems, because of the *direction* of power. A bottom-driven company has supervisors, but they answer to the employees.

Traditional colleges (and many hospitals) have some characteristics of bottom-driven companies if they still hold to the tradition of having faculty members appoint chairpersons and deans from among their numbers, and if the faculty determine education policy (production), which was a popular tradition throughout the United States in the first half of the twentieth century. Chairs and deans are the equivalent of the middle ranks of a typical management structure. Other examples abound in small businesses controlled by the employees who founded and maintain the businesses. Food cooperatives and similar organizations are also bottom-driven in some measure. In Berkeley and Seattle there are supermarkets that are "owned" by the membership. These members are voters and can be elected to the boards that control the supermarkets. However, most members are shoppers, not supermarket workers. The true bottom-driven company is not usually very large.

How do you fit into the picture? You might look for bottom-driven organizations if you have the interest that such a system requires: involvement. Businesses of this type are likely to be exciting, and possibly risky. They are often socially involved. They may demand much of your time. If you want to punch in and punch out, these organizations are not for you. They are, however, thought to be highly rewarding by the employee owners.

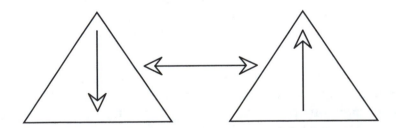

Another popular form of bottom-driven organization is the union. Unlike the true model, which drives the company from the bottom, the union coexists with the employer in a checks-and-balances configuration. **Unions developed as an adjunct or counterforce to**

balance the power of an existing top-down business. The function of any union is to check the power of managers by having an opposing power in the union—usually exerted by sheer numbers. Managers exert ownership and financial control; organized labor exerts production control and labor standards (quality control). The system of pitting downward power against upward power is complex, and the relationship between the two forces is often, by its very nature, confrontational or adversarial. The "strike" strategy should come as no surprise to anyone in a corporation with unions. The strike represents impasse, which happens occasionally in union and management negotiations.

Unions have had a healthy influence on the evolution of American industry from the worker's point of view. The unions fought for wages, benefits, the forty-hour week, time-and-a-half overtime, safe working conditions, and other important concerns of labor. In recent decades organized labor has declined somewhat for many reasons. Unions were once very strong in heavy industries that have themselves declined; American steel production is a well-known example. Since unions exist *inside* the industry, if the industry fails, the union will decline.

Unions must also grow if companies grow. As corporations become larger and larger, so too, the unions grow larger to maintain the checks-and-balances system. The resulting union giants seem to alarm many young graduates coming into the marketplace today. They complain that the unions are just another big corporation. In other words, they seem to feel that the bottom-driven tradition has become a top-down power structure. The larger unions have tried to deal with this problem by giving the local units varying degrees of control over their local affairs and finances. If you find yourself in an occupation that is unionized, you will have to consider these issues carefully.

The union system was designed to counter the corporate system in other ways that will affect you also. In order to control promotions, pay scales, and other shop-floor issues, many unions have endorsed the concept of "seniority." The virtues of a system that is based on years of service are issues for you to think out. The concept was based on the bottom-driven system and the basic idea that workers care about and devote themselves to their industries. The union will argue that seniority protects the new employee as well as the older employee; in an all-for-one and one-for-all organization, the younger worker can depend on a secure future, a good wage, proper benefits, and other matters *because* of the union. The opposing point of view is that talent cannot be rewarded properly, and employee success is therefore in the hands of the union.

If you are considering the union sectors, examine your traits and work habits and seek your own interests—by recognizing *your* best interest. If you think you will be looking at a union more closely, examine it as you would a company. After all, in a sense you will be joining not one company, but two.

New Ideas in American Industry

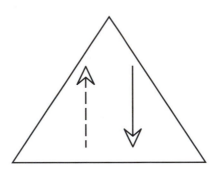

We have looked at the three most likely power structures you will encounter and we have observed how they work and how they view the employee. All three systems have a management tree or hierarchy to delegate power in some sort of chain of command. Nonetheless, the guiding principles of the three systems are quite different. We might add a fourth "influence" to the discussion (it isn't exactly a "system") because of current trends in international business.

As recent developments point more and more in the direction of global economies, increasing competition is coming from nations that have their own ideas about how to run corporations. The extraordinary success of Japan, in particular, has made American executives sit up and take notice, and hardly anybody, from executive suites to the boiler room, fails to discuss these trends—usually with alarm. The executives are arguing about "Z theory," and the steam fitters are complaining because the wrenches are "cheap imports" or because the best power tools are now Japanese, not American. The success of Japanese industry has helped encourage American corporations to experiment with new management ideas and new corporate practices. Evidently, worker input and worker-manager cooperation are the keynotes to this interest. Critics argue that there is nothing new under the rising sun and insist that all Japanese business practices are borrowed from the United States, right down to the company softball team. The arguments do not matter to you and me, but the outcomes, such as "quality circles," "profit sharing," and the like may have an impact on you.

In seeking ways to increase quality and production and innovation—not to mention communication—companies have encouraged the development of open dialog with employees.

This may seem natural to you, but it is not. The three systems we have looked at clearly indicate that power is directional, and it moves up or down. In contrast, horizontal power is the more likely outcome of quality circles and similar discussion group practices. Although the idea of teamwork or team spirit would seem to be quite old, you may find yourself in a company setting that is placing a renewed emphasis on quality-circle discussion groups, shareholder investment plans, and all manner of devices to involve the employee in the fate of the company by trying to get employees to looking at the company as *his* or *her* company, at least in part.

The trend is obviously healthy and may provide you with the opportunity to contribute to the success of the business—and the success of your future. If you are part of a weekly coffee klatch with your coworkers and supervisors, you may have a new and different feeling about a company, and that company may develop a very different perception of you—to your advantage. Besides, you will have the valuable opportunity to be seen—and heard.

The Power of Market Trends

It is obvious that the *structure* of a company can affect you. We have looked at power as the defining force that determines configurations inside a company. Every college-educated manager, every MBA, knows this discussion backward and forward. Employees should understand these principles also. These concepts are hardly guarded secrets, but employees tend to look at companies as the source of a wage and never think to examine the management organizational chart—or the annual report.

Some authorities argue that the final power is the marketplace and argue that real power is a force exerted in environments where the corporation struggles to survive. You might not realize it, but most businesses, including most of our corporate giants, operate on remarkably trim profit margins—much thinner than your own. If you exceed 5% in your profits over last year's gross, because of a 5% pay raise, you are probably ahead of IBM and GMC. Most corporations, regardless of their billions of dollars, must operate competitively and rarely take home more than 5%; usually the figure is an amazingly risky 3% or less.

When you are considering a company as an employer, bear in mind the idea that the corporate gate may look sturdy and elegant, but it could be auctioned off for debt payments in two years. Since markets do exert a type of ultimate power, there is no doubt that you should look at a company in terms of product demand or service demand. Also look at the wave-and-trough effect of swing cycles to see what is the likelihood that you will be *steadily* employed. Some corporations look at the worker as the safety hatch. If they start to sink, the worker starts to swim. The layoffs begin. Other corporations assume considerable civic responsibility.

The Boeing corporation in Seattle is a case in point. One of the stars among the many U.S. corporate success stories, Boeing is a technological giant in the global economy. It is also committed to the Seattle community, as best as it can be considering its global reach. In the late 1960s and early 1970s the company fell into a slump that resulted in massive layoffs. In the process of rebuilding, the company decided to make some effort not to overhire, because in a downturn in the national or corporate economy, a policy that overhires will also become a policy that overfires. Recognizing its commitment to the stability of the local economy of Seattle, Boeing's policy has been to do its best to maintain relatively constant employee levels. Even with such a policy, the aircraft giant dropped thousands of employees in 1992, on the tail of a recession. Then the unforeseen global disturbances of the Asian stock market problems hit the corporation in 1998 and 1999 and new layoffs— in the thousands—loomed once again. Here is Brian's perspective on the problem. He was a machinist at Boeing who decided it was time to go back to college and redirect his interests toward computer technology.

> *I'm currently working at Boeing at the Renton Plant in the flight test department. I have worked for Boeing for three years. Boeing had a lot of airplane orders for the last three years, but things changed because of the world economy. Competition from Airbus also took away many orders from Boeing. In fact, last year the company lost many airplane orders to Airbus because of its production troubles. Even long-term loyal customers like UPS and British Airways went to Airbus because of production problems. Because of slow sales to airlines, Boeing decided to lay off thousands of employees over the next two years. Many workers have started to worry about their future here.*

Watch the market if you can. It is not always predictable. Everyone knew CDs would someday take over the recording industry, but no one could predict that vinyl records and the record player would vanish in two years' time! And no one in the wallpaper industry back in the 1940s could have anticipated the destruction that would result from one humble little invention: the paint roller. Look at government regulation in the timber industry. Look at deregulation in the airline industry. And then there is the fickle taste of the consumer. The venetian blinds we used to see haven't been manufactured in forty years. Overnight nobody wanted them anymore. Yet amazingly, in the 1980s everybody decided to buy Levalors, the skinny current trend in venetian blinds. Smoking is out but cigars are in. Who knows why?

Company Size

We have looked at power. It is directional. It affects the destiny of a company and has an impact on all the employees. It has an effect on you, and you might even have an effect on the structure of power yourself. Power also assumes shape, of course. It is probably fair to say that it is always a pyramid due to a basic law of efficiency or some similar factor. Whether the driver is at the bottom or the top, the result usually will be a ladder of some kind, usually wide at the bottom and narrow at the top. Be aware that the pyramid itself (not the power) may be of interest to you.

Let's assume a company has 5000 employees. All is going well and the employees, including the executives, are living off the fat of the land. Given the luxury to do so, ranks in the ladder of power increase. This may appear to contradict the desire to hold on to power, but people in power like days off just like you and I. They will design levels of management to work on the ladder rung below them so that those new managers can delegate and control power over some number of other managers or groups of employees. The resulting multiplication of managers will increase the steps of the ladder. The result changes the shape of the pyramid. If the pyramid represents the volume of employees, and the height represents the number of managers, then our pyramid, in good times, may be a tall, narrow one. This is the familiar "bureaucracy."

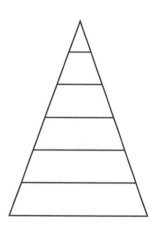

Now suppose that the company decides to downsize or cut its fat. Suppose that suddenly there are no buyers for its product. The "lean and mean" corporation will usually plunder its own management hierarchy in hard times. There may be employee layoffs, but public-sector investors know that removing a lot of expensive managers may be less harmful to company stability than removing a critical base of skilled employees. (Bureaucracies in the government sectors are a different matter, and they are much more difficult to trim.)

Like other employees, executives are just as easily upset by major policy changes that alter their expectations. David was in charge of Research Project Development for his company. He was responsible for forty employees.

In a serious downturn, we lost 20% of our market share in one year. Ron was fired and his entire technical staff was put in my charge. That was about thirty-five people at the time. They didn't offer me a dime for the extra responsibility. What could I do—except perform two jobs for the price of one? Some of my people were none too happy because they depend on me. Others didn't care. Some probably liked the situation because I wasn't available as much. Lloyd and Barbara actually joked about finally getting me out of their hair.

Once a company realigns itself in the lean-and-mean mold, the pyramid may look vastly different. It may be almost as flat as a pancake, since the volume of employees may have remained fairly constant while management ranks were greatly thinned out. In other words, an employee working for this company two years ago had an immediate supervisor, and perhaps six more above. Now she has a new boss who was a former "level four" supervisor, and only two upper levels of management.

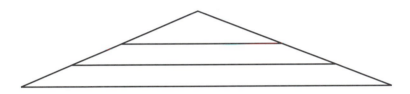

Executives argue the pros and cons of these practices and the management-to-employee ratios that result. For you, as an employee, it may help to know how the labor is organized if you are looking at a company as an employment prospect. How many employees are responsible to your immediate supervisor? Five? Fifty? Five hundred? It matters. If you are easygoing and want to mind your business and do your job and go home, you might enjoy a very high employee-to-manager ratio where there is not someone "looking over your shoulder." On the other hand you may want to go places and be seen—and heard. Success-oriented employees would probably favor a much smaller ratio.

There is another way to look at the management ladder, however. If you are many, many layers from the mountain top, your chances of climbing may be difficult. This is exactly the opposite of what we would expect from the low manager-to-worker ratio. We get the ratio we want but look at all the layers above! At times, perhaps, a higher ratio of employees to managers is a better prospect if there are only a few management levels.

Summary

- A company is a society and exhibits social characteristics similar to those of other societies:

  Divisions of labor

  Divisions of profit

  Social roles

  Normal or accepted practices

  Justice

  Education and history

  A company "language"

- Each level of a company structure is given certain powers.

- Top-down management structures control production and costs, partly by defining the roles and expectations they assign to each employee.

- Employees run bottom-driven organizations.

- Unions are bottom-driven in theory, but large unions share many of the social characteristics of corporations.

- Unions attempt to achieve a balance between the bottom-up needs and expectations of workers and the top-down needs and expectations of management.

- Market trends have highly pronounced effects on employment trends.

- Company size is an important employee consideration.

- There is a social contract, largely unspoken, that guides the complicated life of a group that shares the goals of a company.

- Employees can fulfill their personal interests by selecting a company where they will be satisfied with their role in the context of the group.

Activities Chapter 1

Present a memo to your instructor that develops a discussion of a work experience that relates to the chapter. The memo should be 500 to 1000 words, typed. In it you can explain how you have related the text or the lecture discussions to some event you recall. This exercise will give you a better understanding of the material because you will explain incidents in terms of your perceptions of workplace structure and communication.

Select one of the following suggestions and develop an analysis that involves your current employment or a former position.

- *Discuss the basic social structure you see in the company where you work. Consider language, law, education, norms, and so on.*

- *Relate your experiences with top-down managers. Describe the attitudes of other employees in relation to these managers.*

- *If you have worked in an environment where the leadership was weak or chaotic for some reason, recall the events and explain the problems.*

- *Identify a bottom-up issue that had significance to you or other employees. Union issues are typical, but favoritism and other concerns are common also. Present and discuss the sequence of events related to the issue.*

- *If you have experienced any changes in power structure—a new boss, a merger—explain what happened. Discuss employee attitudes toward change.*

- *If you worked in a company that fired fellow employees or went through a serious "downsizing," explain the situation and the attitudes of the parties involved.*

- *If you clearly see the organizational goals of the company where you work, identify the ways in which those goals are in conflict or agreement with employee attitudes.*

- *If you have had any experience with foreign corporate values, indicate the noticeable features and compare the apparent values and interests with those of your U.S. counterparts.*

- *Explain an instance of production difficulties in terms of input, production, and output problems.*

Note: *Feel free to change the names of the people involved in your story. You may use a fictional name for the company also if you prefer to do so. You may use fictitious names in subsequent memos also.*

Work in Progress

2. Team Players

Many people have the mistaken idea that to be self-employed in a contracting business means that you are a "loner" who has a hard time being a team player or working in a team environment. While that may be true for some, it has definitely not been true in my case. I could feasibly do most of the network installation contracts my company gets. I have found, though, that having other people working on the project not only makes the installation faster but allows us to each check each other's work and catch mistakes that a person working alone might overlook.

My job as president of NetWorks NorthWest is to be both my team's coach and a team member during the installation, guiding my employees but allowing them enough latitude to make practical decisions on their own. It's hard sometimes for me to let go of total control on a project, but the increase in self-confidence my crew gets knowing that I have confidence in them is well worth any concern I might have. They are not shy about pointing out ideas they have about how the project might be done faster or cheaper, and they know I'll listen because sometimes they come up with a winner!

Sometimes it feels like quite a balancing act to manage both an installation and employees, but stressful though that might be, I wouldn't trade it for the company politics of a corporation. One of the best things about being self-employed is that there isn't any company politicking or schmoozing. True, I have to field complaints and squabbles with my two employees, but, for the most part, if something needs to be changed, I change it without any drama.

Perhaps this is a good time to "introduce" my two employees, Randy and Cloe. I'll start with Randy. Randy is a hardware maniac. He has a full home network complete with a four-port hub and an old PC he uses as a file server. He admits he doesn't need to have a home network but he insists that being able to check your e-mail from every room in the house is "too cool" not to do—so he does it. Randy has an incredible sense of humor and an amazing amount of energy. He is the sparkplug for our team and a fountain of good ideas, although he can get frustrated easily if he encounters a problem he can't quickly fix.

Cloe is the quiet, serious type who can stay with a problem and remain fully focused until she finds a solution. She originally started work in computer programming but soon discovered she loved the three-dimensional problem solving that network engineering presented her. Cloe doesn't tend to get thrown off-balance when problems come up but instead seems to light up at the chance to figure them out. She is sometimes reluctant to offer suggestions and make constructive criticisms, although this appears to be getting easier after each job.

J.Q.C.

Predict-able Paths of Company Communi-cation

Our look at pyramids of power and directions of authority has more or less been oriented toward explaining company environments in an effort to understand why they are designed the way they are, and why it matters to you. Perhaps you can select a company on the basis of these perspectives, since company policies will serve you best if they happen to fit your ambitions. With an understanding of company power structures, we can also look at other areas of interest. One concern is communication systems. Power communicates. It has to—otherwise, it is powerless.

Since the structure of everything in a company is very directional, it is no surprise to see that communication is directional also. The output of memos, telephone messages, hallway conversations, seminars, correspondence, or any other vehicle for spoken or written communication is highly directional *and* conforms to company policy. All you have to do is violate the practices of your company to see the sparks fly. An anonymous letter criticizing someone in your company for a bad performance is a good example. I have seen such letters placed in hundreds of mailboxes on Sundays while the offices were empty, and sometimes the documents were signed! The defiance we see in "speaking out" fits our discussion perfectly because it violates normal paths of acceptable communication, which are usually, of course, designed by powerful controlling forces inside the company.

The acceptable, reliable, everyday paths of communication within a company have been called "circuits" by some authorities. Imagine an electronic component and all its circuits; think of a company setting as a similar situation. Circuits are predictable patterns. They are expected—and respected. The paths are as regular as a 9:00 Monday morning kickoff meeting for executives who are preparing for the week. The paths are as respected as the CEO's gold-embossed stationery that speaks power even if the sheet is blank. The systems of communication convey information, and in fact *are* information. Uh-oh, a letter from the boss; you don't even have to read it to sense the authority. A circuit is as predictable as the phone memo (in pink) left on your lab table—a low-authority document, unless it says "call your supervisor." It is important for us to understand and respect circuits. Trouble abounds if we do not.

Once we identify the typical circuits, you will realize that you already know them quite well. What is interesting is to see the pattern, the directionality, the "tradition," and the realization that we usually abide by them: "We don't do that here." "It's not done." "We do it differently."

The circuits can be grouped by direction of flow. **Typical "downward" circuits are the communications from an immediate supervisor.** She is likely to control your role in the workplace. She may have written your job description, and she certainly determined the requirements for your day-to-day activities. She will also be responsible for any work changes, such as new instructions, for example. She must apply company policy and explain it to you, and she must appraise your work performance. These are communication activities that are often written, often spoken, often both. Written performance appraisals have become a popular corporate practice. They may be annual documents or even quarterly documents. Fortunately, this particular type of paperwork is an office practice that often stops on the doorstep as you enter the labs and shop floors; that is, evaluations of individual employees and employee self-evaluations are uncommon at that level.

Of course, not all communication circuits travel up and down an invisible ladder. As an employee, most of your conversation will be among your coworkers. **In terms of directionality, all your communication in the office, lab, or shop is commonly "horizontal."** These work-related conversations are a valuable dimension of company communication. All discussion directed toward getting the job done involves time shared among coworkers in job-related communication. Much of the conversation conducted among friends and workers is also going to be friendly.

Some supervisors may not encourage casual conversation on the grounds that it is a distraction, or a waste of time. Horizontal communication is where this problem occurs. You and I may both know that a job may even be helped by distraction; conversation can keep us from being bored. However, I think you will find that you don't want to have to defend this point of view with supervisors. Of course, if the work is interesting, the shoptalk is directed at the tasks in a very positive or constructive fashion that would please any manager.

Beyond sharing work and sharing a social relationship with our fellow employees, we also share our discontents. We have all experienced the gripes of our coworkers who seek our ear, if not our sympathy. If the gripe is directed at a supervisor during a conversation with a coworker, the conversation has been given the interesting name of "powerchecking." If your boss criticizes you and you then promptly return to the job and complain to your friends, you are powerchecking.

> John is always on my case. Shelby and Bob were over their quotas and I wasn't. Big deal. Let him look at last month. You remember who went over the top last month don't you?

Obviously this can be healthy, unless an employee is *always* grumbling. Healthy or not, this is horizontal conversation to guard with care, since word could get back to the supervisor. Words can also be overheard. A supervisor will not be likely to see how good the ventilation of gripes is for employee morale.

There will be "upward" communication also. You do not work in a vacuum and your supervisors do need to hear from you. You can look at all quality circle activities as upward communications. Your effort to develop or discuss new ideas or other innovations is a very productive form of communication directed toward supervisory levels, at least in the end. So too, **all your questions about how to perform your job or any specific job-related task are the most common type of upward-directed communication.** Personal considerations also exist in bottom-up relationships because an employee may have any number of concerns of a nature that could include the employee's family, the employee's health, or other difficulties that might affect job performance.

All the examples of circuits are obvious once we think about them. Because they are normal and acceptable, they are safe (except for powerchecking). You can trust the direction, the intent, and the tradition of circuits. For example, there is usually only one way to ask your supervisor for a raise, and yet, if you have a union, the traditional private discussion about a raise may be a taboo. There will be a circuit for inquiring about a pay raise, and it will be a clearly marked path of communication. We can conclude, then, that the way to conduct ourselves is probably quite apparent if we operate in the normal channels of our company practices. We usually recognize the risks of defiance and we operate in circuits because they are usually a convenience to us.

If for some reason we abuse circuits, there will be trouble. Consider this example of a misunderstanding of circuits.

At one time, Jeff worked for a group of upper-level supervisors who held an open-door policy, meaning they invited anyone to drop in at any time. Jeff was welcome at *three* levels of supervision. For a number of reasons, all three levels were swiftly replaced in about two years. Retirements were the primary cause. It happens that Jeff's secretary was having difficulties with other personnel in the secretarial pool at the time. He thought nothing of directing her to the uppermost supervisor of the three *new* supervisors because he thought the issue had a possible significance with which the top manager would want to be familiar. Well, poor Suzanne took quite a shaving. She was badly redressed and it was made clear to her that her problem was much too minor for this new top executive. Jeff was to blame. He did not realize that there had been a change in the circuits.

> *Worse yet, when the door closed on Suzanne, it closed on me too. I sent her over there and the new CEO knew it. That team lasted about five years. I was shut out because I sent a secretary to speak to the new president. When Blake was president, anybody was welcome.*

You might think that this new supervisor was simply arrogant. Perhaps he was, and possibly Jeff should have guessed as much, but misreading the attitude of the new supervisor was not his big error. Jeff's error was misreading the new circuits. You see, it was now a taboo among the new leaders to operate outside of their chain of command. The doors had closed, and Jeff did not know it! His error was to address a minor issue to a major supervisor. In fact, his error was to address *any* issue to the president without first going through the channels of the other two levels of supervision.

Networks

Having noted that employees are expected to operate in channels of communication that are the *usual* communication paths at work, we now turn to another type of communication that is present. All the normal paths of communication—the circuits—are designed for the basic stability and efficiency of the company. They have little to do with another powerful force of human behavior: politics. Put a few people together for any length of time and persuasion, allegiance, defiance, and all the usual complications of human nature will surface. The calm look of company communications is only the *appearance* of life aboard ship, as every employee knows.

Communication that takes the shape of politicking is often called *networking,* a word that has become quite popular in recent years. Networking certainly isn't new. Perhaps the open acceptance of it as a type of powerful activity is new, but the behavior is probably as old as the human community. **Networks are political structures or communications *outside* the normal channels, usually with the specific intent of exerting influences.** The network is often very directional: it is power (or, at least, influence) directed upward. Since most companies are pyramids of downward-directed power, you can sense the political nature of the networking idea.

Networking is the *unofficial* channel of power. In fact, the effects are so strong that some authorities argue that networking is often the *true* path of power and the *true* path of communication in a world that appears to be run in circuits. This may sound a little abstract, but networking simply means that a round of golf may achieve levels of agreement or decision making that endless rounds of meetings failed to accomplish. Networking simply means that the new policy adopted in the boardroom was actually decided on over drinks at the club the day before. You see, the board meeting is the official work location—the circuit. The club dining room is where the networking was done. The task was really completed as a networking activity and that is why we refer to the practice as "true" power and "true" communication. Honest, open control of authority is often in the hands of the network process.

Admittedly, you are not likely to find yourself at the boss's country club, much less at a "power lunch" among your company's executives at a swank hotel. True, but there is another face to networking. Networking is not only politics. If people come together because of shared interests, the motive does not have to be political. In other words, there is a second kind of networking that is not very political at all: the company softball team. If you and your supervisor happen to be on the team together, you begin to cultivate a shared interest that increases your acquaintance beyond and outside the work place. You may even find that your daughter is in a scout troop with the daughter of your boss's boss.

It might be easy to overlook the relevance of the softball team, because a chance connection is not necessarily a political tool, so you need to understand how it is relevant. Recall our discussion of the ratio of employees to supervisors. The significance of the ratio is that you can decide whether you want to be seen and recognized for your role in a company. The softball team leaps all the hurdles at once. It is not particularly political; it is simply a chance for you and your supervisor to get to know each other. The rewards are the trust that can be found in turning a working relationship between a superior and a subordinate into one of equality, or into two pretty ho-hum ballplayers laughing together. Interestingly, this shared interest is perhaps the truest of networks and can be very powerful, since trust and sincerity are important to everyone. Who knows, the next promotion could be yours—as long as you have a good throwing arm. Workplace logic can get a little odd at times.

Be warned, the most innocent network can be viewed as a tremendous threat. If you get to know your boss's boss for some reason, be aware of the implications of this. Your boss will be wary. It is also not going to make your coworkers too happy if your promotion seems to them to be based on networking, even if it is softball and scout troops. Innocent, shared interests will not protect you from the spurned employees who also wanted the promotion you got. They will instantly gossip and powercheck you. Whether political interest or mutual interest generates the networking, you should be aware of the maze of complex situations that are involved.

With regard to networks, affirmative action hiring practices have proved to be a curious and complicated case. Several decades of affirmative action guidelines have been directed at the "old-boy" networks, but legal recourse cannot legislate control of networks because networks have a very unofficial existence. They exist almost *because* they are unofficial. As a result, affirmative action practices are directed at controlling *standard* hiring procedures, which are circuits. The cure may not quite have addressed the ailment, since it is difficult for official control to regulate unofficial practices, but affirmative action did bring new rules to an old game. Affirmative action redefined a circuit or normal practice. This has angered employees not because they have racial prejudices but because they have circuit expectations. The idea of hiring the most qualified applicant is a time-honored expectation. It is the way it is (or was). It isn't right; it isn't wrong. It is the expectation. If you are a recently hired member of a minority or a recently promoted woman, for example, be prepared; you may be powerchecked. You may be viewed with suspicion not because of who your are but because employees may think you were hired or promoted outside the rules of a traditional circuit. Worse, employees will feel that you have exercised privilege (or that the law has exercised privilege on your behalf). Alas, privilege is the game of *networking* and that is the problem. It is very ironic. The reaction will be predictable, even though affirmative action is not a likely path to networking (or privilege). If employees *think* they see networking, there can be an unpleasant reaction.

Gift giving is often seen as networking. Many kinds of extravagance are characteristic of networking, and government agencies are often concerned about how to interpret the donations of special interest groups. Caps are sometimes placed on the donations, and donors are usually required to report donations. A reported donation is, therefore, acceptable, and unreported donations are taboo. This type of policy attempts to control networking with guidelines and official channels. In other words the network is forced into a circuit—in theory.

Extravagance is another typical element that is often criticized in company environments, if only in whispers. We have all heard union members complain about "high-rolling" big spenders among their union leaders, for example, but their union lobbyist will ask what kind of lobbying is done over hamburgers and a milkshake? To wine and dine is an operational fact of networking life in the political world. Circuits appear conservative by comparison, which makes extravagant networking appear inappropriate to anyone who trusts in the circuits. Consequently, networking is often going to be in conflict with circuits.

Compared with networks, the clean and neat channels of circuits look more practical and effective in day-to-day life. However, politics is a natural force in any society. Doing your job in an arena of circuits and networks is often complicated.

Being a Team Member

The word *conformity* has a conflicting twist to it in the minds of Americans. In this great land of individualism, the need to conform is always looked at with mixed feelings. On the one hand we are an amazing sports culture. In all the team sports we watch with such enthusiasm, teamwork is the keynote. We *believe* in teams. On the other hand, as an employee, you probably would have second thoughts about a company flag, or company housing. Perhaps you don't even like the idea of a company uniform.

We believe in individual rights, yet even the smallest lumberyard will proudly distribute those polished satin jackets and ballcaps with some sort of large logo, and all the hearty individualists will wear them. Individualists are joiners also. There is a delicate balance between *team* spirit and the *individual's* role on the team. We all sense these two conflicting values, and our democracy is designed around the two forces at work here, forces that are reflected in phrases such as "checks and balances" or "majority rule–minority right." It is a balance we know well, and the company setting achieves the same balance. A company expects skills, but a company also expects loyalty.

The work environment of a company is a community. A company may not hang out a company flag such as you might see in Japan, but in a sense you, as an employee, become a "citizen" of the company for which you work. There is a contract between the two of you—a psychological bond. You depend on each other, and expectations are shared. The bond is mutual. For you the contract means work, wages, retirement-building, and so on. For the company you represent dependability, service, and production. The simplest bond is that you profit from each other, even though your role may look minor if the company is a large one.

If we think about the training strategies utilized by companies, we begin to realize that the employee is clearly treated as a member of a community. **The organization provides the employee with training to mold his or her role in the company, and to encourage the employee's loyalty.** The shaping of the employee involves not just professional skills but company culture as well. The extent to which companies train employees for the company values depends on their status level in the corporation and the corporate policies.

For example, Mike, an MBA, interviewed with a major U.S. wine producer and found that their training for executives involved ten weeks in isolation (without family) and three years of being stationed outside the city where the executive resides. Mike expected training, but not boot camp. He felt very uneasy about the corporation and its perception of its executives. He decided at the interview that company policy would clearly conflict with personal style—at least in his case.

> *It took me no time whatever to realize I wasn't the guy they needed. I don't know how easily they enlisted married people, or executives with children. My wife has no plans for moving, and if we had kids, the problem would be even more complicated. Maybe they were primarily interested in single executives. I don't know.*

This is a fairly extreme example perhaps, but it suggests the extent to which companies take the matter of molding team players. Mike obviously decided he was not the team player the company was looking for.

Every company has expectations of employee behavior; the problem is that the expectations are seldom in writing. Corporate training procedures only suggest the presence of team-player concerns. The *selection* of an employee, for example, is a very serious process for most companies because they will seek out *their* kind of employee at the outset. Once hired, an employee encounters a variety of traditional learning processes, all of which "socialize" the employee. *We are not just hired. We must also join.* We join by becoming part of the spirit of the community. This is the "socializing" process. Corporations have a host of teaching processes you have encountered, or will encounter, in the near future. *Training seminars* (of the classroom or committee room type) and *apprenticeships* (in which you work with someone) are the two conspicuous tools used by companies that can afford them. Learning on your own through *trial and error* is another practice companies may have to depend on. *Failures* are also a strong learning tool for us all, since we are criticized by superiors and hazed by coworkers. Larger companies are not, likely to encourage learning-by-calamity, of course, but small companies may intimidate an employee during the rookie period of a new job.

All these learning processes appear to be focused on skills, but they are part of the employee socialization process as well. The effect of forming the mutual bond is the company community, and all the training techniques occur fairly rapidly in order to train and socialize employees. The experience of going to college is somewhat similar in that college is a community. When students arrive at the college where I work, the men and women, usually in their twenties and thirties, are very timid about being here "with all the kids." They see themselves as mainstream American workers who do not belong here. Very quickly they acquire backpacks and briefcases, wear jeans and sneakers, hang around the labs, enjoy the library, and pal around in the cafeteria. What happened? Socialization. The technical programs offered the courses to the students but the community or social bond is a little tougher to explain. Nonetheless, I see it happen. We all live in these communities.

Perhaps the most relevant perception is the earlier comment. "We are not just hired. We join." As instructors bid good-bye to the students at the college, they know that when they see them again, the students will have changed yet again. Once the students are hired, they leave the college community and enter or join the corporate community. Another image then emerges.

Rebellion and Creativity

One situation that a company really does not want is *rebellion*. The rebels are weeded out because their behavior not only fails to conform but also because they may be defiant. Defiance is a red flag being waved in front of anybody. It would be a mistake, however, to assume simply that conformity is the alternative to rebellion. The appropriate word for the desired behavior is *creativity*. It is likely that you are in your chosen profession because you enjoy a career field where you find some satisfaction and perhaps some self-expression, partly because what you do is done very well. In other words, you seek out creativity. So does any company. **Getting the job done conforms. A job *well done* is something a little more desirable.** Creativity is an intense level of commitment, and creativity will get the job done with that extra touch. No company will discourage you.

Perhaps you then say to me that last year you had a supervisor who discouraged creativity. He squelched every extra effort you made. Possibly so, or possibly he was just afraid of you. Maybe the company did want your contributions, but you may have had an easily threatened supervisor. Perhaps he was looking out for number one, not for his company. Sooner or later he will be found out. But the self-interest of this manager is valuable to us in this instance because he has a lesson to teach. He may not be the only one who will worry about creativity. We all have heard coworkers snipe about "the new kid." Especially if he is the only draftsman in the office who knows how to handle a new CAD system and some of the drafters still refuse to sit in front of a computer. It does not take much to rock the boat. The social community is delicate.

Obviously, there is a catch to creativity. Companies want imagination and energy, but there is a risk. Coworkers don't always admire achievement. What is it about the college student who excels that makes her classmates have a grudge against her? Success or fear of her success is probably the answer. But is it her fault? Probably not—unless she is also networking. You must guard your creativity and channel it with care so that you do not threaten those around you. You do not want to lose friends for the sake of your success, and you do not want to be held back either. What do you do?

We can divide creativity into categories so you can see the way people look at each other's skills. If you hear students say, with frustration, that student Smith is "brilliant," it is not spoken in the same spirit as a comment of annoyance directed at student Jones who "goes overboard on everything." I think you will always sense a distinction between talent that demands respect and talent that can be criticized. How can this be so? Networking. Coworkers resent inside tracks, and they resent excess. If you wash the boss's car, there will be talk. If you stay late at the office too faithfully . . . . If you always work overtime on weekends . . . . Never mind that no one else wants to. You must see that *your* logic does not correspond to *coworker* logic. Which is the correct view? This is an interesting problem, particularly if, for example, you really need the money and have no other motive.

You can be creative in a company setting and not feel the pressure of being feared. The trick is to simply divide all your tasks into three possible categories. In the first category—which we will call "priority activity"—list all the *essential* tasks you do for the company. In the second category (call it the list of "relevant or useful tasks"), list all the other *usual* and *desirable* tasks that you do or could do. Create a third category for the *apples,* and think of anything that is not necessary. One authority calls this third category "peripheral behavior." The student who is reluctantly respected because she is "brilliant" cannot be attacked. She operates in the priority skills category. As you descend through the categories the risk of gaining disrespect from coworkers becomes apparent.

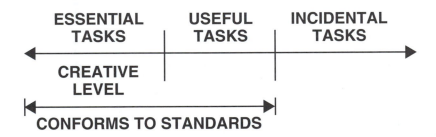

If you get a promotion through the essential skills of your company, you will be a good foreman because your former coworkers will respect you. But woe to the overachiever whose promotion is not looked at as fair play by his or her fellow employees.

Alice was a very dedicated employee—and a staunch unionist. She devoted a great deal of free time to union work and gained the respect of all her coworkers. During contract negotiations, Alice fought hard to get a dozen shop stewards promoted to foreman positions at company cost. She succeeded. The contract was a success. But Alice's brother was one of the shop stewards who received a promotion as a result of her negotiations. Of course, he deserved the position. Sure he fought long and hard battles for all the employees. Just the same, there was a lot of talk about the fact that the promotion looked questionable and many people—including union members—saw it as a bite out of an apple. Alice's brother didn't anticipate the gossip:

> *It didn't last long fortunately, but I heard there was talk. If a union works for its members, everybody wins. The negotiating team didn't figure on any politics from our own people, but by promoting our twelve stewards we got more than a say in management. The higher salaries bothered some of our members and they looked at all my union work as though I was planning on this. And because Alice was on the team that didn't look good either.*

Summary

- There are predictable paths of communication within a company. These *circuits* are directed downward to employees and upward to supervisors. Probably the greatest volume of employee communication is horizontal talk with coworkers.

- Downward communication defines important directives from managers.

- Upward communication allows employees to address their interests as members of the company community.

- Horizontal communication among coworkers resolves the day-to-day needs of production and creates a supportive social context.

- Networks are the unofficial channels of power, and the popular use of the word *networking* indicates that there are few employees who do not understand the concept.

- Circuit communications follow clearly defined patterns, but networks are happenstance and can be subject to criticism.

- From the company viewpoint, every employee represents dependability, service quality, and appropriate production standards.

- Any employee who fails to meet these expectations is considered an uncooperative risk by supervisors *and* by coworkers.

- An employee must strive to meet company expectations and must recognize any conflicts if he or she cannot match those demands.

- An employee should deal with personal conflicts to his or her satisfaction, either by resolving to adapt to the expectations or by fulfilling personal expectations in some other setting.

Activities Chapter 2

Present a memo to your instructor that develops a discussion of a work experience that relates to the chapter. The memo should be 500 to 1000 words, typed. In it you can explain how you have related the text or the lecture discussions to some event you recall. This exercise will give you a better understanding of the material because you will explain incidents in terms of your perceptions of workplace structure and communication.

Select one of the following suggestions and develop an analysis that involves your current employment or a former position.

- *Explain the proper channels of communication where you work. Identify how these channels relate to the power structure. Discuss any memorable violations of these practices. Include an organizational chart from the company or develop your own. (Military, police, and fire department environments are interesting.)*

- *Recall a conflict in information flow, either downward or upward, or horizontally. Discuss the events and the outcomes.*

- *If you have ever used networking to achieve a personal goal in the workplace, explain the event.*

- *If you have ever been outdone by someone else's networking success, explain what happened.*

- *Explain the training methods of various former employers, and explain why the training was significant. Include informal methods, since small businesses do not use seminar training and similar training procedures.*

- *Identify the events involved in a situation in which an employee became rebellious. Discuss the problem. Explain the compliance difficulties and the final outcome.*

- *If you have ever been either an overachiever or an underachiever in the workplace, discuss the manager's reaction and the reactions of your fellow employees. You may also discuss this issue with regard to another employee's ambitions or lack of incentive.*

- *Discuss an adaptation problem in which an employee had difficulty coming to terms with the expectations of the employer or fellow employees. Examine the motives of the various parties and indicate the outcome.*

A Reminder: In this memo, and in all subsequent exercises, please feel free to use fictitious names for employees and companies.

Work in Progress

3. The Public World

Something you quickly learn when you own a small business is that basically you are the business in a very real way. Customers may need to have networks installed or upgraded, and there are dozens of companies who can do the job, but why do they choose my business over the others? I think they choose my company because they are choosing me. They are buying the image that I project about the business I own. The actual service I provide is another matter. If I was referred to them by former clients of mine, I am confident they will take my offer seriously. Since image counts, I can plan accordingly and tailor my business image to be whatever I want it to be. What you see is what you get!

I am most comfortable in a casual and relaxed, yet efficient, environment, and that is what I project as our NetWorks NorthWest company image when I sell our services to prospective clients. My goal is to come across as the confident professional who can take any job in stride without making an emergency out of problems that might crop up. My message to a client is that if they hire my company they will have work done by people who are thoroughly experienced, easy to work with, and committed to making the installation or upgrade as painless a process as possible. When I am bidding a job, I do my best to look the part as well as act the part. Rather like an actor wearing a costume, I wear casual clothes that are neat and tailored, and I forego any fancy jewelry or frills. I want my clothes to reinforce the image of a casual but serious pro with an eye for detail. These are small things, but taken together they add up to the picture my company makes in the mind of a potential customer.

Owning a business like mine, where I have to be the salesperson and "executive director," is complicated because I do the actual work of an installer, which requires me to wear a number of hats. This can get interesting. When I am in a white-collar mode I wear one set of clothes, and when I am grubbing around in crawl spaces laying cable I wear another. The difference can be disconcerting to the people who have hired me, but if I have done my job of setting up my image as a "can-do" kind of person, the problem is minimized. I also take into consideration when the job will be done. If it is scheduled during their working hours (definitely not an ideal situation), my employees and I will wear work shirts with the company logo on them and nice jeans. If it is a weekend or off-hours job, we wear whatever work clothes we want.

I have found over the years that all the thought and effort I put into my business's image have paid off. People will see what they want to see and will make whatever conclusions they will about you and the work you do. However, if you are conscious about the subtle messages you are communicating, you can at least stack the deck in your favor.

J.Q.C.

S Perception

Social interaction is a remarkably complex phenomenon. We walk a precise and narrow path if we make a concerted effort to gain the respect of both managers and coworkers. Since their values and perspectives differ, it is a challenge to maintain the balance that has to be maintained in the workplace. Top-down forces, horizontal forces, and bottom-up forces exert strong influences at work. Beyond these rather one-dimensional roles, however, there is the far more complicated matter of community. A workplace contains a social group that involves all the complex interactions of any community. We must interpret those around us. We interpret their behavior; we interpret what they say. Always, we assume that our perceptions are true and that only our judgments could occasionally be at fault. This isn't quite the case.

If we remove power from the discussion of communication, we can remove the directionality of much of the interaction as well. Even then, the complexity that remains is unchanged. If anything, power and direction help define types of communication. If we take away these elements, the communication is no less involved, because all the actors remain on the stage. Much of our daily interaction in the workplace can be shaped by a great many other conditions. We must look at certain *basic* categories of these conditions to understand why there is enormous complexity in workplace interaction. In particular, we need to look into the sources of communication *conflict*. Why do we understand each other? Why do we misunderstand each other?

If you were supposed to meet a network technician at the computer lab at 1:00 P.M. and you arrive at 2:00 P.M., there is a fundamental problem. This error is incidental in that it is so commonplace, but it is frustrating for both of you to have missed the connection, and it is ultimately costly. Miscommunication in the workplace is an overlooked expense in an accountant's bookkeeping. If we analyze the event, we realize that you both understood part of the message. The technician was in the lab at 1:00 sharp. You showed up late, but obviously you both understood where to meet. The misunderstanding was the hour. Perhaps someone distracted you as you listened to the technician. Maybe there was a lot of noise in the hallway. Possibly you depended on remembering the time rather than writing it down.

In one sense this small event suggests details concerning interaction that are quite mechanical. What did she say? What did you hear? Was the event loud and clear? At this basic level there are many considerations to examine. Communication between two people depends on *transmission* and *reception*. It is worth our attention to look at the forces at work in this fundamental output and input transaction. There are ways to improve the transmission and the reception of what we communicate.

At the same time, any communication between two people is marked by the most complex human behavior imaginable. Our interactions are not simple matters. There is a transmission and there is a reception, but as our simple lab appointment demonstrates, our ability to be confused by even the simplest and most straightforward comments is baffling at times. At least you are probably a native speaker. That eliminates not one but many complex issues that can cause miscommunication. Perhaps you run late most of the time; this habit is guided by very complex inner workings of the mind. Perhaps you wanted to meet John, the other technician, and gave less importance to meeting Shirley, who was available at 1:00. Perhaps you are a male and thought less of Shirley's skills for some reason that is not explainable.

It is easy to muddy the waters of workplace communication if we introduce the social, historical, emotional, and even global forces that are at work around us. Although we do not want to lose sight of the company matrix in which all our communication is suspended, good communication is an unquestionably larger matter than company policy. Our every contact is a complication based on our human community. Although that community is a workplace, it consists of its members, each of whom has individualized perceptions that are influenced by realities beyond the workplace. We will have to examine a host of these realities. If we perceive the complexity, we will find one very important path to understanding: *awareness*. Awareness is perhaps the simplest route to improved communication.

First, we need a baseline for these many communication difficulties. Perhaps the most practical place to begin is at the level of perception. We must stop. We must look. We must listen. Let's assume that you are absolutely certain you were to meet Shirley at 2:00. She said so. She in turn insists it was a 1:00 appointment—and that it was your suggestion. Well below all the intersecting social forces of community are the realities of what we think we see and what we think we hear. We are, in some sense, victims of a very faulty system of tools that make up our awareness of our world. We might not experience the truth of that world at all times. We are supposed to assume that either you or Shirley got the time confused. We can take another perspective and assume that you are both correct. This may be the best way to acknowledge our basic human condition. At a basic level, the issue is not even community. The basic issue is what we perceive to be true. The truth then means that both of you, alas, are correct.

Coming Alive

Basic person-to-person communication is a vital area of service skills, and conversation is one key to your success. "Talking" says it all, and talking is actually a very complex process. **The private world of the mind is the point of origin for any thought we present in the public world of communication.** We must look at this fundamental human condition first because individual isolation is so much in contrast with the community of busy humanity that is constantly building societies around us. We seem to understand what goes on "between" us, but life for us is not really so simple.

The "expression" of ourselves is an absolute condition of our humanity, even without reference to communication. Death is an appropriate counterpoint; it is expressionless. There is a mental illness that results in a total lack of animation: catatonic schizophrenia. The catatonic patient is rather like someone who is in a coma. There is no one there for us to understand or respond to. Both of these conditions take on the aspect of death in that there is no inner self that is being constructed for us to see. The lack of animation is a wall to our understanding. A body in a coma is as great a puzzle as the silent Sphinx sitting in the desert. In these circumstances, the personality ceases to exist for us, and regrettably, perhaps the victims cease to exist for us as well. In this respect our animation is "who" we really are, or what we really are.

Our existence depends on more than three dimensionality; we exist largely in the pleasure of the fourth dimension. Time is movement. We exist in activity. We laugh, we cry, we live, we die—and we work. Animation is the substance of meaning for us. We are actors acting out our lives. No matter how we define someone, the definition will be based on the *activities* of that person for the most part. If I ask you who you are, you will find the question peculiar. The immediate response is usually a name, which does not mean much. In casual conversation, talk has a predictable tendency to quickly shift to what you *do*. We discuss your *work*. Conversation may later reveal that you are also a wife, a husband, a parent, a mother, a student, a bowler, and so on. All these aspects of ourselves are largely projections of *activities*—of parenting and fishing and so on.

Who you are is also what you are willing to *share* with people. We are in great measure what people experience of us—at least for them. We play out roles that they perceive or experience. This creates a peculiarly lopsided set of "selves" for us at times. Your coworkers are not likely to see you as anything other than a coworker. If you grow roses and belong to a rose club, the members of the club will see you as a rose grower. Both aspects of yourself are elements of you, but the perceptions of the people around you can become skewed. At times, indeed, people may be shocked to find that they are speaking about the same person, since two people can experience a third person in very different ways. We wear many hats and alter our behavior somewhat to fit the circumstances. Those around us then see us in different lights. We become a series of intersecting realities, some few of which are shared by everyone who knows us. You are the sum of your activities but only you ever see the total.

We are always *more* than appearances. Appearances frankly do not give us much reality. "I know him to see him" is the sum total of our knowledge of a person we have not met— a person we have not experienced—because our perception of people is a perception of activity. Put another way, we always exceed the sum of our parts as others see us. What we *do* with our body mass is exactly what we come to be for those around us. Our activity is the key to the way people will see and define us. We exist in a world like spinning toy tops. We arrive in a whirl of animated dust and never stop—until we do.

In perceiving other people, we depend on their animation to define them. We go so far as to give life to inanimate objects. The play activity of a child who has a doll or a teddy bear is a method of play acting in which the child gives life to the doll or bear. The child animates the inanimate. Otherwise, the little friend has little meaning for a child. And when we, as adults, hold long conversations with our dogs and cats, we are conducting a similar activity. Pets are quite nonhuman, but we add human activities to the relationship. Talking to a dog assumes that the pet is going to understand even if the animal does not talk back. No matter, we *project* to them our moods and perceptions, and hope to see in them the desired responses, even if we have to create the response as well. We build dialog where there is none—or perhaps a little, to give our pets their due.

In sum, we want somehow to emerge from ourselves. We are trapped inside our minds. We struggle to construct some sort of self to send out. It is all so obvious that we may not realize that the way we express mental activity is the way we create exhibits to "show" thought. We think of ourselves as human beings, as *objects*. Rather, we are humans *being*. We are people *performing*. And we see each other in terms of *action*. What you give us is all we get. Your public self is you as far as anyone is concerned. The words, the voicing, the body language, the image, and everything else you give us to experience is the only known result.

This concert of articulations, this maze of expressions, makes up your behavior as you project it. It is you, always emerging for us to see, always developing. Of course, it is obvious then that all you give us is all we can get. Your public self is you as far as anyone is concerned. If you want your boss to see that you are a wonder worker, you had better work wonders.

We show ourselves for what we are, but we also show ourselves for what we are not. We manipulate the image to create impressions. We try to look relaxed when we are nervous. We try to look richly dressed even though we shop at the discount stores. We wear baggy clothing rather than diet. Obviously, it is in our self-interest and for our self-satisfaction to mold our image as we project it for others. We design it as we go, with a perfectly healthy interest in looking our best.

I introduce these very basic comments so that you can understand the role of your spoken language in "acting out" your everyday life. You speak what you have on your mind. Speaking is your thought—inside out. Speaking is usually the most precise animation of your thoughts. Of course, speaking with words should not be viewed as your only public activity. It may not even be your dominant public activity. Certainly it is only one of many vehicles you have for expressing thought. The activity of speech is only one of the activities of your animation. You can speak thoughts and you can perform thoughts. Thought involves a vast arena of components and processes that are not necessarily handled by the mind in words, or in an articulated structure of words. Similarly, your speaking activities range far beyond your need for expression in words and word structures.

However, **"expression" is only half of the act of communication: sending the message is only half of the intent of expression. Receiving the message is another matter.**

A World of Illusion

Before we look at whether we accurately understand people during communication, we should look at how accurately we understand anything at all. When we think, we create a world of illusion and call it reality. The first variable we should examine is the scientific reality of perception. We know we bias most of what we take in during communication with another person, but the question is whether we are perceiving reality in the first place. Our ability to perceive is limited, and perception errors are very common. You have seen X-ray photographs. When you look at an X-ray, you are seeing something we do not see and never will see. Is the photo real? Of course it is, but we cannot "see" what the X-ray machine can see. We operate in such a narrow band of light that it is amazing we know the world as well as we do. We cleverly invent devices to present us with perceptions of the broader spectrum of light in perceptions we can understand—or see. We invent infrared cameras, for example, to show us an infrared world. But if the light spectrum is so vast, then obviously our understanding of the world around us is fractional. We operate on a hint of the actual fullness of things.

The extent to which we can know reality is a long-standing debate. The entire science of psychology dwells on the question. For our purposes here we can state that our awareness is greatly restricted by our physical limitations. Fortunately, the most significant objects of our attention, our fellow human beings, are operating in the same way.

The first issue really is not frequency response—as in the X-ray example. True, if the frequency of a sound is over 20,000 kHz you will not hear it, but we are overlooking the obvious: first you have to be around to hear. We usually are not. We are very small fry in the scheme of things. Even at work, if you have fifty employees working eight-hour days, sixty minutes to the hour, and you calculate the sum of how often all the employees speak to one another, your part in perceiving all this reality is staggeringly small. Much of your perception of what goes on is indirect and consists of what I call "echoes." "Hey, did you hear that Ray got canned?" You are told what happened, what is happening, what will happen. Almost all of it is elsewhere. In sum, the slice of reality you see is quite small, and much of that is echoed as secondary comments that you experience.

What we know of the world depends on the limits of our sensory capabilities, our amount of exposure, and the trust we put in echo information. We are prone to errors in our perceptions for other reasons also. Consider Beth. Beth did not see the new stop sign on her street corner one morning and drove her Bronco right through the intersection into another car. The stop sign is bright red, a warning color in our culture. The big letters of all roadway signs are the result of studies that show us how big to make the letters to be effective. And they are always predictably located—to the right of your lane for example. How could Beth not have seen the *new* sign if it is designed for our perceptions? Well, it was not there. Of course, she cannot tell the police officer that it was not there, so she explains that she did not see it. There is little difference. It still was not there, for her, because of patterned behavior, predictability, call it what you will.

Our perceptions, primarily our visual perceptions, can be in error for many reasons. Simply the fact of forgetting a pair of glasses can dramatically alter the perceived world, but there are other problems that have little to do with 20/20 vision.

Illusion

One consideration is that we may not see what we are seeing. Mirage is the legendary optical illusion of the desert sands. Such genuine errors in what the eye perceives should be the first consideration on our list of perception problems, not because such situations are common, but because they are so real. World War II pilots who ignored their altimeters often flew to their deaths in overcast weather because of a combination of sensory responses that made up seem down and down seem up. Who can resist the temptation to keep a foot on the brake at the car wash? The water sprays and the brushes create the illusion that the car is moving. Point your two index fingers at each other one foot in front of your eyes. Let them touch. Now look over the tops of the fingers and focus your eye on an object several feet farther away. Notice that the fingers now create an odd illusion, a little sausage of a finger that hangs in the air suspended between the index fingers.

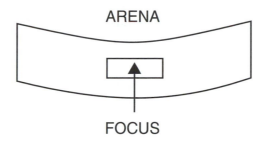

ARENA

FOCUS

Selectivity

The second problem is that we are limited by the focus of the eye. Of the full width of our vision, only a small area in the center is in focus, and *is* the focus. The eye moves by what is called "psychomotor" control. Let's simply say that the mind determines what deserves attention. The attention area gets focused and centered. All else moves to the larger field of vision.

One possible explanation for Beth's accident is that the usual focus in driving is the immediate path of the vehicle. To *really* get the stop sign into her mind's attention it would have to be moving. Think of the speed with which you respond—let's say by ducking—when your eye perceives a motion on the periphery. It is an interesting demonstration of reflex action. The perceived motion is very important; otherwise we favor only the central focus of our attention—the front center of our field of vision. All else moves to the larger field of vision. We respond instinctively to our flanks but otherwise ignore the fields at the edges of our attention.

Expectation

We often create patterns by conditioning. We see what we know should be there. The following illustration is well known, and you have probably seen it before.

You probably will not see the mistakes in the figures initially, which means you are making a visual/mental error. You see the graphic incorrectly because you are mistaken. The perception was "miss taken," or taken incorrectly. Conditioning caused the error. You see what you know *should* be there.

The easiest place for you to experience this phenomenon is not in conversation, but in writing. As a college student you probably have written a number of projects. You probably expected to have a few errors in the projects, but you may not have noticed that careless errors are among the most common errors. However, "careless" means you were sloppy in your preparation, which is fairly uncommon. You will see errors you would *never* make, such as omitted words, for example. The cause is memorization. You saw what *should* be on the page rather than what *was* on the page.

You memorize the document as a result of working on it too much. You want it to be perfect, but there is a force fighting against you. It takes time to perfect your project, but the time you put into the project increases your memorization, and then you do not *see* the errors because you scan the text rather than read it. And Beth? She certainly do not *expect* to see a stop sign. Her conditioning *knew* there wasn't one, so why look? A blinking light would have saved her Bronco because movement would have made her aware of it. Hazardous intersections use blinking lights to encourage our awareness of the stop signs.

Organization

One of our great gifts as human beings is organization. Human intelligence can be viewed as achieved organization. If you see a tree, you think "conifer" or maybe "fir." You organize the tree. You put it in a box of kindred trees, all of which have needles and do not shed greenery in the winter, and so on. Entire philosophies such as Zen Buddhism are based on the efforts to "see" without systems such as organization. Organization is an obsessive activity for us. It is play. It is work. It is art. It is science.

From childhood on we teach organization. As an infant, my daughter had a set of three or four round plastic rings of different sizes that were shaped like donuts so a child could harmlessly taste them, touch them, and put them on or take them off a little post that looked somewhat like a candlestick. It must have been fun to try to see what bright green tastes like, or find out how yellow feels, but infants quickly learn that colors do not taste or feel. But the rings could be *organized* on the post from largest to smallest in a pyramid, or from smallest to largest, in a reverse pyramid, or randomly. Kelly never liked the random look. She *always* made a pyramid! Look at the ten lines below and ask yourself what you see.

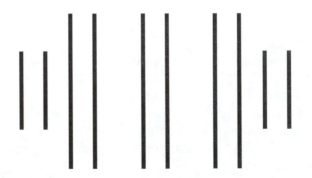

You organize by habit. Look at the following boxes and ask yourself what you see. Look at the sample on the left and focus on the square composed of four circles at the upper left of the pattern. The square includes the empty circle in the corner and the three adjacent black circles. It is *not* the way you would probably see the various relationships you perceive. It is almost difficult to see because the four circles seem asymmetrical as a group.

Historical experiences

The largest single factor in our perception of what we experience is the psychological factor. We will dwell on this aspect of perception extensively, so here we will consider only one basic illustration. Everything you see will have color. It is thought that there may be some *universality* to color perception. Light greens are supposed to be relaxing for example. Then there are *cultural* perceptions. Americans seldom appreciate the traditional black and red decor of Chinese restaurants. Black and red are not happy colors for us. Color values are different in China. Finally, there is the historical shaping of our *personal* responses to color. We tend to like or dislike certain colors. We will usually wear our preferences right on our backs. In sum, if you and I look at a color together, we will see the same color more or less, but the reality of that color stops there. Beyond is the great land of the mind's interpretation of the color. My perception will not be yours. Your perception will not be mine.

I have used visual perception as the basis for these comments about what we perceive and know of the world around us, but our perception is also based on all our senses: what we touch, smell, taste, hear, and see. The other senses share similar personal and historical influences. We have to somewhat reverse priorities to "see" the other senses in perspective because we are visual at the expense of all other sense responses. Nonetheless, our other senses are very influenced by historical elements. For example, I do not like to touch sticky stuff. To me, to my historical experience, jelly between my fingers is unpleasant. Of course, touch can also be universal; nobody has a pet slug. Sliminess seems to be particularly universal as an unpleasant sensation. Okra and oysters are the brunt of plenty of humor. Other sense preferences are also cultural. Countries either cook with garlic or do not, and never the two shall meet, as we all know. Cumin and other unique flavorings can stimulate one appetite and repel another. Perfume and cologne are far more personal matters. A man or woman will wear a favorite scent with the idea that other people will enjoy the scent. Certainly no one wears a scent to anger their friends and coworkers, but oddly enough, people fail to realize that a scent can perfectly well trigger annoyance—not to mention allergies.

In the end what is real is mostly what is real for each of us, and we bias the reality we experience. Since this reality is not entirely shared with anybody, and since anyone else's understanding of the world is also colored by historical experience, then mutual understanding between us is quite a challenge. Mutual understanding is the focus of our interaction as people. Everything we do is, in some sense, a disclosure of ourselves for others to interpret.

The work of understanding each other is played out in our behavior and communication efforts. We talk to share. What we share are our perceptions. We then see our differences. We can be wrong in what we perceive. We often are—partly because people tend to think that a shared experience is understood in the same way by all the people involved, which is just one more example of illusion!

Summary

- Your public self is the only image of you that most workplace acquaintances have.

- Every person manipulates the self-image that he or she projects.

- In the workplace, an employee will usually use a great deal of control because the projected "image" is an investment in employment security.

- Because the employee is busy with specific tasks in an enclosed environment (usually), the employee will have little direct experience of other areas of the company setting.

- As observers in a workplace setting, employees will know surprisingly little about most acquaintances. The information is often indirect because they have been told information about other people.

- Second-hand information is highly prone to distortion and misunderstanding.

- Employees closest to a task or task environment will be the group that has the most authentic understanding of one another. They relate more seriously, more often, and with direct information about each other.

- An employee does not see his or her work experience in quite the way others will see it. Life experience becomes part of the work experience in that each person sees the work world from his or her own perceptions. For the most part, those perceptions were not shaped by the workplace.

- Error and illusion complicate the seemingly simple world of work relations. What we think is true is not always correct. What we see is not always accurate. What we hear is often distorted. And all of us do our part to manipulate what we show others.

Activities Chapter 3

Present a memo to your instructor that develops a discussion of a work experience that relates to the chapter. The memo should be 500 to 1000 words, typed. In it you can explain how you have related the text or the lecture discussions to some event you recall. This exercise will give you a better understanding of the material because you will explain incidents in terms of your perceptions of workplace structure and communication.

Select one of the following suggestions and develop an analysis that involves your current employment or a former position.

- *Explain a professional misunderstanding of blueprints or schematics or similar technical graphics. It happens often. For example, the new classroom building across the patio from me had a concrete wall on the north side that was the result of misreading the blueprints. It took two days of jackhammering to remove the 10-inch-thick concrete misunderstanding.*

- *Discuss an accident that was caused by a visual miscalculation.*

- *Study several employees: one you hardly know, one who is an acquaintance, one who is an old friend. Why is the first one-dimensional? Why is the second two-dimensional? Why is the third three- or four-dimensional?*

- *Think about the differences between a very popular employee and an employee whom nobody seems to notice. Explain the dynamics of the way they project themselves and the way they are seen.*

The Mechanics of Talk

4

Work in Progress

4. Conversation Confidence

When I started my company, I never thought that I would spend as much time being an amateur psychologist as I do being a network engineer. Offering a quality service and competent people to do the job certainly goes a long way to creating business for my company, but this alone would not make the company grow or build our reputation. For that, I have to rely on my skills at figuring out what makes a potential client "tick" at the very beginning of the client relationship. I try to deal person to person and not just contractor to client. The interactions I have with this potential customer will follow most of the same lines as the nonbusiness interactions I have with other people, and so I focus on the fact that clients are people first and foremost. There is more to a contract than a business agreement.

When I go in to bid a job I keep in mind that I am meeting with a person (sometimes several), and that person will be reading me as much as I am reading him or her. This is where the psychologist part comes in. I know that the client in front of me will be trying to judge me to get more of a "sense" of my value than the bottom line on my proposal. Do I sound shaky or unsure, or do I come across relaxed and confident? Is my presentation canned, or do I really know what I am talking about? Can I clearly answer any questions that pop up without being defensive? Since I know these ideas will probably be running through a client's mind, I can prepare myself accordingly.

If there are details that I might muddle or forget, I go over them before a meeting and write them down to take in with me. If I notice that clients are looking the least bit confused or skeptical, I stop and ask if what I have said so far is clear and if they have any questions. If I don't immediately know the answer, I am honest and tell them I don't but would be happy to do some research and get back to them. I want the customer to trust that I will not go ahead with something that I am not really sure about and that I will be confident enough to get more information.

I also try to tone down the technical jargon as much as I can because it can get in the way of communication with clients without a technical background who might feel too intimidated to ask questions. These unanswered questions oftentimes are the very things that come back to haunt you later when you are in the middle of the job, and a client says, "But, I thought you meant . . . ." That is definitely not the time to try to figure out what your customer thought you said. Clearing things up at a late stage of the game can mean hard feelings on both sides, and that is not productive for anybody.

Finally, I use an old technique to make sure my client and I stay on the same wavelength and have a clear understanding of what is needed and what is offered. I periodically repeat the customer's concerns in my own words. "So, what I hear you say about the second-floor computers is . . ." This way, if I totally missed what the client meant, the person can correct me right away rather that after the job has started.

J .Q. C.

58

4

The Communica- tion Loop

The last chapter suggested that we are, each and all, islands unto ourselves. We try to communicate to overcome our isolation. We are stranded observers trying to understand the world beyond us. At the same time, we seem to want to be citizens of a human community, which is probably why we try to understand all the other members of our fold. Our method of meeting the challenge is fairly obvious: conversation. It is communication that creates community.

We can look at the tools of understanding and communication in very basic terms. Speaking activities consist of two skills: wordcraft and stagecraft. One entire structure of the act of communication depends on language, or language shaping, or wordcraft. Talk is the application of language. Word symbols are the tools of our trade. Another entire structure depends on acting out the language that we communicate.

When we talk to one another, there are very distinct processes at work that make the communication happen—and they are very predictable. There are six distinct stages involved in my simply saying "Hi" to a coworker in the lab. The first three stages involve me as the speaker. The other three stages involve the other person as a listener. The stages are easy to experience and understand, but they occur so quickly so many hundreds of times a day that you must pay close attention to catch them in action. A normal nerve in your body will conduct a message a distance of fifty-five meters in one second. You will perceive and translate the world "Hi" just as quickly. We have to put ourselves in slow motion to see the event properly.

Paul meets Candace in the CAD lab and they start to discuss the merits of electronic design systems

Paul:	"Pads is the industry standard in the field of PCB design."
Candace:	"Maybe, but for volume sales Pro/Engineer is the big system."
Paul:	"And AutoCAD?"
Candace:	"AutoCAD started it all, and they are still competitive."
Paul:	"How is it that Solidworks is so popular?"
Candace:	"Invitive command structures."

In the brief diaglog above there are six comments. You would think that the structure is simple enough, but notice that four of the comments are judgments. Two are questions. Furthermore, each of the six comments goes through a process involving stages of construction and delivery and interpretation. The brief dialog goes through thirty-six processes—all in the 20 seconds it takes to complete the dialog.

If you say even one word to someone, the communication process is initiated and executed. It is composed of the six stages of the communication loop, which is called a *dyad*. A dyad (think of dialog) is the basic communication level between two people. The process is illustrated on the right.

The Loop: The Basic Stages of Communication

SPEAKER

A. *THE THOUGHT PROCESS OF THE SENDER*

First, your thought decides on a comment.

B. *THE MESSAGE-SHAPING PROCESS*

Second, you turn ideas into symbols such as words. This is called encoding.

C. *THE DELIVERY PROCESS*

You channel the message by way of sound, visibility, contact, or some combination of sensory signals.

LISTENER

A. *THE RECEIVING PROCESS*

Some percentage of your signals are received by the senses of your listener.

B. *THE MESSAGE-TRANSLATING PROCESS*

The person you spoke to also structures the signals into symbols. This is called decoding.

C. *THE THOUGHT PROCESS OF THE RECEIVER*

The person then works on the symbols to understand your ideas.

These six stages are being used at an amazing clip when we speak to one another. It takes less than a second to say Yes or No, but the stages must be handled, one by one, so all six phases happen in an instant. There is plenty of room for error, as you can see. Speed is a classic difficulty in itself. When we use the expression "you put your foot in your mouth," we are overlooking the speed at which it got there! **There may be many reasons for saying exactly the wrong thing, but the usual culprit that lets you do it is velocity.**

Apart from rapidity, of course, we must look at each of the six phases of the activity of a dialog to understand how each phase can create problems. All six phases are important to communication, and all six are problem areas for understanding even the simplest comments between two people. Each of the six elements of dialog helps to shape (and alter) meaning.

You can look at the concept in simple graphic terms:

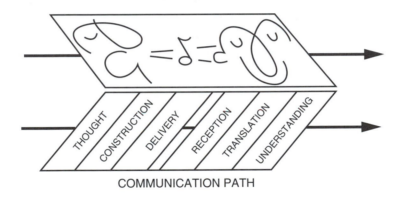

COMMUNICATION PATH

The Thought Process of the Sender

No two people look alike. We do not think alike either. Every life experience is a separate one. Chris grew up in Pennsylvania, and his neighbor grew up in Kansas. Both have lived in Seattle for many years and Chris has always enjoyed the city. It is a fairly hilly city. Pennsylvania was hilly also. His Kansas friend has never liked it here. "Can't see anything" is his only comment. You have to think about the comment to understand it—unless you are from Kansas. Or perhaps it is an issue of personality: the hills can be seen from the top or from the bottom. Some people see the glass as half full and others see it as half empty. We are conditioned by an infinite variety of life's circumstances, and those circumstances condition and shape our thought. Everything behind us shapes the way we see everything in front of us. Most natives of the northwest love the cool, overcast climate. Sharon is from a sunny, arid climate. A perfect day for Seattle natives does not come close to her preferences. The locals do not understand her complaints.

The Message-Shaping Process

Shaping the message can be a problem. As we turn ideas into symbols we do so on the assumption that all the symbols will be shared. Obviously, if someone does not understand your terrible Spanish in a cafe in Albuquerque, your efforts to keep the chili peppers out of the dish you have ordered could be a calamity. And if you wink at someone, will he or she take the symbol the way you intended? As much as any force at work in the communication process, our "vocabulary" is a prominent stumbling block—probably because we have a habit of assuming that everybody shares our vocabulary. This can be a huge problem if you are a field service representative for an industry that depends on selling technical applications to nontechnical businesses. It is hard to "sell" something people do not understand.

The Delivery Process

Most of what we send as a message is shaped in sounds and sights. We use our voice to designate and define the symbolized ideas, and we add many visual elements also. This mixture can get very confusing for some specific and peculiar reasons. Some authorities observe that we are as much as 80% visual. This does not leave much of a focus on sound, and yet sound happens to be our most organized and complex method of communication because we speak and share a language! This contradiction probably says a lot about why communicators have trouble communicating. Of the six levels we are discussing, the one you probably thought you could count on would be the strength of your voice—the delivery process. **We need to learn that people do not so much hear what you say as much as they *see* what you say.**

You can easily test the visual dominance of our focus of attention. When you have friends over for the evening, be rude and leave the television on. You can turn the volume down, since the sound probably will not make any difference. Now, watch what happens. And watch you will. The evening will be difficult at best and probably disastrous for conversation—unless you talk about what all of you end up watching on the television, even without the sound. It is too difficult to resist the visual cues of the movements perceived by the eyes. We depend on our eyes for everything. Only in the depths of a black night will we really turn to our ears for help.*

The Receiving Process

At this point we turn to the process involved in the *listener's* activity. The listener does not simply take it all in. The television experiment demonstrates that it is not that simple. Up

---

* As an interesting footnote do not forget that we humans do not share the same sensory focus that we find in, for example, our pets. At night, a cat's eyes have three times the acuity of ours. Authorities argue about the exact power of perception of a dog's nose, but there is no question that a dog's nose is dozens of times more aware ("more sensitive"), and possibly hundreds of times more aware, than ours. We cannot really imagine such a world. And dogs, by the way, often have poor vision and have monochromatic vision (no color).

Since the dog is not a beast of burden or part of our diet, we have to ask ourselves how this particular mammal became our best friend. The answer is likely to be the purpose a dog served, and still serves, in being the ears and nose we do not have. They protect us. In return, with our eyes, we round out a fairly convenient world for the two of us. We protect them. If the dog scented the prey or heard danger, our ancestors responded because they had the keen eyes to use their bow or their spear. (All projectile weapons are sight dependent.) At a very fundamental level we share our skills with the dog for survival: he is fed; we are safe.

to 70% of all perceived sounds are ignored. Suppose several people are talking at once. Suppose there is noise, industrial noise perhaps. Suppose someone has a hearing problem or forgot her glasses. Problems abound if there is any loss in the total available signals reaching our senses. On the other hand, a surplus of sound, particularly in the case of noise, can be just as bad.

The message-translating Process

This phase represents the effort to properly understand the speaker. The decoding process is what might be called "first-phase understanding" in that a listener will try to determine what the speaker meant before trying to do anything with it. This is more or less the issue of getting the symbols straight. If we misunderstand the vocabulary, we can misunderstand the entire intended communication. If we misunderstand, we will not understand instructions, explanations, commands, details, and so on. "Tommy, get me the pliers." Tommy goes down the ladder and fetches pliers for his dad. He climbs all the way back up to the roof and hands his father the pliers. His father grits his teeth. "Not those pliers. I wanted the blue handles."

The Thought Process of the Receiver

To suggest the difference between simply *translating* and then *processing,* recall the comment about "first-phase understanding." I call this final level "second-phase understanding" or the processing stage we use to analyze the message.

First, the symbols are looked at more or less without interpretation. The second phase is the moment of *interpretation* or the moment at which we bias the meaning. It is the point at which you decide what a comment means to *you.* For our discussion, we are forcing the dyad sequence into slow motion, but your neurons are wired for speed. You are reading this sentence without much sense of a first-phase process. It happens too fast to be obvious, but the moment you do not understand or do not agree, the distinction between phase one (translation) and phase two (understanding) becomes quite clear.

For example, Holly's husband tends not to spend much time with her. He loves her. He is a very contented guy, but she wants to spend more time with him. If Holly says "We don't spend enough time together," there is a problem. If Scott shared Holly's *concept* of time spent together, she wouldn't ever have to bring up the issue. Her comment that they don't spend *enough time* together won't make sense to Scott because Scott obviously spends plenty of time with Holly—as he sees it. Holly has one concept of time—and probably affection and love and so on. Scott has another. The vocabulary can hardly work, since both of them will operate at a personal level of interpretation to explain the expression "enough time." The first phase of understanding is the translation, but the second phase, obviously, takes over, since the term will *assume specific meaning* only by interpretation. Scott responds,

Holly we spend plenty of time together, and you know it.

No, she doesn't know it.

The entire process of communication is a matter of mind over matter. Errors occur in the process. We need to see the communication activity in its parts: three distinct parts of a process in delivery, and three more in receiving.

These six elements of the communication loop are what a conversation is all about. You need to be keenly aware of these six stages because everyone depends on them as speakers and as listeners. The stages are likely to be reliable, but they are just as likely to be variables that shade or color an interpretation of communication.

The dilemma of speaking is essentially the challenge to construct ideas and translate them for others. It is a construction problem. Imagine a species of beings for whom everything that goes on, goes on in their *brains,* and nobody else can know what is going on there. It sounds like science fiction because we ignore the fact that it is the human condition.

The Drama of Communication

Up to this point the communication process looks fairly analytical, but a tidy six-stage process is not the stuff from which we are made. The world *is* a stage, and we are out there for all to behold. The communication pattern is real enough, but the players on our center stage are complicated, as we know. We should now look at the actors a little more closely.

There was a time when theater acting was a stylized discipline. Not just anybody could act out a stage drama because the gestures and voicing had to be visible at quite a distance. When silent movies came along, the early movie actors and actresses came to the film studios with stage training. The result is the comical acting we see in the old silent movies. Of course, the exaggeration of early movie acting helped audiences understand the plots, since there was no sound track. The huge gestures of the face and body were no longer needed for visibility in a theater but were used as a replacement for *speech* in the silent movies. The early motion pictures were an odd situation for an actor. The casts consisted of actors without voices.

Once sound came along in the early 1930s, acting slowly lost its pantomime or exaggerated quality, although you will still see the old tricks when Fay Wray sees the original King Kong. Rather than scream at the top of her lungs, she will show her horror by turning the back of her hand to her lips in a grand sweeping gesture you could almost mistake for a polite yawn. But in its day, such gestures *showed* real terror in a symbolic fashion. Sound was not the only reason for the change in acting styles, however. Cinema technology advanced on other fronts. You might not notice that silent movies did not use close-ups extensively. The camera was a rigid device that did not move, and close-ups of Valentino kissing the heroine had to be spliced into the footage. In the thirties the camera was given more portability. First, the old tripod mounts were replaced with tracks and wheels, later the boom was added, and along the way optic technology led to the development of zoom lenses. The moment the close-up became a practical working reality on the screen, acting reverted to the world of everyday behavior. Actors and actresses ceased to be stage performers.

We take for granted the way in which a film is a mixture of shots representing amazing changes in physical scale. Clint Eastwood approaches on his horse. He stops just outside of town. Cut. Now his face, thirty feet wide, shows up on the screen, and we see every hair of his eyelashes as he suspiciously squints through the blinding Western sunlight to see if any gunmen are on the rooftops. The film goes on in an endless confusion of long shots, close-ups, and everything in between. This activity *should* be very confusing for the viewer, but it isn't. Why? Because optic technology has reached a point where it can imitate the eye, and the mind, and the resulting film editing imitates our own habit: we rapidly focus and refocus and scan as part of our nonstop process of understanding our world.

Since the camera can suddenly place the face of an actress within inches of our own, we no longer need the exaggerated acting of old. Because the camera zooms in on a face, just as our eyes zoom in on the face of someone we are talking to, we do not need stage tricks on the screen. For this reason, movie actors often do not seem to be "acting" at all. The old craft went out of fashion along with all the greasepaint and stage makeup.

The kinship between cinematography and our habits of perception demonstrates two points. **First, what we call *speaking* is largely a matter of *acting*.** Words play a fairly limited if indispensable role in the activity—whether it is a movie or real-life events. One reason complex conversation is difficult is because most of us are not great speakers of words alone. We *are*, however, gifted at acting. Conversation does not act out meaning because spoken words alone speak what we mean but because we add voicing and animation to shape the truth of our messages.

Second, we are highly visual by preference. And we sense that everyone else is also, which governs our methods of communication. Our performance is visual because *we* are visual. We read what we see in people and their behavior with great skill. We zoom in and out constantly. In fact, we do so with such skill that we are inclined to call it "intuition" at times. In truth, if you arrive at a party and, within moments, have the intuition that the party will be a bore, you have *not* had an intuition. Your eye has seen that one fellow is pulling up his socks for something to do, a woman is staring at the landscape in the darkness outside the window, small groups are clustered in muffled conversation, and you do not see a smile or hear a laugh. Your scanning eye will gather all this while you say hello at the front door and *almost* remove your coat. Indeed, your perceptiveness is as brilliant as your gift for acting. We do not need intuition, auras, electromagnetism, psychic skills, or other such explanations for our obvious acting and perception skills. We need only to realize how sensitive and subtle these abilities can be.

Speaking without Words

We have observed that we are primarily visual in our orientation and that this conflicts with the idea that communication is primarily oral. More specifically, our visual orientation conflicts with our notion of communication as spoken words. We will get around this conflict if we view dialog as what we *see* spoken (forgetting telephones for now). The communication path of the dyad illustration explains what we say, but it also explains what we act out. The dyad can be heard, or seen, or both. When we *see* a conversation *spoken,* we experience the entire dialog correctly, or fully.

I am imposing the idea of "seeing" conversation onto the usual idea of hearing a conversation. People often use the expression "do you see?" We also have the popular expression "I hear you." The idea of both expressions is that seeing and hearing mean understanding, but we need a term that bridges the hearing and seeing separation. Think of "seeing" conversation: *Do you see what I am saying?*

All the dimensions that add up to good communication, *excluding* words, are popularly called *nonverbal* communication or nonverbal behavior. Often this category excludes "voicing" (the way you phrase what you say), from the nonverbal category, but I want to include voicing here on the grounds that voicing is not directly part of the word designation and word ordering process. I prefer to include voicing with all the other dimensions of communication that *enhance* communication through anything other than the symbolic transfer of word significance. Even though voicing is verbal, let's start with this issue and work farther and farther away from spoken words.

Voicing

Voicing (or *paralanguage* as the specialists call it) is what we hear, apart from the direct symbolic shaping of a word; in other words, pitch, tone, and rate are inevitable parts of speech patterns, but they operate beyond the mere shaping of words. *Volume* goes up and down with every comment we make. We usually whisper affection and yell our discontents. *Pitch* tends to follow suit. Your pitch or tone will follow the emotional color you want a comment to have. *Rate* is another aspect of voicing. You might speed up a story to give a tale some excitement. You might speed up a comment because you yourself are excited.

Listeners translate the voicing of a comment as quickly as they translate the words. A fast talker will be perceived as an anxious or a nervous personality. A slow speaker will be seen very differently. There are people who are "loud," and there are whisperers also.

Pitch is equally variable. All these variables carefully shape language, and they can shape an image of a speaker also. Voicing can have quite distinct characteristics that reflect our personalities in many ways. You "recognize" your friends on the telephone. The voice is an image. Sound is the picture.

It goes without saying that we need to correctly interpret voicing to complete the meaning of what we hear. If someone snarls as she says "yes," you know she means no. In fact, since the person got hostile about it, she means an emphatic NO. Much of our understanding of volume, pitch, and tone is probably universal, but to the extent to which there may be cultural differences, there will be confusion between cultural groups. Americans seem "loud" to some cultural groups, for example. In other words we seem to speak loudly enough to seem a little rude, or graceless, or even vulgar. We will not notice this effect until the decibels go up a lot higher, relative to what we feel is our normal volume. Oddly enough, we will also think that loud people are graceless and rough, but first, they will have to be louder than we are!

Pitch can easily be misunderstood in certain cultural situations. Chinese is a unique language (actually, a number of languages that share a common writing system) in that the language depends on four tones to change the meaning of a sound into four different words. If you say "blue," you mean one of several ideas—color or mood for example—and we use the sentence around it to figure out what you mean by "blue." In Chinese, a word, let's say "blue," can be given four different meanings by changing *tone*. As a result, Americans have always said that the Chinese language sounds "singsong" or song-like. True enough, it is. In a case such as this, it can be difficult for a Chinese speaker and an American speaker to depend on each other to communicate with tones that the other will understand. We do not use their tones, and to some extent they may not use ours either. This complicates communication with Pacific rim cultures. It is one thing to struggle with a shared vocabulary. To make matters difficult, the voicing structures might not match either!

Noise Talk

Beside coloring words with voicing, we are also prone to simply making a lot of *noises,* which become *another* feature of speaking. At what point these noises become vocabulary words of the sort we find in a dictionary does not really matter, because noises function meaningfully anyway. Such sounds as "yeah, "yuk," "ugh" and "uh-uh" are obviously words, if rather unofficial ones, when compared with "yes" or "no" or the now universal "okay." Others are not easily defined as vocabulary in our word system but are obviously some sort of vocabulary of their own that we understand.

If a child is naughty and you gently wag a finger at him, you might simply make three noises we call "tisks." Both the wag and the tisk have meanings as though they were words. If you are on your bicycle and have a near collision with the little silver bulldog on the front of a semi, you might wipe your brow and sound an exhaled breath to mean "close call." Again the gesture and the noise are symbolic and have meaning. Whew!

Some sounds we utter may seem more like noise than words—"oops," "ouch"—but they are socially acquired, although we may not notice it. They are words, because no Spanish child is going to say "ouch" if she burns her finger on the stove. She will say "Tuy!" (pronounced too-ee). And she will not use quite the same physical gesture either. The American child will use the American gesture of shaking the hand quickly up and down with a very loose wrist. The Spanish child will keep the wrist firm and make strong downward strokes of the hand to make the fingers snap together in an audible sound (takes practice). All manner of nonwords and noises become a key element in our language affairs, and many have evolved into what I call "noise words."

Our use of words, nonwords, noises, and gesture, all with the same meaning, can be seen in a number of examples. If you want to tell someone to be quiet you have a choice of languages. For example,

1. You can say "Please be quiet."   (words)

2. You can say "Shush."   (a noise word)

3. You can make a shushing sound.   (a noise)

**4. You can tap your index finger on your lips
(with or without the shushing sound).**   (a gestured word)

5. You can scowl.   (a performed word)

In the same sense that there are noises that mean something, there are sounds and words that mean nothing! We all speak with a certain amount of meaningless filler, particularly the sound "uh." The result will be a sentence that reads "I uh think I uh can go." Some people will insert the words "You know" or "like" with this result: "You know, I think, like I can go." The filler words mean nothing in the context of the sentence logic. We seem to want to keep the sentence moving, with noise if necessary.

If we add voicing and noise talk to the language system, we could say we are fine-tuning the language communication. From another point of view, we could say that we are increasing the number of communication systems in the process. If we look at language as any complex set of symbols that are used to communicate, obviously we speak in several languages at once.

Animating Words

So far, we have three major components working together to express our ideas when we speak: words, the voicing of the tones we give to the words, and the culturally acquired noise words (the *uh-huhs*). We are looking at three language systems that we use together constantly. There is much more. What we have to this point is a telephone conversation. We have another major way of speaking that we add to the use of words and voicing: body language. You are probably familiar with the term, but you may not realize the extent to which body language shapes and colors your way of speaking.

We should look at body language briefly if only to realize the extent of this "vocabulary." There really is no mystery about this skill we all master with such dedication, although the literature on the subject may be mysterious. The notion that you can learn to "read" body language and understand the secret realities behind your friends and foes seems to make secret this very public stuff. Everybody seems fascinated by the topic of body language (or *kinesics,* the study of body motion), but we are already quite expert in the field of acting out conversations and in interpreting the meaning of the way other people act out conversations for our behalf.

Perhaps people are fascinated because they have seen books or articles on how to *really* understand a person's "secret" motives or feelings. This is occasionally true, but most of us already wear our hearts on our sleeves when we talk: we *want* our feelings to be understood. We seldom truly mask feelings. When you grumble "Nothing is wrong!" you are not masking a thing. True, a shaking voice reveals the real fear you might want to hide in a nervous moment, but looking for "concealed" feelings is probably not the way to look at body language. Instead we simply need to realize we are all million-dollar movie stars. We are very public and we all have tremendous acting skills.

Each of us is also extraordinarily sensitive to what he or she perceives in the body behavior of others. We do not need much literature on the subject. We spend our lives reading other people, and creating behavior for them to read in turn. Our behavior is a combination of exposure and design. It is true that we might expose inner feelings or thoughts from time to time. However, unintended expressions of body movement can be hard to interpret because the meaning of the behavior is not being projected in a clear way with the *intention* that you will understand it. On the other hand, the minute we project ourselves through *intentional* body language, no one will mistake our intent.

Face Language

Basically we act out body language. The fundamental areas of the drama are neck-up and neck-down. With your face and your body you gesture an incredible spin on words. Try to say "really?!" with a lot of pitch to show that the comment is both a question and a show of amazement. In other words, your best friend just won the lottery. Now say "Really?!" with lots of pitch, but with little volume. What you will most likely also do is tilt your head up and forward and raise your eyebrows. The body language is part of the message.

The area of the eyes is central to one-to-one dialog. Eyebrows are particularly important, by the way, even though the idea may seem comical. We raise our eyebrows for amazement. We pinch them to the center for annoyance or anger. We raise one and drop the other for doubt. Then we open our eyes wide or squint to add more to the image. Your facial movements easily define or enhance the meanings of words. Smiles will do the job. So will frowns.

In the process of holding conversations, we make conscious efforts to use our faces and our bodies. The gestures of the face are a language unto themselves as they weave into the voicing of words. Some elements of the language—the smile or the frown—are universal. Other elements such as a wink may have regional meanings, as though such elements were English, or French, or Japanese. Still others are the personal touches you lend to the projection of your thoughts.

Perhaps the face is the focus of much of our body language communication for reason of the proximity of the brain, the eyes, the ears, and the mouth. Attention seems to fix on the head, since all the back-and-forth channels of communication are centered, indeed attached, to our minds! There is little to dwell on here because we are experts at using face language and we are experts at reading face language. You are as gifted at it as everyone else. All we need to remember is to be sensitive to what a face says. The signals do not always correspond to words.

Body Language

What we call body language is, in terms of contributions to our spoken communications, largely a matter of facial expressions. Conversations have a neck-up focus. However, neck-down body language is a part of the physical activity of body language also. The example of timidity demonstrates the fact. A person who booms with self-confidence is not likely to diminish his or her body mass. Quite the opposite, the largeness of the personality will probably be accompanied by confident erect poise and much-used arms and hands. The confident personality can almost seem to exceed his or her size, like the cat that will often meet its enemy broadside.

Body language of the neck-down sort is a matter we handle expertly. We control our projected meanings with skill. Consider the examples of confidence and timidity. You can project confidence by standing tall, being sure to keep a full frontal position in front of anyone to whom you are speaking, and perhaps by cocking your arms akimbo by putting your hands on your waist. What you are actually doing is maximizing your body mass.

Now, if you want to be timid, you want to minimize your body mass. First, pull in your neck. Then stoop your shoulders. Hide first one hand in one pocket. Put the other hand in the other pocket. Turn diagonally to the person in front of you. There, now you are a much smaller target aren't you? You can also stare at the floor, because the performance of the face in the move from confidence (100% eye contact) to timidity (100% eye aversion) is partly a downward curve of diminishing use of the eyes for conversation. It is rather comical at times, especially when I ask for volunteers in a group discussion and 100% gung-ho eye contact suddenly flattens out at zero, and the coffee stains on the table suddenly seem to become very interesting to everyone.

The anthropology of body language, if there were such a science, would be interesting. Consider a cat's first line of defense: bluff. A dog chases a cat only to a certain point because of cat strategy. Cats are on the small side in dog terms, but a threatened cat will do a phenomenal thing: it will double its size! The cat often turns broadside to the dog. This is a dangerous strategy until you realize that it allows the cat to look larger. Then the cat will arch its back and, in fear, its hair will stand on end. A smart dog will then call it quits as the competition suddenly looks a lot tougher to deal with—specifically, a *lot bigger*. What is more interesting is that we can do this too. When we get goose bumps out of fear, we are trying to make our hair stand on end. Body language evidently has some very old origins, indeed.

We can also see body language function on several levels. Consider the simple gesture of the handshake. The basic symbol is ritualized and has symbolic meaning as a greeting. But how you interpret a handshake will be based on how you read the delivery, which will run from the crusher to grandmother, and every subtlety in between, not to mention temperature and humidity readings. Perhaps there are even handshakes specific to politicians. The handshake is also cultural, since a woman is unlikely to bear down on your fingers the way a man might. And if a man is light in the hand, the gesture may seem insecure or timid. The handshake, precisely because it is an introduction, is also a cue. We cue people with such subtleties as the pressure and vigor and style of such a seemingly incidental activity.

The fascination with body language probably has to do with the way in which we control it and in the way it will betray us. We usually handle the craft with skill and use it to shape our meaning in conversation—and to shape the overall perception we want people to have of us. It is this second point that is a little fragile because it can involve concealment. For example, you can hardly *appear* calm and collected if you are really a bundle of nerves. Even if no one sees you bite your nails, they do not have to. Your hands are right there with you, and the results of the nervousness are visible even without the gesture. I do not know that we can "dress for success" with body language if it goes against our nature. It is difficult to be something we are not. To *defy* ourselves in body language is not usually effective. In the workplace, use body language to *be* yourself. There should be no reason to *hide* yourself.

Other Communicators

Time

One unique feature of nonverbal communication overlaps with the other features in the makeup of communication. Time is a nonverbal communication form in that *when* you do something is very different from doing it. Time points to a space and moment. The use of the moment is a strong communicator. Marcia's boss tells her to be sure to type an important report for Monday morning. If the report is ready on Monday, Marcia has said something by being timely. If it is ready the next day, she is saying something about being a little late. If it is done next Friday, well, Marcia is obviously fed up with all these projects isn't she? Maybe her boss never even got it. The time distance between point A and point B is communication and speaks *loudly,* though without words. It is a very powerful communicator. Being punctual is one of the easiest paths to success. **The proper use of time is as effective as the abuse of time.** I would hope that it is more effective, at least in terms of positive signals.

In the workplace setting, quality can frequently conflict with punctuality. Both the quality of a project and the timely completion of a project are important considerations in any business. What are you to do? Strike a balance. Be sure you know which one is preferred if time is running out. Do not get into too many situations where you never get time for quality. Some businesses operate by minimizing overhead by maximizing labor effort. If you sense you are being asked for 130% while your pay remains the same, act in your best interest. Besides, if the quality is poor, the work is no fun anyway.

However, if you stall on deadlines that other employees easily meet, the signals are quite different. Be wary of pokiness, and also be wary of perfectionism—a more dangerous problem. Perfectionism is more dangerous because it is fairly self-righteous, and the results are often late. The perfectionist *thinks* he or she is doing the right thing, *always.* Yes, quality is an asset, *unless* it conflicts with other realities. If my role in a project puts other people's involvement behind, no one is going to thank me for the quality of my work if they are at the office nights to catch up because of me. Perfect and late are a common couple, and an explosive match.

Another concern about time as a communicator is that the business world lives by the clock anyway. Everything is a matter of timelines. The workplace clocks everything. Time is the key to profit and loss in ways other than the obvious employee issues such as eight-hour days, ten-minute breaks, forty-hour weeks, and so on. *Everything* is time managed. In some businesses the length of time it takes an employee to walk across a room matters, and some well-known industries, such as UPS, do not hesitate to use the stopwatch.

Now, if an employee happens to be habitually late about tasks, he or she is taking a serious risk, not because the employee will be fired—which is likely—but because the employee is likely not to know *why,* or to *accept* why. People who are habitually late are like perfectionists in that there is a "superior" motive for their behavior. The perfectionist rationalizes the rightness of quality. People who are habitually late usually rationalize the rightness of being late. They *always* have an excuse. It is never, however, an "excuse" to them. It is an explanation. The communications barrier is never resolved, and explanations turn into no job.

Sandy was always late because she had to drive all the way from Butler County. Then she had to drop off her daughter at a day-care center. She was a single parent. What Sandy never realized was that even her coworkers were not sympathetic with her being late. The boss "let her go," but the employees let the boss know it was okay to let her go by making certain comments. "Look, I live twice as far away." "Lots of us are single parents." "My kids are in day care also. Big deal." "We all have problems." Sandy never realized that her number-one critic will always be her fellow employees. Employees resent double standards.

That would ruin my day, that business of her coming in late. She just reminded me of my brother. My brother is never on time for anything. I won't invite him to dinner anymore because he doesn't appreciate it enough to arrive on time. Sandy always showed up puffing and apologizing. Why puffing? She didn't walk to work! Were we supposed to think she was making an effort to be on time? What a joke.

Since habitually late people are self-indulgent and always rationalize their lateness, they may *never* really learn.

Somehow we never stop to think that most of our fellow employees are much like ourselves, particularly in terms of the strain of everyday challenges. They are not very sympathetic with the "exceptional" case, because the exceptional case rarely *really* is exceptional, especially in light of statistics that reveal that true exceptions—the physically handicapped—will be there, and they will be on time.

We can never take time lightly in the workplace. It is a very strong communicator, and there is only one appropriate behavior pattern we can accept if we want to maintain a positive professional image.

Distance

There are still other considerations in nonverbal communication. The actual distance between two people is a very significant communicator—and partly a culturally controlled factor. The study of space relationships of this sort (proxemics) is intriguing. In general, anyone closer to you than arm's length is usually a family member. The length of your arm is significant because the close proximity means you can and are likely to *touch* a person in that space. It is family space. Friends are also welcome; they are "close" friends. Beyond this intimate space is the more casual space you use for conversation. It will not be too far—perhaps from an outstretched arm to the other end of a typical dinner table.

This space relationship between people is a constant. If someone invades your space you will feel discomfort and be very suspicious. If someone stays much beyond appropriate space you will wonder if you had too much onion on that burger. Distance is a curious and subtle communicator. When you think about it, there is nothing there. It is the most nonverbal of communicators. It is simply space. But it speaks.

Different cultures deal with space differently. It is not necessarily as universal as the smile. It will be uncommon for Americans to violate their own space traditions. Beware, however, of foreigners! They can get "suspiciously" close. If a culture is fond of frequent touching, patting, and other daily contacts of this sort, they *must* stand close. This is common in the Mediterranean region for example. The result is that you, as a tourist, will be confused by the "overly friendly" behavior of strangers. It will not be the smiles that worry you. It will be the space—or too little of it—and what's worse—being touched.

The meanings we give to space vary from culture to culture, which once again suggests a peculiar similarity to "languages" where there is no language. A lunch table that seats five or six is an ideal experiment. Go to the table and ask if you may sit down. Sit beside the person or at least across the table, directly facing the person. The results are predictable, maybe even risky, because the polite convention, in the United States, is to sit *diagonally* more or less *opposite* at a *maximum* distance from the fellow student or fellow employee—unless you know them well. But in other nations such behavior would be an offense, and all the tables in a cafe may sit empty except the one where people gather. Space evidently speaks many languages.

Flags

We have a physical reality and we manipulate it a great deal. Many of the communication methods we are looking at pivot on animation, but there is at least one static or fixed constant that we use to communicate: dress. We project an image of ourselves in the way we decide to cover our bodies. Clothing speaks.

What we communicate and the way we will be interpreted will include clothing, hair style, and the general concept of the *physical* image we create every morning before we leave the house for work. Whether the message is subtle or conspicuous, the fact of our physical presence is as obvious as a waving flag. Once we are done up for work we may forget about it, but the presentation on our backs is one that projects us all day.

We manipulate our image. This is most obvious when we want to strike a certain appearance, say for a job interview, but it is equally true of us on a day-in, day-out basis. We need not dwell on the attention we give our physical appearance as a communicator. It is an enormous industry—from weight clubs to suntan parlors to clothing companies and cosmetics—and it is worth billions. There are body shops everywhere—for us!

Most people probably dress for comfort, and they probably dress in accordance with their own self-image. Saturday is probably a good day to see our dress-for-comfort style and our personal self-image at work. We tend to dress in a workday image from Monday through Friday. Within that context only those who do not conform stand out. Eccentric dress, unpressed clothing, soiled clothing, and being too well dressed are all topics of discussion in any business.

You have heard the expression "dress for success." You may have also heard the somewhat contrasting view "dress for self-confidence." Both ideas are smart ones. They work, but since you probably know they work, let's look further at another idea: *dress for trust*. If a police car signals you to pull over, the lights and the logo on the car speak with authority and you pull over. Now the officer comes up to your car window. Just suppose he is dressed correctly in his stiffly pressed state-police uniform but has a three-day beard. You will be very nervous. The police car tells you to expect a certain image. **The image must be the one you expect or you lose confidence; you lose trust. Dress codes build trust. Image builds trust.**

Trust in an image is not always reliable, of course, but developing an image is an effective *tool* to use. Your mail carrier, the bus driver, and thousands of corporate service areas rely on an immediate visual image for identity and for *public relations* and *public trust*. Doctors do not really need to wear white smocks around the office. Those who do not will usually have a stethoscope hanging from their neck. They do not *really* need the stethoscope either.

When I bid a job, I wear a flannel shirt, chinos, and work shoes, and I keep a trusty Stanley measuring tape on my belt (which I rarely use). It is my stethoscope. Some time ago the dean of my college asked me to teach a home remodeling course as a community service for our area. I thought it would be fun and I prepared the course and offered it for a few years. I never told the class I was a faculty member. I just showed up in my flannel shirts and work shoes, always with my trusty Stanley tape. On one opening night of a new course, I was busy getting out tools and equipment for the evening lecture, and two older women came in and sat down. Obviously they did not know quite what to expect from the course, so they scrutinized *me* while I set up the gear. One turned to the other and smiled and said, "Well I don't know, but he sure *looks* like a contractor." I always dress for trust. You should too.

Spaces

In the workplace there is one more communicator we might mention. Environment. Workspace. Whether environment is only the leather briefcase of a computer service technician or the workstation of a local drafter, the van of an on-call service employee or the entire lab being used by an ultrasound tech, the space communicates, from little details in a briefcase to entire laboratories. *Use* of space and space *images* matter. Do I mean decor? Sort of.

Consider Carl's veterinary clinic. He had an architect draw up the plans for the space. The architect specialized in veterinary clinics and understood the "ergonomics" (meaning, more or less, the study of "used" space) of vets. The designs, from the point of view of utility, were fine. Carl's big concern was the front office and the exam rooms. These were the work areas most seen by clients and their pets. The space had to have the ability to communicate. He worked out every detail.

I didn't want that country kennel flavor in the air. I wanted a clinic that equaled "people medicine" to give people the right trust in "pet medicine." Double doors kept the disinfectant and clinical odors away. That was the first measure. I added sophisticated background music. I had to balance usefulness with the "people's medicine" image I wanted. I didn't want linoleum floors, but sick pets can be a cleanup problem. It took a lot of shopping to find carpet that was just right. Super heavy duty stuff.

Carl carefully placed all his medical certifications in one corner, out of the way but impossible to miss. On and on he went, detail after detail. Sometimes the ideas were comical. He learned that large potted plants look great in clinics for people's medicine but get a little ragged around male dogs. In the end he created one of the most instantly successful clinics in the city. The space *spoke*. The issue is one we discussed earlier: building *trust*.

We should not overlook the fact that a lot of our space is plain for all to see. Neat, messy, plush or plain, simple, complex, bright or dull—space is usually interpreted by an observer and leaves a message. Be alert to the fact. You may not pay nearly the attention to space that Carl did, but you need to control your space, if only so the boss or your client gets the "right" idea. Your public may be an entire service area, or only a few coworkers, but people will see and interpret your environment. As Carl realized, success is a result of trust in a service's business. *Image* builds *trust*. Your lab, your office, your van—whatever your "space"—is part of your image and you should construct it to do what you want it to do. Mold it to mean what you think it should mean—or say what it should say.

People even value empty space. It too is a communicator. Space functions as a reward or status. As you know, the executives of a corporation get the top floors with the view. However, open space is an even more precious reward. It is seldom observed in the studies of the American system of perks that office "space" increases as the status of an employee increases. People place considerable value in the cubic volume of nothing. In fact, it has status! For example, at the college where I teach the deans' offices are three times larger than mine, and the president's office is three time larger than theirs. I do not grumble because up to six part-time instructors are huddled in offices the size of mine.

The Challenge

Your ability to understand the intentions of people around you is best served by your awareness of the many levels they use to speak to you. There is the primary "speaking" loop in which words run between people in the basic dyad we first discussed. Beyond the dyad of words are all the other dyads. Add a face dyad, a body dyad, voicing, space, noise, silence.

Even *noises* and *behavior patterns* are significant communicators. If you arrive at your home and close the front door with a little bit of a bang, the dog will go hide and your spouse will quickly arrive to see what is wrong. The noise spoke. Patterns also speak. People expect us to behave as we always behave. If you suddenly change a conventional pattern of activity, you will communicate that something is amiss.

We need to be keenly aware of the big picture, particularly because of the tendency we have to say one thing (words) and mean something else (other projected cues). We can be very contradictory. If you ask your friend to go to a concert and he says "Sure," there is no problem. But if he pauses, neutralizes his voice, looks away from you, and says "To see what?" a complicated comment is constructed. These complications are created all day everyday. We need to understand them to understand what people really mean. The words often are in reverse! People often mean the opposite.

We have a responsibility to project ourselves with clarity. Understanding the complexity of our coworkers is another responsibility, and we want to be clear in the meanings we project to them!

Summary

- The communication loop consists of six discrete areas that deserve attention.

  - ✔ A "speaker" has a thought.

  - ✔ The person shapes the message.

  - ✔ It is delivered with the voice and other features.

  - ✔ It is received as sight and sound and other sense responses.

  - ✔ The "listener" translates the message.

  - ✔ The person interprets the meaning.

- The loop is extremely fast; it is used repeatedly, and it can be a problem in any one or more areas in any one or more uses.

- A short conversation can create a hundred loops in a matter of minutes. Any detail—even one word—can trigger a problem in one or more of the six areas.

- The manner in which a person speaks—tone, volume, speed—has a pronounced effect on the interpretation of the message.

- Speech is animated by the body, which adds meaning to what is spoken.

- Body language is controlled but can be quite spontaneous as well. Control allows a person to shape the way a message is perceived. Spontaneous responses allow the observer to understand but such messages have far less intentional control.

- Time is a communicator. It is best used by respecting other people's expectations of time rather than your own.

- Dress communicates also. Workplace clothing communicates a shared behavior pattern. A highly distinct concept of clothing will violate the shared behavior.

- Workspace is a projection of an employee, whether it is a toolbox or an office or a service vehicle.

Activities Chapter 4

Present a memo to your instructor that develops a discussion of a work experience that relates to the chapter. The memo should be 500 to 1000 words, typed. In it you can explain how you have related the text or the lecture discussions to some event you have experienced. This exercise will give you a better understanding of the material because you will explain incidents in terms of your perceptions of workplace structure and communication.

Select one of the following suggestions and develop an analysis that involves your current employment or a former position.

- *Select a skilled communicator at work. Discuss the person's speaking skills, body language, supportive behavior, and listening skills.*

- *If you can recall an event where listening skills were either a success or a calamity, explain the event in detail.*

- *Employees often hold back and repress rather than speak up. The restraint may be typical of the person or a response to a specific situation. Identify such a situation in your work experience.*

- *If you have traveled widely, discuss cultural difference in speaking and one-to-one communication habits. Include touch and distance in the discussion.*

- *Consider your communication patterns with employees. Is there a meaning to the frequency with which you see them in passing, or speak to them, or phone them? Who gets high contact? Who gets low? Is it all business? Is there a friendship link involved with some parties? Who visits whose space? Who gets personal distance? Does anyone shake hands or pat you on the back and vice versa?*

- *Discuss your sense of space: your office, your car, the right image for your business.*

- *Explain any company programs you have had concerning employee image training. What expectations were identified? Discuss the training sessions and employee responses.*

- *If you have had experience with dressing for a role, explain the situation and the expectations of the employer. Did the image create some kind of understanding with the public?*

Listen Well to Speak Well

5

Work in Progress

5. Speaking and Listening

Through the remainder of my sketches in this book, I will present an actual contract from the initial contact to final conclusion and explain all the usual difficulties that occur along the way.

I first heard about Tempo Storage Systems through a friend of mine who runs an adjacent business in the same office park. He had heard that Tempo was thinking about upgrading their local area network and gave the president, Tim Simmons, my business card. Along with the card, he gave Tim an earful about how great my company is. He also mentioned how pleased another company was with our work after we upgraded the company network.

Because Tim heard about me through a known and trusted source, he was open to giving my company serious consideration. I have found that this kind of word-of-mouth referral is absolutely the best form of advertising a small company like mine can have, and I always make it a point to thank the people who refer me. A simple thank-you note for a referral goes a long way toward ensuring that that person will keep giving me referrals as new opportunities pop up.

Simmons called me and we had what I think of as a phone interview. This is the first of many make-or-break points for getting or not getting a contract. I encouraged him to explain the situation and tell me the details of the job as he saw it. I did't rush him or interrupt except to help him with terminology or to clarify technical points. This is the customer's chance to lay out what he or she needs, and it is my chance to get a sense about both the type and size of the job and the type of people for whom I will potentially be working. I do not often turn down jobs, but there have been a few, and this is when I make that decision. Either the type of work was not what my company did or the person I was "interviewing" did not seem like someone with whom we could work. I find it is better to be clear at the outset and recognize a contract that isn't meant for us.

During my conversation with Tim, I jotted down all the key points he brought up as well as the questions and concerns he had. It was hard not to push him or "sell" him a little on Net-Works NorthWest, but this wasn't the time. Selling would come later; for now, I wanted him to know that I will listen carefully to what he has to say. I repeated his main points back to him toward the end of the conversation and asked if it sounded clear. I also encouraged him to call and leave a message if he thought of anything else.

He expressed his interest in having my company develop a bid, we set up a time for me to do a walk-through of the site, and I told him I would fax him the information he had asked for before we had our first meeting.

J. Q. C.

Looking for Control

Next, our discussion will deal with communication skills you should develop and communication problems you should avoid. I noted at the outset that conversation will be infinitely varied and that there will be endless situations with which you will have to deal. The next chapters are devoted to the problems of these varied and vast communication situations. We will look for the problems, develop tactics, look at samples, and talk out what happens when people communicate. There are strategies, but you must realize that conversation is highly conditioned by circumstances. Unlike a committee meeting or an interview, every encounter you have in the cafeteria is likely to be different from any other encounter of the day. Of course, lunch is as likely to establish patterns as any other social behavior. People often gather in small groups. Gatherings that are repeated by the same people for the same purpose take on the elements of ritual. In this case the events are predictable. Then the patterns of communication are more predictable as well.

Whether the communication is predictable or unpredictable, you need to be able to influence the conditions of the encounter. Because you will either contribute to or in some way control conversation situations, we will both look at theory and then investigate practical circumstances so that you see the conditions involved.

Let's start with practical rules for good listening and practical rules for good speaking; These are two key areas of concern that greatly affect every communication encounter. There are basic strategies you can apply to any situation.

Rules for Good Listening

One key to good communication is to keep quiet. This contradiction is an important consideration that communicators need to understand about good conversation. Of course, people must *engage* in conversation and speak. The more accurate comment, then, is to say that we must stop, we must look, and we must listen.

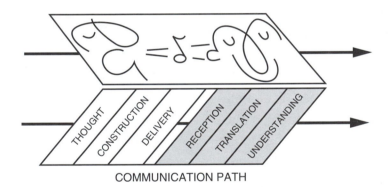

COMMUNICATION PATH

Stopping

To really understand someone you must be prepared to listen. If you are doing a task of some sort at the same time you are listening to someone, you are not going to pay attention. This is perhaps the single most frequent barrier to accurate communication at work: the "uh-uh treatment." It is also probably the single biggest barrier to accurate communication in general. If you talk to someone who is doing something else, he will usually keep saying "uh-huh" because he has no other response in mind. His mind is focused on the other task. The lack of eye contact will tell you he is really not part of the conversation. You could probably say "Give me a hundred dollars" or "Your wastebasket is on fire" and you would still get the "uh-huhs." Well known studies on parent-child relationships (such as *Between Parent and Child*) reveal much the same communication barrier. At certain ages, a child may be given very little quality attention and not be taken seriously. The problem is that a child's world is not our world, and the very important hamster that Suzy saw at school is given the uh-huh because Mom is doing the dishes. Not only, of course, is the uh-huh treatment a disaster in communication, it is also insulting. The solution is simple: focus.

Looking

It takes only one major tool to focus on someone who is talking to you: attention. Think of *looking* as attention because you can almost measure your attention by the strength of your eye contact. If you do not take your eyes off someone, it is because your attention is quite focused. As attention declines so will the visual intensity.

You might think that stopping and looking are very obvious practices, but people abuse these basic good manners. Why? Well, the reason is fairly clear to Suzy. She will give up and go tell her girlfriend about Fred the hamster. She will seek out someone who takes her seriously, who sees the importance she sees. If a conversation is important to one person and unimportant to someone else (at least at the outset) we have the cause of most stopping and looking failures. The clash is sad because we do not identify someone else's worries or fears, or joys, or excitement at *exactly* the moment they are trying to share those interests with us.

Listening

The art of listening is the art of humility. We must be willing to put down our task, set our ego aside, and give importance to what other people are trying to tell us. Listening is a craft, it isn't that difficult to practice, and the rules are easy to follow.

1) **Understand by listening.** First, do not play on your own ego. It is very common for listeners to be imagining their own next line or their next joke. Do not be thinking of what you are going to say when the person is done talking. And do not barge in and cut off a thought—or worse yet, finish sentences for people. The very real problem is that if you start to hear yourself think, the voice you are hearing from the other person will get a lot less attention, if any. The next time you are in a class or at a meeting, observe what happens when you raise your hand to speak. You will hear very little that is said until you are allowed to speak because you have something "on *your* mind." Conversely, if you raise your hand to speak, refocus on what is being said by *others,* and then have the opportunity to speak, it is likely that you will *forget* what you were going to say.

2) **Listen by understanding.** Understanding is not just hearing. You know from seeing the many levels of the dyad that can cause problems, that hearing is *not* understanding. Understanding is the hard work of listening. You must use all your thinking skills and your perceptions to piece together what someone is trying to explain. The issue may be robotics or autoimmunity or taxes or hamsters but you must *understand;* this is largely the job of *your* mind, not the task of the speaker. The speaker gives you all the tools you need: words, pitch, tone, facial movements, body language, space, time, and so on. You must piece it together. We observed that listeners have to quit thinking about themselves or their ideas, and that is because they must spend that time *thinking about what they are hearing.*

3) **Do not judge.** Neutrality is always a safe haven. In conversation it is an excellent tool. Maintaining neutrality in a conversation for political reasons is not the central issue. The point is, you should not hastily evaluate what you are hearing. You do not want to judge, nor judge in haste. In conversations, if you maintain neutrality you will hear more. First, you will not be shaping your defense. You will keep the ego out of the picture. Second, if you jump into a dialog by agreeing with someone, you will shape the rest of the person's commentary. Third, if you charge in by disagreeing, you may cut off the conversation and never hear what could have been said.

4) **Support a speaker.** If you are supportive, you help people along. It is a simple kindness, but being supportive builds trust. You do not even need words. Eye contact, a sense of concern, or a smile will achieve a great deal. In addition, saying "yes" or "right" or "for sure" or simply being agreeable will enhance a very desirable condition of most honest conversations: openness.

It goes without saying that you can never betray the confidence you build with supportiveness. Confidentiality is a top-level condition of openness and trust. It is also an honor of sorts—and a responsibility. For people to confide in us, they must have confidence in us.

5) **Maintain clarity.** It is almost certain that small or large chunks of a conversation will be somewhere between vague and incomprehensible. The problem with asking questions is that questions shape answers if you are not careful. The car salesman pretends to be busy while you walk around the showroom. After you show your true interest in the car you came to see, the salesman will happen over to the car. "It's a beaut,' eh?" The answer came first; the question came second. The sale comes third. In conversation, however, you do not want to sell the conversation by shaping

questions. The best tools, then, are neutral "checks" to simply make sure you are understanding:

- **Paraphrase** You might restate the ideas you hear to be sure you got the message. The classic expressions that indicate paraphrases are "Do you mean. . .?" or "In other words. . . ." Observe that one expression is a question, and the other is not.

- **Interpret** You can do a lot more than restate ideas. You can interpret the conversation to see if it is something you understand from your point of view. "Something similar happened to me last year," you might say, and then tell a personal story. Or you might use the comment "so what you really wanted to tell her was. . . ." You can search for the meaning with an interpretation.

- **Ask clarification questions** Use simple questions that do not point at answers. Ask for the meanings of words, ask about a feeling, ask for background, ask for anything you do not understand, but in simple brief questions. Avoid cutting off a conversation with questions that interrupt what you are being told. Wait and remember them. Avoid the expression "I don't understand," but ask questions until you do. *Showing* confusion gets in the way. It will frustrate a speaker and build self-doubt. Let the questions do the work. Confess to not understanding only if you find it necessary.

Contradictory Behavior

I have assumed all along that people say what they mean. Life, as you know, is not so simple. People often say one thing and mean another. In such a case the spoken language conveys one message, and all the nonverbal systems carry the other message, which, as often as not, can be exactly the opposite of the first message. Jody wants to go to a late movie with her girlfriends. Her mother does not want her to go. Jody insists. Mom and daughter have gone around and around about this before. Finally Mom explodes, "Go ahead and go!" and throws down the garden gloves and marches out of the garden. This is *doublespeak,* the craft of saying one thing and meaning something else.

People use doublespeak a great deal. Usually the doublespeak is clear. If someone at the office comes up to you and says, "Howard really deserved that promotion didn't he?" and you say "Oh, sure," you will not necessarily notice that both comments can be structured to mean exactly the opposite of what is said. First, the question can be said with a smirk, which is doublespeak in this case. Then you roll your eyes and say in disgust, "Oh, sure."

Since the doublespeak can shift from obvious to subtle, you will often have to ask people what they mean if you suspect a doublespeak situation.

The reasons for doublespeaking are easy enough to see in daily environments. The process is a clever way to state objections or moods that conflict with the desired response that someone else is looking for. The idea of agreeing and not really agreeing means there are *two* ideas that have to be projected at the same time. Basically, if you agree and reject at the same time, you are simply expressing concessions. Concession is an important behavior among us, a behavior so fundamental that it is a pivotal element in modern psychology. Concession is the price of society. It is the glue that keeps us voting rather than fighting.

In the workplace, the issue of doublespeak is sometimes a little more subtle, since the conditions of the workplace depend on courtesy and compliance—courtesy toward clients and coworkers, and compliance toward superiors and contract commitments. We discussed the problems of rebellion earlier in the text. There cannot be much defiance in the society of the workplace. The workplace social makeup depends on teamwork. Polite doublespeak is probably the most convenient method of protest in the office or on the shop floor—although true agreement is less frustrating, and honest, levelheaded objections are sometimes valuable.

A good tool for keeping up with conflicts between words and behavior, or words and nonverbal signals of all kinds, is what is called a *perception check*. This is a valuable concept to remember. The term basically refers to any question that asks if someone means what he or she is saying when the person's behavior is saying something else.

Harold falls asleep during the first act of the opera. This is sincere behavior. Brenda tells her husband not to be so dull. During the second act he remains poised and taps her hand to the rhythms. But the second act is even worse! His upper eyelids hang over his pupils, trying to put him out of his misery again. Brenda looks over and sees precisely the closing eyelids and sees right through Harold's taps and his smiles and his poise.

> *You really don't like this do you?*

Perception check.

The next day at the office Harold is giving a few employees free baseball tickets. He does this every year just to be a good boss. He gives Dawn four great tickets on the first base line "Oh that's very nice of you," she says as though she doesn't know what to do with them. Harold has a funny feeling but can't put his finger on it. Later he remembers that he saw Dawn at the opera the night before. Click. He goes to Dawn's desk.

> *Would you rather have some opera tickets I would love to give away?*

Perception check.

As a final point, be constructive. Try to avoid criticizing doublespeak, since everyone does it. You can make it clear that you understand someone's mixed feelings, but try not to be upset by doublespeak. You use the device as much as anyone else. Everyone depends on it.

Good Speaking Skills

If we think of good listening as a group of "receiving skills," then we should think of good speaking as a group of "sending skills." We should place tremendous emphasis on listening because conversation is reciprocal. There is a give and take. Conversation has a remarkable back-and-forth pattern that shapes the meanings of what we are saying. Often the progress is slow because the back-and-forth movement, usually prompted by questions, serves as "correction," just as though we were reorienting a flight path. A skillful speaker can shorten the length of a communication by speaking correctly in the first place.

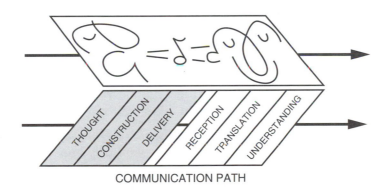

COMMUNICATION PATH

There are a number of basic rules that will build a practical speaking strategy. Talk is more a matter of design and control than an issue of words.

Sharing

Speak in words that will communicate and in a manner that will communicate. The vocabulary is determined by your judgment of the capability of the listener. The circumstance or situation determines the manner. Think of conversation as sharing; try to *share* words. This task is effortless for you and for me. We do it all the time. We adapt like chameleons. When the family all comes together for the big New Year's party, you speak in any number of different manners with differing vocabularies. There is a vocabulary for the younger grandchildren. There is a different way of speaking to teenagers. All your brothers and sisters get one or more variations of your constantly fluctuating word selection and manner.

Someone gets slang. Someone get shop talk. Someone gets sports jargon. Someone gets a few profanities. And the grandparents get all the respect and dignity they deserve.

At work you need to apply this skill of adaptability. You can safely judge the need for three specific levels of sharing: *downward, horizontal,* and *upward.* You will recall that we discussed the types of company communications in these three levels. When you speak to subordinates, you probably are in a position of some specific kind of authority, and you may have to speak in simplified terms if the subject is highly technical, but in normal terms if the subject is clearly understood. In a horizontal situation, you speak as an equal and speak in a shared vocabulary, even with regard to technical issues, because you know the vocabulary is shared. *But,* when you speak upward, beware of tripping on your vocabulary. Do not speak in a vocabulary greater than yourself; flexing your vocabulary can create trouble. Never snow. Snowing is usually transparent whether it is up, down, or sideways. More often than not, the only person who is bluffed is the person who builds the bluff.

There is a pitfall to speaking downward that you must guard against. Nothing will stop workplace communications faster than condescension. When you think of speaking simply, you must not speak simple-mindedly. Share. Always respect the person you are talking to.

You have to be careful with the idea of "sharing," because it can offend people. Our adaptability can, and does, get quite inappropriate at times and we do not see it. Parents who "baby talk" to children seem not to realize that children do not "goo goo" and "ga ga" for long. In this case the parents' adaptability is a fiction. The child seldom speaks in baby talk; baby talkers talk in baby talk. Fortunately, the voicing redeems the intent because the affectation is usually quite obvious in this situation.

Similarly, at work we must guard against subtleties such as talking loudly to foreign people in order to be understood. Foreign-born workers obviously hear as well as we do. The trick is to speak without slang, without contractions ("don't," "won't," "can't") and without idioms (thousands of expressions that say one thing but mean something else such as "get a load of that," "a shot in the arm," "keep an eye out," "catch the late show"). Sometimes people will also turn on an ethnic button and try to give their speech pattern a certain style because they are talking to someone who has just such a speech pattern, but the copycat simply creates a very embarrassing situation. The intent may be honest enough, but the result is not likely to be what was intended: sharing. In fact, if someone thinks they are being put on, the outcome can be disastrous. I might be flattered by a "gimme five" handshake from a black friend, but it might not have the same clear meaning coming from me to him.

What are the specific devices you can use to speak effectively? Precision and understanding. You must speak with accuracy, and you must understand what the receiver needs to achieve a maximum understanding of your intent. There are a number of more specific tools that will help you achieve precision and understanding:

Clarity

The usual approach to discussing good speaking is to dwell on precision. Know the facts, be specific, name names, use proper nouns, refer to brands, identify products by number and types, date activities by explaining when things happen, be clear in your feelings. Do not be abstract; keep ideas very simple. Every book on speech communications will speak of being "specific and concrete." The point is to think in numbers, names, events, and clearly phrased suggestions.

Numbers

Speak in numbers whenever possible. There is little likelihood that you will bury your speech pattern in statistics since most people speak in a dramatically opposite manner; they speak like their best recipes—without a single measure of anything. Think in numbers, from quantities, to ages, to parts numbers, to dates, to dollars, to math. The impact will not be extreme but the result can be very helpful because your language will gain precision.

Events

Every sentence is an event because it contains a subject and a verb. Most conversation is uncomplicated and consists of descriptions of actions, usually one per sentence. The trick is to speak about events in a responsible manner. This is particularly important at work. You must distinguish between events in much the way a scientist determines experiments. You must think of events as either true or questionable. By "true event" I mean you are constructing a conversation based on firsthand experience. Focus conversation on true events, on primary experience. Relate other people's experiences with caution.

An event is questionable if it is an *echo*. When other people's experiences reach you in the form of "echoes," you are the last to hear it. "Jones said that Brown said that Johnson said that Smith got a promotion to Anchorage." Beyond a certain point you should omit this sort of distortion from your conversations. Rumors are common in workplaces. Every child has played the game of whispering a sentence down a row of children to see what comes out the far end. The results are great fun—unless your union is on strike and you have a hotline number to call day and night to be certain you get the truth. Rumors fly in any crisis. Keep "near" events in your conversation. Speak of what you know to be true. Consider Ruth's comments, which she regretted:

Ruth's husband was a real popular guy where he worked. Everybody knew him and respected him at the plant. Shortly after his death, Ruth remarried. Ruth also worked at the same plant, but she had not told anyone about her remarriage. Her friend Beth was one of the first people at the plant to hear about the remarriage. Allan, another employee, has profound interest in such matters and Beth mentioned it to him somehow. She kicked herself for weeks about that because the circumstances of short bereavements are always treated as scuttlebutt. Allan was *too* interested and wanted to know more. Beth knew he would talk it up. Beth could hear the echo already! She had made a big mistake in mentioning the marriage. She felt it was her fault.

> *The problem with office talk isn't that anybody is gossiping; it is just*
> *that you always talk about people if you aren't talking about work. It*
> *is hard to be constantly on the alert about what I should or should*
> *not say. It's hard for anybody. But Allan was all ears and I didn't like*
> *that. I try to keep my conversations with him focused on business.*

Names

Names—names of anything—become real if an experience is real. If you are the voice of experience, name names: people, streets, devices, parts, positions—there is no end.

Contributions

Put yourself in a conversation with the full expectation that you will be asked to explain your ideas thoroughly. Accept contributions. Reciprocate with restated ideas. Listeners can usually contribute to your ideas. Accept help to bring your ideas to life. The reciprocal nature of dialog will do the job. It is one task to *describe* events or ideas; it is another matter to *evaluate* the events or originate ideas. Fortunately, you can quickly place most workplace dialog in one category or the other. Comments are frequently descriptive. They can also be evaluative. You may or may not have to add much elaboration to a dialog in which you simply explain what you *saw,* for example. The difficulty begins when you explain what you *think* about what you saw. Then the attention shifts from description to evaluation.

You may often find yourself sharing your technical specialty with a client. The dialog will depend on equal contributions from the two of you. For example, in residential architecture it is very apparent that people have trouble articulating space ideas when they imagine a design for a home. The architect's role is to take hazy ideas and define them. Ideas can always be handled by speaking rather than drawing or writing them, but "explaining ideas," developing thoughts, is the hardest level of spoken communication. The architect can build what the client can only imagine, but the architect must first allow the client to talk out concepts and ideas. Drawings will test the architect's ability to capture the client's intentions, and dialog continues until the concepts are clear. People often need help to express ideas.

Openness

A final note concerns receptiveness. Our earlier discussion of listening indicated that you should save your breath if someone is not listening. Build conversations when the environment is receptive. Do not insist on conversations, but wait for the moment. Whether you are listening or speaking, keep the conversation receptive. Be positive, and avoid strong or long negative comments, even though the comments may not be directed at anyone present. Everyone tires of complaining, and openness should not be confused with candid ventilation that can damage a dialog.

That there may not be a receptive condition for conversation suggests a new set of problems. We have looked at the conditions of good listening and the conditions of good speaking. If all goes well, these conditions can be managed. They become tools. But circumstances are often less than ideal, and you will have to use these suggestions in the presence of some mighty challenges. The next chapter will identify the problems you are most likely to encounter.

Summary

- Good listening is half of the craft of communication.

- The pace of life and the pace of the workplace create communication problems because it is hard to follow rule number one: stop and listen.

- If you stop and listen, you will then be attentive, which is rule number two: look and focus.

- With a serious effort to listen and focus you are on the way to rule number three: listen to understand.

- There is a craft to listening and there are rules to practice:

    1) Listen with complete attention.

    2) Listen by trying to understand.

    3) Do not judge what you hear.

    4) Be supportive.

    5) Ask for clarifications, paraphrase, or interpret ideas.

    6) Watch for doublespeak (contradictions between words and actions).

- Good speaking is the other half of the craft.

- You must speak with respect for the listener. Choose attitudes and vocabulary with the listener in mind.

- Precise details add to precise communication. Be precise about ideas or abstractions to the extent possible. Also be precise about numbers, any reference to events, and the names of things or places or people.

- Leave events untold that relate to other people. Secondhand information about people is prone to distortion.

- Openness builds trust. If you show openness and are receptive to others, you will be respected.

Activities Chapter 5

Present a memo to your instructor that develops a discussion of a work experience that relates to the chapter. The memo should be 500 to 1000 words, typed. In it you can explain how you have related the text or the lecture discussions to some event you have experienced. This exercise will give you a better understanding of the material because you will explain incidents in terms of your perceptions of workplace structure and communication.

Select one of the following suggestions and develop an analysis that involves your current employment or a former position.

- *Explore an incident that involved a misunderstanding of technology. What was the role of spoken communication? Did it create the problem? Did it resolve the problem?*

- *If you have been challenged by the difficulty of explaining your technology, explain the situation and how you handled the matter. This can be discussed as an ongoing issue if you are involved in sales or client relations.*

- *Relate an event that involved contradictory behavior or "doublespeak." Was the concession accepted or was the issue talked out?*

- *Test the stop-look-listen strategy at work. Be extremely attentive and supportive. Report the results and discuss.*

- *Discuss being mistaken about a supervisor's orders or similar work-related difficulty that resulted because you were not listening. Or discuss an employee who misunderstood what you told him or her to do.*

Barriers to Communication

6

Work in Progress

6. Office Politics

There are "rough spots" with any job, and my experience with Tempo was typical. As with most jobs, the first problems (or potential problems) to become obvious centered around the major characters.

Enter Tim's wife, Sharon. She is the vice president of Tempo and a serious designer. She designs most of the systems and has won awards for many of her ideas. Initially, I was tempted to not take Sharon's role in the decision-making process very seriously. During our first meeting, when I was getting a fairly noisy and distracting tour of Tempo, I noticed that she seemed bored when Tim and I delved into the more technical aspects of the network installation, and she wanted to know only if production would have to slow down or stop while the network was being installed. I was concerned. She is second in command with full power to approve or veto any company decisions. I began to ask Sharon if she would like any specific ideas or vocabulary clarified. She did, and I also agreed to fax her a copy of our company's "network primer," which we give to our nontechnical customers. Sharon told me later how much she appreciated having it. I would always ask her if she had any questions, and that showed that I took her seriously.

Enter Tim. The president of Tempo is the opposite of his wife when it comes to technical matters. Although he knows very little about networking, Tim Simmons is passionate about electronic devices and tinkering with them. During my tour, while Sharon kept quiet, Tim talked nonstop and had a million ideas and suggestions. He said he wanted to be a part of the installation and had tools I could borrow if I did the work. Tim second-guessed some of my ideas, and it was a treat to deal with someone who knows the jargon and has technical knowledge. He was also easy to "read," since he is a very animated guy.

On the other hand, he worried outloud about the cost of hiring someone to do the network when he and Steve (I'll get to him in a minute) could do it themselves. The hitch! A client like this can be a problem. I answered questions, gave short explanations on the points he questioned but was clear about the fact that, due to professional ethics and insurance, we wouldn't use his tools and couldn't let him be part of the actual work. In truth, as the president of my own company, I couldn't have another president at my side. That could get awkward.

The final actor was Tim's young nephew. Steve is a computer whiz. He can program in most of the popular computer languages as well as troubleshoot hardware, and he has been hired by Tempo to keep their PCs up and running. He doesn't, however, know much about networking or network equipment. His uncle Tim is justifiably proud of Steve's abilities and was probably right in thinking they together could eventually get some kind of upgrade installed at Tempo.

I saw my challenge as finding a diplomatic way to show Tim and Steve that although they could do the installation, they neither had the time to learn the skills nor the right tools to do the job cost-effectively.

J. Q. C.

Workplace Blocks and Office Politics

6

Most of our discussion has been directed at pinpointing locations where the speed of conversation or the complexity of language seem to create dilemmas. We assume goodwill as a basic condition of communication. This may not be the case. It's a jungle out there. It is not that malice lurks behind every watercooler or that every assembly plant floor is a snake pit. It is simply that our discussion has been directed at how to make conversations go right, and many conditions will make them go wrong.

What do you do if the shop is too noisy? What do you do if you do not understand a technical conversation? Or if you are being talked down to because you are a woman? Or if someone is just plain angry? You cannot assume that a good conversation is going to look like a controlled experiment conducted in a quiet classroom. When you are at work, you have no choice but to roll up your sleeves and dig in. There will be communication barriers everywhere. You must anticipate the usual difficulties so that you can recognize them when you see them.

The growls and howls and hisses of the jungle are sometimes very subtle and clever in the office or out in the plant, but you can meet them head on if you understand them. You can tame or temper the problems. You have to. The work must go on—production, wages, bosses, clients. You have to *manage* communication to make the entire system move onward. It is not supposed to be a snake pit. The solution begins with you. However, you cannot always expect predictable formulas to solve the endless variety of unpredictable people problems.

The dilemma inherent in communication suggestions is that formulas on the order of how-to-do-it suggestions are not difficult to come up with, but they may be difficult to apply because of the infinite variations involved in communications. In a given day you may have *hundreds* of "conversations," ranging from hallway hellos to intense conference meetings. These events will, however, follow patterns; we can group the communications into clusters using any of dozens of criteria to define the divisions. The *barriers* to good communication are predictable, but they are subject to infinite variation. No two conversations you will ever have can ever really be the same, and the barriers you encounter will equally be subject to variation. Nonetheless, barriers to good communications fall into boxes or types, even though you cannot expect a "textbook" case when you actually confront people problems. Obvious problems such as shop noise or what is called *information overload* are identifiable. They "block" communication. Other barriers are often subtle, including all cultural barriers. For example, the sheer agreeability of many immigrant people who try to speak English, but for whom English is not a native language, may mean that none of the smiles and yes's and nods of acknowledgment can be relied on. These are blocks to communication if you do not recognize the confusion behind the smiles of agreement.

I will use the communication path illustration to encourage you to analyze the *sources* of different communication problems. Sources suggest solutions or strategies that, in turn, help remedy the problems. Let's begin our discussion of barriers with a very basic problem that can occur in the workplace: noise. Noise is one of the major communication barriers, or blocks.

Noise

A common and obvious communication block is noise. Many workplace situations are too disturbing because of the noise. Leave a noisy area when you want to speak to someone. Job safety for you and your coworkers may depend on getting the message across.

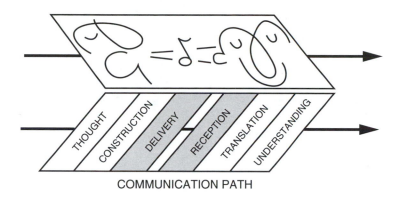

COMMUNICATION PATH

Also be aware that some people have more difficulty hearing in a noisy situation than others. You may notice that older employees hear less. The problem is not age but too much noise for too many years. Damage to hearing is a national problem in the workplace.

Tolerating noise is simply *enduring* noise. The ear *does* suffer. The result is what is known as "peripheral hearing loss." The symptoms are easy to diagnose. Brian spent thirty years working with power tools. Somewhere in his early thirties he noticed that conversations in his van were difficult for him to follow. He thought the problem was the noisy van, and it was, except for one thing: *other* people in the van *understood* the conversations. Only *he* couldn't follow the talk.

> *Actually, it got pretty funny. I would always be yelling "What?!" and that noisy old van had us all yelling. But at some point I realized I was the only one who didn't get the message. After that, I could tell when other people had the same problem because they did exactly what I did and kept asking people to repeat things.*

Peripheral hearing loss is the failure to distinguish and separate discrete sounds (such as a voice) if it is mixed with other noise. The cause: noise.

If *you* hear but realize that someone else is having difficulty because of noise, leave the noisy area. And start to protect your ears, even when you mow the lawn.

What is the point of origin for this problem? In this case the problem is a mix-up between delivery and reception that is caused by unwanted sound (see the preceding illustration).* The mix of delivery and perception becomes jumbled.

Word Walls

Vocabulary must be included in the discussion of communication blocks because people in business and industry are highly specialized nowadays, and their vocabularies set up barriers. The barriers fall into two types: unintentional blocks caused by technical jargon and concepts, and intentional blocks caused buy a desire *to look* impressive. Let's deal with the second issue first: "muscle words" and "buzzwords."

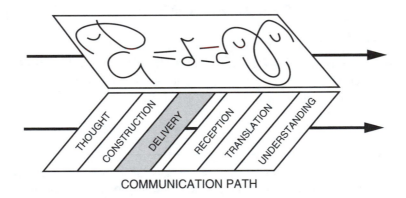

COMMUNICATION PATH

People who talk over your head are probably not trying to make you look stupid. In fact there is the possibility that you will be flattered because someone may easily assume you *do* understand. Nonetheless, some people are not very concerned one way or the other;

---

* Pages 61and 64 explained the communication path illustrated on the preceding page. The shaded areas indicate the location of the problem under discussion.

they are more or less chest-thumping. In effect they do not much care about what listeners think, or what *they themselves* think for that matter. For them the medium is the message. Words become power.

If vocabulary is self-serving, it does not communicate. *Muscle words* are the technical terms that you might slip into conversations to leave a good impression of your professional knowhow. *Buzzwords* are the current trendy words in business and industry. Such words as *viability* and *interface* come and go much like slang rises and falls. *Interface,* of course, like most such corporate jargon, evolved from technological terminology, except that it can mean "power" breakfasts or coffee for two in its trendy use. These terms are highly unnecessary. Buzzwords are language junk anyone can identify and stop using. Are you "comfortable" with that? See if you can "access" a few examples yourself. They "impact" your ability to understand.

The problem of being understood in a technological field can be a very difficult matter and a very serious one indeed: it can spell profit and loss. A sales team must be trained with great skill to overcome the tendency to *flex* words. Perhaps service is second only to sales for its importance in the outreach of industry. There must be no chest thumping in either field. It is very difficult to speak the language, not of your company or business, but of the person you are assisting or serving.

The old adage "the customer is always right" means so much more than customer privilege. It should mean customer *understanding:* what the customer thinks or understands or knows *is* what is right, not the knowledge of a salesperson or service tech. You must make the effort to speak "across" and explain your knowledge in terms that can be understood (but never be condescending, or sales will be lost and contracts will be canceled).

Spell communication with the word *patience.* Every word was explained. The transaction—Sandra's first house—got so complex that Pat, the real estate agent, had to call in a partner of hers who specialized in "creative financing," which meant putting together a mortgage with imagination and unconventional practices. Pat humbly admitted that she needed help even though she had years of experience in real estate. When the expert arrived, Pat explained every word *he* used also. Sandra was very impressed with the real estate agent and appreciated the explanations. Pat was a good teacher. Teaching may not "sell," but patience does. Patience will bridge. Pat's profit was five thousand dollars.

> *I got Pat lots of additional business. She was so helpful. There was so much real estate jargon and endless papers to file. I sent all my friends to her and she handled my second house also—and that was seven years later!*

You can build bridges over communication barriers. Joan recently called her telephone company for a business phone in her home. She wanted to know the rates for installation and so on. The employee informed her that the "systems interface for the new phone line would be installed." Joan was puzzled and said "What does that mean?" The phone company representative said, "the guys have to come out and put a phone jack on the outside of the house." This situation suggested that the phone company employee wanted to talk in jargon but could perfectly well speak in lay terms also. Why waste time? Buzzing does not impress anyone, and it does not explain the bill either. **Speak to communicate, not to impress. Speak to bridge.** In fact, technical sales services probably should *teach* in order to sell. Service serves best if the client is *taught*.

Technology can be difficult to sell because of the vocabulary that usually is an inseparable aspect of a technology. Have you ever gotten the rush in a primo stereo store, the kind with ambient lighting and plush rugs and free coffee? Here is the ideal setting for educating a client, but it takes a sensitive employee to use communication instead of abusing it. Few people *really* understand audiophonics and the latest in the stereo electronics component systems. Perhaps people will buy a sound, or maybe a brand, or sometimes a "look" in equipment. Few of us are prepared to discuss frequency rollover response, crossover networks, integrated frequency locking, mixing a Dolby B FM take, or whatever. Although a salesperson may be flexing muscle by talking such technical stuff, the motive is to sell the *technology* of the components—hence the technobabble. But what does someone care who wants to play his or her favorite Willie Nelson records? Selling technology is another version of selling horsepower and cubic centimeters ("Hey guy, 400 watts per channel! Blow your windows out!"). The pitch speaks, yet it says nothing. "Feel the power," as the saying goes in the well-known stereo speaker ad.

The origin of this problem is obvious: words. We would call this a delivery problem in our illustration of locations for communication barriers. The motives play a role in this practice, but the real problem is vocabulary.

Overload

For years, there has been ongoing lobbying by the secretaries at the Aronson Corporation to change the personnel policy to create a "level" of secretary to function as a receptionist, someone to handle all the inquiries from walk-in and phone sources. Probably for lack of funds, the receptionist idea has never been adopted, so the secretaries, being a resourceful group of employees, try to see to it that the entry-level rookie handles all these conversational details. The problem is that Aronson, like most companies, puts the secretarial staff in *very* public spaces. The result is that they are constantly on the phone answering an inquiry, at the same time they are waiting on people, at the same time they are trying to get typing done, at the same time they are preparing for a meeting, at the same time . . . .

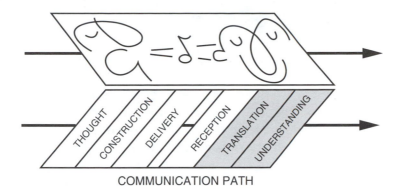

COMMUNICATION PATH

This is "overload," a typical communication block. It is familiar to you if you have young children and invite your friends, who also have young children, over for a relaxed evening dinner. It is also a problem for most community college students in technical and engineering programs. They rarely are the eighteen-year olds you thought they would be. They are in their twenties, thirties, and forties. They have jobs—often a full-time position or two part-time jobs—and they have families. Then they add fifteen or twenty credits of college work. The result: overload.

Too much input causes confusion. In the instance of your own personal overload, you must be alert to your performance risk. Overload spells inefficiency, and your job performance or your grades will show the result. Even your safety can be a risk if you are overtired or confused, not to mention whether you will still accomplish other important tasks at home. Always try to help overloaded workers by telling them you see a problem. They need that recognition. They are hardworking people who need to know when enough is enough. You do not have to say much, but it helps the overloaded employee to have someone see the problem. Often a solution will follow.

Not only can you be understanding, however; you can also remove yourself from the load. Consider the example of secretaries. If you go to the secretaries and they are busy, perhaps you should leave. You can also avoid their services by being as efficient as possible so that you can limit the work you require from them. The opposite tactic would be to barge in while a secretary is trying to wait on someone because of a must-have-ASAP project. Self-importance can blind all of us at times.

What is the point of origin? The messages are received, so the first half of the communication is completed. The muddle occurs when too many messages are translated and the mind has to try to deal with them—more or less all at once. The problem is the receiver's ability to cope. The understanding of the message will be confused.

Dialog Cutoffs

A few barriers are nothing more than mechanical blocks in effect. If you converse with someone who interrupts you frequently, you will experience some sort of cutoff with each intrusion. **Frequent interruptions (often questions) are annoying and can destroy a conversation as easily as a noisy truck outside the office window.** The problem is that a speaker is stymied and does not get anywhere, and may lose either the control of thought or the interest in the effort the conversation is taking. You might think that both speaker and listener would control interruptions because the problem is obvious. The interruptions are not only conspicuous, they are annoying. The speaker needs to be assertive: "Hey, let me finish!"

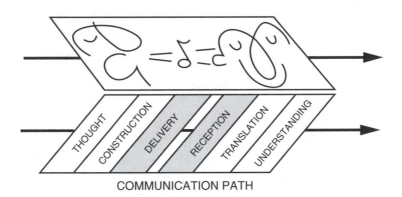

COMMUNICATION PATH

Frequent questions are one problem, but a number of cutoffs that destroy communication are *not* caused by questions. A common cutoff is one we might call the *completion*. It is that habit some people have of completing our sentences for us. We start a sentence, and then out of nervousness, impatience, or excitement or some such cause, suddenly we hear "and then . . ." and find the story written for us. If you relate a sports event to someone, be on your guard if someone else is standing with you who also saw the game!

Another cutoff is the *interjection*. People often cannot resist adding a point, or a bit of wit. For every detail of your ninth inning story, the other guy throws in his two cents worth. Even worse than the two-cent cutoff is yet another trick, what might be called "the cheap shot." Your story is interrupted not to add details but to tell you that you are wrong! Even if you are, he could have told you about it later rather than cut you off with *corrections*. And in front of your friends, no less! The reason we have these slang expressions for such interruptions is because they are well known as rude annoyances.

All these interruptions are conspicuous, and you can control yourself to readily avoid them. Awareness will limit your tendency to use interruptive blocks. The greater problem is controlling your listener's bad habits. You must politely and with coolness explain to the listener what he or she is doing. If it happens to be your boss, use great coolness. At times, it would be helpful to tell people to shut up and listen. It is more effective to explain the problem to them.

> *I can't explain if you don't let me explain.*
>
> *No, that isn't what I was going to say.*
>
> *Why don't you tell the story.*

Consider this a mechanical problem, and call it a delivery dilemma in which delivery is blocked, but blocked by the intrusions of the *perceiver!*

Stereotyping

We have to look closely to analyze the usual causes of misunderstandings. A misunderstanding may have to do with vocabulary or some practical matter, but the forces at work are often complex. Unshared meanings can be caused by differences in education, and they are also likely to be caused by differences in space, time, and beliefs. Age matters, nation of origin and even *state* of origin matters and religious persuasion or lack thereof matters. In other words, from race to hometown, variables occur in time and space to differentiate us. One result of differentiation is a tendency to stereotype people as "different."

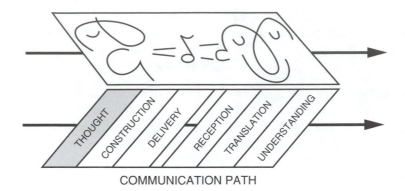

COMMUNICATION PATH

Stereotyping is commonplace. It is hard to avoid generalizing, and generalizing can result in stereotyping. In fact, stereotyping is an effort to be specific, not general. Stereotyping is usually likened to name-calling, and in this respect it tends to point an accusing finger at people "John has an alcohol problem. He is Irish, you know." Usually the stereotype is based on a specific consideration—let's say alcohol—which is then generalized in the comment "all Irish are prone to alcoholism" or "all Irish are drunks." It is a false generalization based on a specific assumption, which is also false in this case. Negative stereotypes are such a cultural taboo nowadays that ethnic jokes are only whispered, and you rarely hear public slurs about social groups anymore. We seem to be learning.

Being politically correct does not mean that we do not see the world in groups. Ethnic stereotyping receives public attention, but the tendency to stereotype people is a much larger issue and often involves no ethnic perceptions at all.

Ray and Dave had been pretty good friends. Dave is a supervisor, and an employee who came up through the ranks. He got along well with his coworkers who still saw him as one of the guys. He did too. During a perfectly normal conversation one day, Ray got very angry about supervisors and ranted away in front of Dave—not at Dave. Dave was very offended because he was, after all, a supervisor. If Ray thinks they are all jerks, Dave reasoned, Ray must think he is a jerk also. He took offense because of the stereotyping Ray had used to blast supervisors.

That really rubbed me the wrong way. He carries on about managers and pretends I'm different—at least in front of me. I didn't have much to say to him after hearing that.

Lumping people together can cost us friends without our ever knowing it. Stereotyping is not just an issue of race relations. It is *very* commonplace. **We generalize about groups far too much. The groups can be women, men, children, workers, doctors, supervisors, carpenters, technicians, clients, or a hundred others.**

Also be wary of what we might call "positive stereotyping." Consider the "natural rhythm" compliment, for example. It may sound flattering to say that a group has a specific attribute, but it is clearly stereotyping and should be avoided. Social groups do what they have to do to succeed. Yes, the Irish did once dominate police departments, but not because of a genetic preference for policing. Yes, the Polish did fill the steel mills with workers in the early decades of the twentieth century, but not because they had strong backs. Sociologists explain that available access to success is the only condition a new immigrant group will look at. And they rarely dominate, for example, a career field. It is the stereotype itself that is dominating, and it is false. Maybe many entertainers *are* African Americans in our time. Thirty years ago they were Italian. Fifty years ago they were Irish. Somewhere in between they were Jewish. These, among others! There seems to be enough talent around that we do not need genetics to explain it.

The point of origin for this block is the mind of the observer who sees the stereotype and translates communication in a judgmental way: "For their age they sure are energetic!" The challenge is to recognize individuality: "Sure he is a politician, but he is one of the good ones!" If you see a group instead of an individual, then you cannot communicate very honestly.

Deceit

To misjudge people is one problem; to be *encouraged* to misjudge them is another matter. Deceptive behavior is often motivated by the hope that a person can be manipulated into a certain perspective. The hidden motive is the problem here. In the workplace the polite expression for the hidden motive is a *hidden agenda*. The term somewhat indicates that it is a conspiracy, and it is.

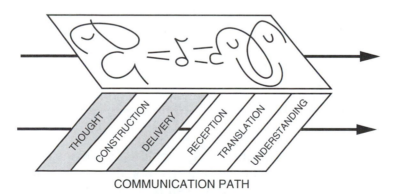

COMMUNICATION PATH

For example, a supervisor might sabotage talented employees who could threaten his position. He might let an employee leave early but report it as though he had not been told of the departure. He might provide faulty directives on purpose, construct impossible goals, or send the employee on all sorts of wild goose chases. He could easily overwork the employee to get rid of her. Always the motive is hidden. Then there is the old trick of "promoting" threatening employees—promoting them to Tierra del Fuego or Timbuktu. Essentially, dishonesty *depends* on misunderstanding. **Innocence is what makes dishonesty effective, because we easily misunderstand people with ulterior motives.** We do not suspect them—at first.

Communication is, after all, self-serving. It is only natural that we will use all our communications to our own ends. We can identify deceit as any communication that will focus on distortion in such a way that the receiver of the communication is duped. Sincere communication does not deceive or dupe. Of course, the reasons for false communication are very complex, but if deceit is suspected, the result will ever thereafter be distrust. It is best for all of us always to be honest.

Jason was a popular supervisor at the midmanagement level. Everybody liked him. He hoped that he would someday receive a promotion into the CEO ranks. He was finally granted a vice presidency. However, his appointment coincided with the appointment of a new president. The new president was not popular with the rank-and-file employees. Jason found that he had to consistently defend the new president against employee criticism. Of course, he was defending his boss, so Jason had to ignore the fact that the president actually was *not* effective. His own job was on the line if he crossed his supervisor. His popularity with employees began to plummet because he was in a dilemma: if he admitted the truth he could be dismissed; if he did not, he would lose support from the ranks. He made his choice—to defend company policy—and the employees saw it as a hidden agenda.

> *It was a disaster. If only Leon had not retired! Tom wasn't a popular president; I know that. But he wasn't some kind of incompetent. I worked with him and he was a talented executive. I just couldn't convince the old-timers.*

Right to the end he defended the president. After the president was, in fact, removed, Jason continued on but not with much of his original rank-and-file support. After all, Tom was removed because he was not the person for the job.

Honesty creates its own frictions, but distrust is not likely to be one of them. Honesty may not always be the best policy, but deceit is likely to be the worst policy. Honesty runs its risks, but people know where you stand. Deceit is another matter. *Deceit cultivates bad faith and lack of trust.*

The point of origin for this problem is in the mind that originates the intent to deceive.

Distrust

Two types of emotional barriers are major trouble spots in communications and set up huge blocks. The first, *distrust,* is usually obvious. Distrust is a problem in communication because the entire premise of conversation—sincerity—is missing. **Usually the distrust is caused either by defensiveness or suspicion.**

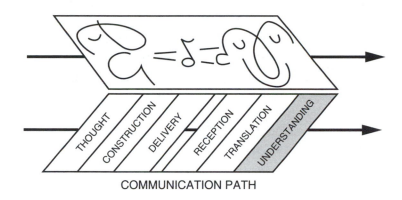

COMMUNICATION PATH

If you have a job that involves "correcting" people, you will find that defensiveness can be a major difficulty from time to time. As long as defenses are up, explanations are meaningless. In my business I evaluate a lot of papers. They are not all "A" papers. The problem is that as the grades get lower, defenses rise. Very often the resistance rises in proportion to the need for a student to understand problems, which makes the task impossible. Any employment that teaches will be corrective in some measure. Any supervisory role has the function of being corrective also. Of course, people might be naturally defensive, but let's assume that your job prompts the problem because you have to tell someone what he or she did wrong. What can you do? Here are a few key strategies:

- Make people realize you are there for their benefit.

- Share the corrections by placing emphasis on your own similar errors, or the commonness of the errors: "Everybody does it."

- Be descriptive and avoid value judgments. Say the job performance is "not adequate." Do not say it was "awful" or a "waste of time" or "junk."

- Suggest solutions, and use the words "maybe" and "perhaps." Use these words to suggest solutions rather than demand them.

- Help people see solutions themselves.

- Be sure to avoid superiority. *Authority* may be appropriate, but *superiority* is a very different posture.

- Speak as the *equal* of the other person. You can be equal and still hope for respect as an authority, but cannot be equal and superior at the same time.

Another typical cause of distrust is *suspicion*. For example, many people react with hostility to any sales pitch. A sales routine of even the briefest sort is something many people do not handle too well. Suspicion is a big challenge to a sales person. Of course, sales involve unique motives we all recognize—but the agenda certainly is not hidden. Because the sales motive is obvious, suspicion is easily aroused.

Suspicion can be an element in a conversation between two people. If suspicion becomes a factor among workers, for example, or a factor between you and a friend, the problem is far worse than a sales pitch, but far easier to solve. Why? Because, again, sincerity is the solution. Many people can never quite get past the falseness of a sales pitch, but they might jump into any problem that looks "suspicious" among friends and coworkers.

What's the matter? You don't seem to believe me. What's wrong?

Sincerity, and the tips outlined previously for dealing with defensiveness, are your best tools.

The origin of this block is the receivers' understanding. If a motive is ulterior, there is a deceptive intent at work, as we noted in our discussion of deceit. The *communicator* creates the barrier. Once that deceit is perceived or suspected at the other end of the communication loop, the other party becomes distrustful or suspicious. Therefore, distrust and suspicion are also barriers, but the *receiver* builds the barriers. The problem leads to a fixed perception of the communication effort, and perfectly honest client relationships can come to be seen, for example, as sales-pitch routines. Communication is then blocked.

Games

We might call the problem of games the high point among barriers. This group of problems results in situations with deep-seated, serious conflict. The problems surface in a thousand ways, but two emotional patterns are generic: *aggression* is one and *surrender* is the other. It takes two players to make the emotional pattern a "game." Your job is to make sure you are not one of the players. You cannot referee the game if you carry the ball.

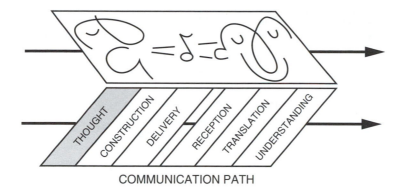

COMMUNICATION PATH

Aggressive behavior is rarely physical, and this text is not equipped to speak about how to handle that sort of problem. The secret is to handle the behavior before you need help. Aggression is occasionally spontaneous. More often, people are frustrated and pent up and boil over at times. **In the presence of aggressive behavior, it helps to establish psychological distance.** Use time (an hour, a day) to remove yourself from the event and look for breathing space. You need *psychological* distance and real distance. Use time and space second—*real* distance. Then draw on a host of resources we will discuss in this book and try to reestablish good communication through all the practices you can utilize.

Not long ago Barbara, who acts as Joan's secretary and who is also the secretary for five other people, got quite angry with Joan. Barbara is tough, outspoken, and proud of it. When she is mean she is very mean, and proud of that too, perhaps. Joan wrote a memo that Barbara felt made her look bad. The memo had nothing to do with Barbara, and Joan would never have even thought of Barbara in connection with the issue of the memo—not even remotely. Barbara yelled, stomped down the hall after Joan, swore at her, banged down papers, got red-faced—the works. Joan was more amazed than annoyed. It was easy to stay calm and cool. The challenge is to *stay out* of any reason to get upset. Joan made a few firm comments to express her dissatisfaction with Barbara's behavior, and left.

> *I had seen her temper flair more than once, only this was the first time that I was the target. You see, I had told accounting that we had to redo a few account histories because there were errors. Barbara thought that she was being singled out because she kept the books on several of them. I wasn't about to get in a shouting match over such stuff. Besides, she calmed down—although it took a week.*

Joan did not get ruffled when Barbara got angry and so she had good control of psychological distance. Real distance helps also. Joan avoided Barbara for a week. She did *not* apologize to Barbara by the way. Barbara was being very annoying and Joan did nothing to hurt her feelings. A week was the correct judgment in this case. Barbara's temper would not have changed in one hour or one day. I recommend removing yourself from the hostility. How long you wait is a difficult question. If you have young children at home and there is ongoing disagreement between you and your spouse, perhaps you cannot wait a week. It might not be fair to the family. Every circumstance will be different.

There might be times when you will have to mediate a situation where someone is openly angry. Technical services often call for people skills and diplomacy. The machines or systems are not the only problems to be dealt with. Engineering technicians can and do run into people problems, and service people can sometimes help when things are not quite right among the clients.

> When Valerie called the service tech because her computer malfunctioned at her office, the service technician arrived promptly and got to work—cleaning up coffee inside the keyboard of the computer! Valerie *denied* that it ever happened. Patience. Her boss, a fairly nervous chap, overheard the conversation with the technician and realized the coffee did not put itself in the computer. There were words, but the service tech did not just slip out the door until he helped resolve the issue in a diplomatic way. Patience.
>
> *Her boss just lost it. I mean this didn't happen in front of twenty people. There were just the three of us in the office and some people in the gallery—but the guy has a temper. I had to stay and calm them down. This doesn't happen every day, but office politics is pretty real and so I have to deal with it sometimes.*

Service must educate. There will never be coffee on that desk again, and the service tech, if he is sharp, will look for coffee rings around the computers each time he makes his rounds in his service area. From time to time he will educate the staff about what could happen. Perhaps he will even tell the Valerie story. People skills and people understanding matter.

If you apologize for something you really did not do—even the basic "Okay, I'm sorry!"—you experience the other deep-seated barrier, that of *surrender*. Surrender and withdrawal are similar issues because emotions cave in. There is no aggression. People give up. The problems are twofold. People who cave in are going to hold resentments. The second problem is that many people decide to accept the victory! If a man pesters his spouse about a new car and she finally, one night, starts to cry and says "Go ahead. I don't care. Buy it!" I hope he resists the temptation. Victory is not sweet when people surrender. **You do not want an employee to surrender or withdraw; you want an employee to *understand*.** Surrender and withdrawal are expressions of resistance regardless of any victory someone might gain.

Because surrender and withdrawal involve doublespeak, you have your choice. You can argue your case, but you must not be thoughtless. In your better moments you realize you have *not* gained what you were really after: understanding. The perceptive tactic is to start all over again, with new communication, if you want honesty. The question of surrender is simple. Do you want to win an argument, or do you want to be understood and be understanding? You should not take the ball and run. Always perception-check a surrender.

Our own thoughts are where we construct hostility and passivity. We all have egos and they can be defensive. Hostility is hard to deal with, but at least it is obvious. Surrender or withdrawal can be subtle issues and call for close attention if we want to resolve the barrier problems of passivity.

This chapter concludes the discussion of company power and communication, the nature of spoken communication, and the problematic areas that create barriers to communication. With an awareness of these conditions, you are in a better position to understand your coworkers and clients, and you are in a better position to be sure that they understand you. These chapters have focused on aspects of communication that are universal and problems that are commonplace. An understanding of these conditions will help you negotiate the day in the ever changing relationships among your coworkers.

There is, however, another dimension to communication that we can investigate in a very different manner. Just as daily communication has endless variables, certain communication practices have few variables and can be studied precisely and practiced with skill. These "fixed" varieties of communication involve the communication activities of committees, spoken presentations, sales, and interviews. In the following chapters, we will look closely at these important features of workplace communication.

Summary

- There are conspicuous communication blocks that can make it difficult for speakers and listeners to understand each other.

- In the workplace, noise is a frequent intrusion. Be sensitive to noise problems and adjust your environment accordingly. Stop the noise or move the communication elsewhere.

- Internal noise consists of those drumming preoccupations that distract a person. Stressors of all types can create enormous blocks that limit communication because they are just as loud, in their own way, as roaring machines. Noise can be industry; noise can be the industry of the mind.

- Choose your vocabulary with care. Engineering technicians are very vulnerable to being misunderstood if they encounter the public sector. Speak to communicate, and never speak to impress.

- Employees often "overload." They are often responding to too many demands on their time and too many demands on their attention. Respect overload, both your own and any overload you perceive among fellow employees. Communication is difficult for an overloaded employee. In such a situation, there is too much confusion, and the focus of attention is weak.

- Listen with respect. Do not cut off a speaker's comments in any way, or the communication will be damaged.

- Avoid stereotyping by being aware that seeing people as groups is a far more commonplace habit than the much discussed issue of cultural stereotypes.

- Be aware that communication can be other than well meaning. Deceit can be intentional. Distrust is also common.

- Do not enter a loop that is aggressive. Keep your calm.

- Avoid false surrenders, and recognize this doublespeak when you see it. Try to resolve the surrender.

Activities Chapter 6

Present a memo to your instructor that develops a discussion of a work experience that relates to the chapter. The memo should be 500 to 1000 words, typed. In it you can explain how you have related the text or the lecture discussions to some event you recall. This exercise will give you a better understanding of the material because you will explain incidents in terms of your perceptions of workplace structure and communication.

Select one of the following suggestions and develop an analysis that involves your current employment or a former position.

- *Explain what your life was like when you worked in a noisy workplace, whether it was a shop environment or an office building.*

- *Identify an information-overload situation. Evaluate the parts of the load and suggest solutions. The employee in question can be you or someone you know. Someone with two jobs and a family might be a good case.*

- *Explore one or more cases of people stereotyping or gender stereotyping at work. This can involve cultural stereotypes as well as age images, union-worker images, manager images, and so on.*

- *If you recall an example of deceit among employees, explain the events and the outcomes.*

- *If you recall an example of distrust among employees, explain the occurrence and how matters were resolved.*

- *Have you ever worked with an employee who became aggressive? What was the background? How was the employee handled?*

- *Have you seen a case of withdrawal that became severe, perhaps severe enough for an employee to seek employment elsewhere. Discuss.*

- *Have you ever lost a friend or acquaintance as a result of barriers between the two of you? Explain.*

Group Decision-Making

Work in Progress

7. Important Meetings

Once I had made the initial contact with Tempo, it was then time to take my findings back to Randy and Cloe (my two employees). Though I am president of NetWorks NorthWest, I do not make the decisions about which jobs to bid or which jobs to pass on without having a meeting with my crew first.

Our meetings are informal and are usually held at a coffee shop, but they are all business. First, I give Randy and Cloe copies of my notes from my walk-through tour. I include layout diagrams and observations about difficulties we might have with the installation. In the case of Tempo, I described the problem of the company having two separate buildings with other businesses sandwiched in between. I also described the situation of an overinvolved president, an underinvolved vice president, and a touchy computer tech who happens to be the president's nephew! Once I presented my findings, I waited and let Randy and Cloe voice their thoughts and concerns. I wrote down the major points.

The concerns Randy and Cloe had about the Tempo installation mirrored my own. The physical problem of the split "campus" could be fixed by using fiber optic cabling as the LAN's backbone. This would be more expensive, but not insurmountable. We decided together that the total capital outlay for NWNW to do this job would be about $30,000 and that our price to Tempo would give us enough profit to make the job worthwhile. Cloe worried that this much outlay would leave us strapped for cash until the installation was complete and thought it might be going too much out on a limb. Randy reminded her that we could arrange the billing clause in the contract to make the first contract payment large enough to cover our initial outlay. This solution satisfied Cloe, and she had no further objections to the bid.

The potential problems with the human element of the job seemed more problematic, and for a time Randy and I locked horns over it. Randy saw the dynamics between the computer tech nephew and any outside computer professionals to be too volatile and not worth getting into. Although I saw his point and told him so, NWNW is not yet in a position to turn down an otherwise ideal job because of a potential problem of this sort. To make the job palatable for Randy (if our proposal was accepted), I assured him that he would not have to deal with the nephew. If any problems with him came up, I would deal with them. With this objection taken care of, Randy seemed satisfied, and we moved on to scheduling the job.

We figured how long the actual work would take, then added two days to the number. The additional time gave us a buffer in case we needed it and would be a great boost for NWNW's reputation if we finished the job in less time than we bid for. The three of us wrangled over dates and times but finally came up with a workable timeframe. Having ironed out the details and the problems together, we were now ready to finalize our proposal to Tempo.

J. Q. C

Group Decisions in Company Settings

In the social fabric of many business activities there is a frequent need for employees, employers, and clients to come together in one combination or another. Meetings are a constant in much of the white-collar work world. A wide variety of these meetings are utilized in larger corporations. Each type of gathering serves one or more practical needs for communication between members of the workplace. The need for group activity of this sort develops in a number of quite regular patterns. Most businesses use one or more of the following types of group gatherings.

Standing committees Certain gatherings are a historical practice or regular event in many business situations. They are as predictable as Monday mornings or the first of the month. The activities may vary from "updating" to any of the other purposes identified in other meetings on this list. Because the meetings are regular, faithful members often form a network. You can tell that the network is in place when the group meets even without an agenda. The regularly scheduled coffee klatch has a similar use and value.

Briefings are another type of very commonplace group gathering. Although a memo often serves as a more precise tool for routing information, it is only at a briefing that a group has the opportunity to ask the questions that are of concern to employees. Obviously, this opportunity serves the purposes of both manager and staff because of the clarification of the information that is presented and discussed. A memo is much more likely to set off rumors than a thorough briefing, particularly a briefing that includes a detailed handout of new information *and* discussion.

Instructional meetings are the legacy of today's technology and today's bureaucracies. The "machinery" of both of these arenas is everchanging and, sad to say, evermore complicated. The turnover in state-of-the-art systems, such as in computer technology, for example, constantly creates the need for training services. (The federal government spends billions yearly on computers that, as you would guess, replace computers. The current budget for this process exceeds $10 billion.) Instructional meetings have become so common that "training services" have become a division of many companies. The training staff offers frequent instructional meetings to familiarize employees with technical changes. Similarly, as new local, state, and national laws and regulatory controls come and go, instructional meetings must be called to keep pace with the paper legalities.

Decision-making meetings are widely held throughout the American corporate world. Executives usually have the power of executive decision, and strongly willed leadership employs many workers. However, enterprising managers in all industrial sectors of today's world are likely to work out decision-making matters through executive teamwork. To act by independent decisions may be to act in the dark. The legal ramifications, and the technical ramifications of today's decisions, are often well beyond the near view of any one or two executives. Decisions are frequently a group response—and not just at the top levels of the company.

Consultation meetings are held to take advantage of the inputs of all members of a group. Although "consultation" may seem new in traditional "union" industries, it is an everyday activity at, let's say, an architect's office or an engineering firm. Teamwork is fundamental to commercial and industrial architectural design, and the shared responsibilities of the team are critical to the success of the business. Large unionized industries—for example, the auto industry—seem to be moving more toward the idea of "consulting" with employees. Certainly they have moved from the "suggestion box" to real activities such as "quality circles," a type of consultation meeting.

Confrontation meetings are a traditional tool for industry. Labor-management agreements are the usual source of this sort of group sparring between the unions and the management teams of corporations. The word *confrontational* is a fairly accurate term because both "sides" truly believe they are right and *know* the "opponent" is wrong. At lower levels of industry the confrontational meeting has the additional purpose of allowing groups to express their concerns, with a defusing effect, it is hoped, that will be of value to all involved.

Brainstorming meetings are used for developing creative responses to business needs by generating new ideas and new ways of looking at issues. Later we will be more specific about how brainstorming is supposed to be done, but most businesses use the word loosely and hold meetings to brainstorm ideas, by which they mean to offer possible solutions to problems. In many instances this sort of meeting is a reaction to an immediate crisis of some sort, and the meeting is held to put out the fire.

As you can see, the variety of the meetings is largely a function of the *utility* the group serves. Although this list identifies only seven types, there are dozens of meetings every week at most companies. Most employees are involved in one or more such gatherings—and that includes everyone from the physical plant area, through the clerical groups, to the executive offices.

The Risks and Rewards of Cooperation

Many men and women who have sat on committees have very specific opinions about the value of the committee concept. Many agree with the old joke that "a camel is a horse designed by a committee." There is plenty of evidence to suggest the truth of group inefficiency. Perhaps the biggest gripe is that committees waste time. Executives think the decision process is too slow. Employees think that, for all the hours involved, the outcomes are not worth their time—especially since their workload is unchanged and all the work back at their desks is still piling up. Few companies provide perks for this sort of activity, although some companies factor committee chores into pay bonuses—with great success!

Another gripe that is common is the idea that committees exist to do undesired jobs, which can be the case. The role of many a committee is defined by a task that is actually going to take the pressure off someone else. We should look at the situation closely, however, because it is not simply a matter of the shrewd boss who wants to say "*we decided* to eliminate coffee breaks." Steamrolling a decision by committee or massaging a committee to make a decision are tactics that are too obvious to be the *real* problem. More likely, the *difficulty* or the *complexity* of the tasks generated the committee.

A more subtle criticism that occurs from time to time concerns the issue of compromise. It goes without saying that a committee—whether it consists of five people or twenty—will have to rely on compromise to make decisions. If you are on a committee, you are as likely to disagree over details as you are to agree. Except under the most ideal circumstances, there will be a need for one or more members of the team to compromise. As you might imagine, no one is going to complain who sees a committee decision go the "right way," but there will always be times when members do not go along. For many members of committees this conflict can result in an unpleasant or undesirable compromise of personal values. Because of a number of employee variables—from job status, to gender, to religion, to personal perceptions—the issue of being "compromised" by committee decisions is common.

On the bright side, two heads are better than one. We do not want the forces around us, from our nation's president to our military leaders, or our supervisors at work or even our teenage children—to act alone. **Group decisions are likely to be superior decisions—even if they are slow.** There is an obvious logic to having more than one resource for decisions. Additional recommendations, friendly counsel, negative differences, professional advice—all the contributions we can gather—should improve the outcomes of our work.

There is, however, an additional value in committee work that is a result of the group activity: two heads are more likely to *disagree* than one. If there is an open environment for discussions, a committee of any number, from three on up, will question the broad spectrum of comments concerning any given issue under discussion. Disagreement will often emerge regardless of how friendly or delicately the members relate to each other. Disagreement can develop as the direct function of constructive dialog. Constructive dialog is something a person cannot do alone. You are not too inclined to disagree with yourself—particularly if *you* created some idea or concept.

Committee members function as "correctives." Disagreements usually help steer the concepts under discussion. This process follows a *dialectic pattern*. The idea of the dialectic procedure is an old one. Simply stated, a *proposal*, let's say one put forward by an employee at a meeting, is called a *thesis*. If it comes under attack at a meeting, the criticism is the *antithesis*. If the members of the team incorporate the solutions to the criticism into the original proposal, they synthesize a solution by compromise. When the group strikes such a balance the compromise is called a *synthesis*. In other words, there is a proposition, opposition, and compromise. In effect there are three possible results, as can be seen in the following diagram.

In many oral communications texts, group consensus (one type of synthesis) is viewed as the democratic and just outcome of a moral and democratic group. The suggestion is probably misleading. The *ideal* committee may function along such lines, but company employees work in a for-profit environment. The idea that "if you are not part of the solution, you are part of the problem" is commonly found. Many work-related committees do not aspire to much discussion; they aspire to efficiency. In other words, committee outcomes may develop in *any* of the three ways that are involved in the dialectic process, and there can be other results as well.

THE DIALECTIC PATTERN

```
   PROPOSITION  <--->   OPPOSITION
      (+)                   (−)
            \            /
             \          /
              COMPROMISE
               (+ / −)
```

Possible Outcomes:
1. Proposition upheld
2. Opposite position upheld
3. Compromise strikes a balance

If a manager sticks to her proposal regardless of, or with the support of, dialog, the "thesis" asserts itself and is the outcome. The supervisor says, "There will be no mileage reimbursements for business trips from now on." Let's suppose everybody agrees. The proposal will be adopted. Perhaps, on the other hand, a proposal is badly bashed by criticism, and the critics build a counterproposal that is accepted by a committee; then an "antithesis" is in control. For example, pay cuts were proposed as a solution for tardiness at Anna's company, but the proposal was replaced, in a rowdy committee meeting, with flextime, to everyone's satisfaction. The problem had really been an issue concerning getting children to school on the way to work. Both of these examples are types of decisions that occur with regularity. At times decisions are unchallenged. At other times the challenge becomes the new decision. The positive side to both of these results is that there *was* dialog. The dialectic process *was* a tool of workplace conversation, or possibly a committee, even if there was a victorious element involved, since one view prevailed. At least there was a contest, even if compromise or synthesis did not prevail.

"Synthesis" or *compromise* might generate a more energetic debate than the first two outcomes. We do not usually disagree, agree, challenge, compromise, support, and criticize without some sense of compromise. The difference is that **when compromise develops, it is much more likely to gain support from *all* the committee members than decisions that are not adapted to the suggestions of the members.** Suppose there is a new state regulation that all employees must be certified in CPR. Who pays? Managers see the law as an employee obligation, but because of strong employee disagreement, they compromise and use company time for the training—synthesis. There is a political side to compromise, of course, if horse trading is really what is happening, and trade-offs may be buy-offs.

Labor-management settlements often incorporate this element out of necessity. Endless other committees adapt their decisions to reflect helpful inputs, and the constructive result is far superior to trading in horses, or plans for the perfect camel.

Apart from the dialectic process there is another factor at work in committee activity that helps explain why group decisions are valuable. Yes, there is the democratic ideal. And, yes there is the efficiency of disagreement that helps define correct decisions. There is also *synergy*, a unique force that develops as a result of group behavior. **Synergy means that the whole is greater than the sum of its parts.** In math, the idea makes no sense, but human behavior is less predictable. The idea is that a group of, let's say, five committee members, can exceed its number—by one. In effect, there are six committee members because more comes out of the group than the sum of the inputs.

If the five members are put in separate rooms, they will not perform beyond their unit value of "one" per member. Working as a team creates something more—the team spirit—which is the missing member. The blend and the outcomes create something beyond the sum of the original parts. All things being equal, what else can a coach look for in a playoff when the best play the best? He looks for team spirit. You will never hear the word, but the magic of the best team performance is the same product found in the best committees: synergy.

How synergy develops is easy to see. First, note that the phenomenon will not occur in a negative environment. A team has to be cooperative. If team members manage cooperation constructively, the elements that encourage synergy are apparent.

> - **There must be an open exchange of all available information.**
> - **Problem solving must be actively and openly discussed.**
> - **Decision making must be openly shared.**
> - **Individual involvement must be honest, and members should have no personal fears.**

If these activities are undertaken energetically, a group will seldom become stuck. Superior decisions (superior to those of individuals) should result because there will be a better and broader base of ideas. In addition, a team that is pleased with itself will develop a positive attitude in support of itself and its decisions, which is valuable morale to cultivate in the workplace.

In sum, the *democratic process,* the *dialectic function,* and the *synergistic effects* are the challenge we can offer to the criticisms of committees, but we have to make it happen.

Committee Basics

What is a committee? You may have noticed that our discussion frequently used the word "group." Let's focus on *decision-making groups* or any group that involves the extensive contributions of its members. Many of the committees of the sort we identified earlier may demand little of our effort. Of more importance is any group involvement that calls for our skills, because then we find ourselves in a highly interactive situation. The decision-making committee, as a task-oriented group, is the most demanding of these situations because of our *roles* in such a group.

A *task group* can be defined as a dedicated number of people who work together to achieve a common goal. In industry, the size of a committee may not be a function of efficiency, but if a group has a specific task, we should look at the number of members to suggest a few ground rules. Five to eight members is often thought of as ideal. Five to eight is probably a matter of numbers and not a complicated social issue. A committee of three or four may not generate a productive volume of dialog. On the other hand, the rate of ideas and the rate of interaction declines with a large number of members. On the high side, an overly large committee serves little purpose.

Once a committee is over a threshold of about twelve, some authorities suggest that the group will start to develop political subgroups. These minor allegiances can change the intent of a committee. Perhaps you have seen this happen. Any large committee—civic, religious, political, corporate, or otherwise—runs this risk. Even more common is the risk of "deadwood." There is little point in developing large committees if the members are not going to be involved. Deadwood is an obstruction to the success of the committee and a burden to the morale of the group.

Notice that our simple definition of a work group calls for "dedication." If you want to know the *number* of members to have, you need to be even more concerned about *which* members to invite aboard. There are several ways to judge this issue. For now, let's consider the matter primarily in terms of team spirit. In an earlier chapter we discussed the socialization process an employee undergoes in a new work setting. A new committee goes through a similar phase. Good committee members stick together and stick to the task. We should select them with these conditions in mind. We judge potential members by looking at three considerations.

- **Mutual tasks**—Who will be dedicated to the task? Who will participate in the work in an energetic way?

- **Mutual care**—Who will be dedicated to the group? Who will find the group attractive and want to share the group's efforts and the group's identity? In a word, who has the same interests?

- **Mutual outcomes**—Who really wants to see the results? Who will be in for the long haul because the end result is of personal interest?

Sample 7.A

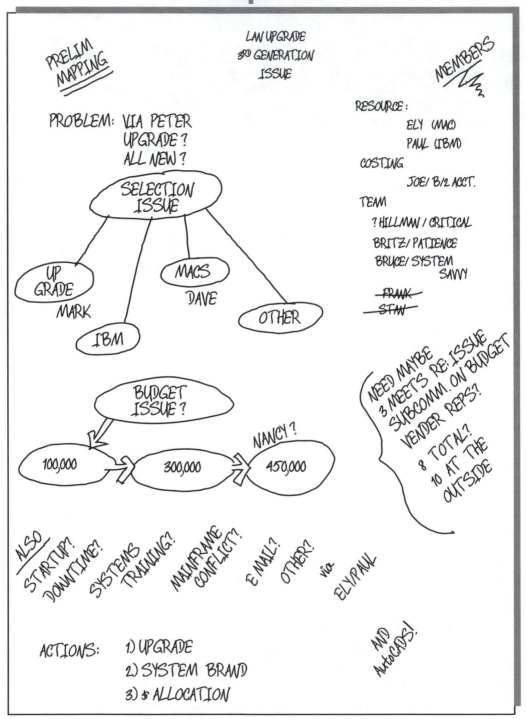

The forces of cohesion are complex. As you can see, the sharing of a committee's task may appear "mutual," but motives may be quite distinct. Some combination of these goals or motives creates dedicated behavior. Committee members may serve many purposes but there must be a mutual concern for the task. The trick is be sure to find members with purpose.

Timing

As we begin to look further at what you can do to help the success of committees and committee meetings, we need to look at different angles that will shed light on the uses and abuses, and the likely successes and failures, of committees. For example, given the opportunity to select a committee (or serve on a committee), you need to look carefully at an issue that has nothing to do with the group members: what is the decision deadline?

Basically, group success can very much depend on an issue that may defy any efforts, superhuman or otherwise: not enough time. Critics observe that poor committees waste time. The truth is that an equal number of committees battle to even come close to meeting deadlines. Cumbersome, tedious committees annoy many company employees, but just as commonly they find their names on hasty outcomes, overwhelmed by the task because there was no time to get the job done properly.

A close look at the available time is a good starting point for decisions about making or joining team efforts to address issues.

---

1) **Establish the available time frame.**

2) **Examine the extent of the work involved.**

3) **Determine whether you will be able to count on adequate participation from group members.**

4) **Determine whether you will be able to attract motivated members.**

5) **Work with all of the first four conditions in mind; try to guesstimate the possibility of completing the job the committee will set out to do.**

If you find yourself appointed as a group leader, as the first order of business, you should sketch your ideas on paper to try to give shape to the committee membership, the tasks, *and* the timelines. Balloon maps are a handy, way to visualize tasks the order of discussion items, and timetables, as in the sketch on page 127 (Sample 7.A).

Committee Decisions and Group Spirit

To observers who do not sit through meetings held by a committee, the only perceptions they will have of the group are ones based on the documents made available to them as a result of committee work. Usually these products are the outcomes of the labor of the committee members working as a team. **The job of the team is, at its most challenging level, decision making.** On the surface, the paperwork seems to be the measure of the committee. This is a misleading method of analysis, of course, since a team effort is then measured in terms of success or failure by outputs.

Although our purpose here is not to keep score, outcomes definitely *do* give us one benchmark we can use to judge committee success. Essentially, a committee is given a job to do, what is sometimes called a "charge." After they wrangle with the issues—whether in a one-hour meeting or twenty meetings at four hours each—they tally their group perceptions in some way. Four outcomes are most likely.

- **Consensus** means total group agreement. This is a difficult achievement. In fact, although a great deal of literature considers consensus to be a goal we should strive for, it is probably more ideal than real. Besides, some observers are skeptical and fear that consensus may be very risky vigilante behavior at times, and milquetoast concessions at other times.

- **Compromise** is a far more prevalent pattern of agreement. **Compromise is agreement by adaptation in which divergent views are incorporated into, and help define, decision making.** Compromise is the more likely outcome of committee work, and this pattern—amending an agreement until it is mutual—is the best possible decision a team can usually make. Its superiority is based on conflict. We can put a fair amount of trust in the outcomes of conflicting interests. Granted, diluted or compromised decisions are also a probable outcome of compromise, but the dialectic process involved in conflicting inputs may also lead to the *best* of outcomes. This is because conflict, if it is analytical and not emotional, can result in the most informed decisions, since conflict will initiate intense and thorough analysis of options in the decision-making process.

- **Forced decisions** Stopgap recommendations are also a common outcome. If we recall the issue of time, obviously any committee will be hampered by inadequate time in which to complete tasks. In addition, poorly informed resources are a disaster, since inferior decisions may result. Many a military committee has made doomed decisions because of inadequate inputs with which to judge the position of their adversaries. Committee members often become resigned or fatalistic if they know they have to decide what to do when they see the risk they have to take. These makeshift decisions are seldom adequate, unless by luck. Companies often refer to these committees as teams that are only "fighting fires."

- **Indecision** Of course, there is the very real possibility that a team will simply fail in the mission of decision making. There are times when no decision is going to emerge. The result is similar to a "hung jury" in which the internal conflicts will not or cannot be resolved. It is also common for indecision to be the result of inadequate information for sensible decisions. Again, lack of time can be a problem. Fear of decision making may also result in indecision.

Although supervisors, and anyone affected by the outcomes of a committee, will often measure the success of a committee by the outcomes listed, the *members* often judge the success of committees in other terms. **Because the behavior of the members and the activities of the team *created* the outcomes, the committee members know that there is more to the effort than simply a final memo.** There is also the process—all those hours!

A committee is sometimes pleased with itself. At other times members are bitter. At still other times the members are hostile and have words. The chemistry of a group is complicated, but we will look at a number of elements of concern to members. When we speak of "chemistry," of course, this is a very vague way of speaking of a group of people and how they did or did not connect to achieve some task as a team. There are a number of perspectives from which we can view the group. All the viewpoints look at the process of teamwork, and that process involves the team members and their activities.

Task Functions

First, we might look at the two basic functions of a committee. It is important to realize that a committee doesn't just *do* a job. A committee *experiences* a job. A committee may have a rational objective, but the task of achieving the objective puts us through the chores of building *community*. It is in this sense that a team has not one but two functions. The obvious function is the achievement of the charge given to the committee. **This *task function* is the "business," or the obligation to get the job done.** Obviously, we will have to discuss how to go about the business of decision making, and we will do so by looking at how to construct an agenda, how to use a few basic parliamentary procedures, how to build outcomes, and so on. Interestingly, most people consider the chores that compose the task function to be the only job they ever thought a committee would set out to achieve.

Social Maintenance Functions

A *second* function of the committee—now commonly called the *social function* or the *maintenance function*—is equally important, even if it is a more subtle force among the group members. The committee, as a group of diverse folks who set about a job, creates a community. The spirit of that community matters—a lot. If the members get along, the committee will handle tasks with much more interest and success. **The maintenance function, then, is the effort to help build unity among the members by constructing good working relationships among them.** The social maintenance function is a *social need*. Anyone who has ever sat on any committee—civic, religious, corporate, or otherwise—realizes the significance of community, but few of us have stopped to look at this other side of group activity. Social maintenance, maintaining positive spirit as a team, is actually *very* important. Ask any coach. What will help you to best serve a committee is an awareness of community. Maintenance of the social spirit is as important as the task itself.

Committee Structure

Pick your favorite team sport and look closely and you will find a committee. The Dallas Cowboys may be a little on the large size by most company standards, but they function as a committee just like the ones we might serve on—with the difference that we deal in paper outcomes, and team sports are concerned exclusively with the location of one ball, and they track outcomes on a scoreboard. The success of their decision-making process is partly determined on the field and partly determined by locker room and athletic club practices. The team is a variation of a decision-making group.

There are three fundamental patterns that characterize groups. First, as we noted much earlier, most business settings are top-down structures. Essentially these are *authoritarian* structures (not in the sense of blind submission to authority). There is authority that controls and guides the activities of business. In this situation, committees are less likely to set policy. Instead, they will work for a group leader who controls the tasks of the committee.

A second model is the *democratic* style, in which we share the burden of the chores but also the burden of leadership. This is the preference of our cultural history if not the preference of office managers. Because most of us are born into a culture of democratic values, the participatory style of committee is obviously quite prevalent, if difficult to manage properly. There is difficulty because we usually have little training in being the citizens of our democracies, large or small. **Democratic process calls for participation, and participation is not just a willingness. Skill enters the picture if the job is to be done properly.** An efficient democratic group depends on some knowledge of all the material of this chapter, and the next.

Without the craft of control, we get the third possible prototype: *anarchy*. The group will be unstructured and out of control if the members are disorganized, or do not have a leader, or do not share the tasks of leadership. The basic options are quite clear: authority, democracy, or anarchy. These patterns of group behavior have structures that can be symbolized in the following drawings.

 **AUTHORITARIAN**

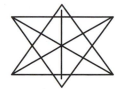 **DEMOCRATIC**

 **UNSTRUCTURED**

The basic pattern of a business (or an athletic club) is likely to be reflected in whatever committees emerge. In other words, a company and its committees will be alike in kind. If an employer is fully in control of the workplace and each employee responds to him or to her, then the model is something like a wheel with the boss as the hub. The pattern of the office communication will look something like this:

 **THE HUB PLACES THE SUPERVISOR IN THE CENTER**

A larger business is forced to begin to develop the usual pyramid of power because chains of command become a necessity. Vice presidents, sergeants, pitching coaches, foremen—assistants of every kind—become part of the structure of business. The communication pattern then begins to involve links such as in the next drawing, where the supervisor has two assistant managers.

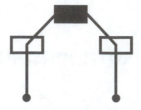

**THE CHAIN OF
COMMAND
USES ASSISTANT
SUPERVISORS**

These basic authority structures suggest that any committees that are formed in these businesses are likely to be in the authority mold, with a leader who is a supervisor. The committee will reflect the larger structure of the business.

Within a professional team such as the architectural firm I mentioned earlier, or a small law firm, the process may be much more democratic. There may be a leader, but the committee is characterized by equal power among the members.

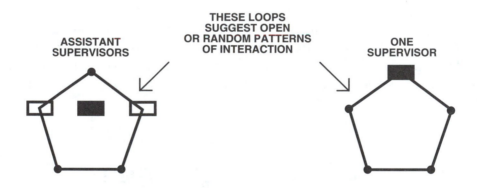

Even with assistant directors involved (on the left), the process can be quite open or true to a democratic model, where equality is a greater force than rank.

These symmetries trickle down. At work we are definitely under the persuasions of these systems. Suppose we have six employees sitting at a table. Can we judge their interactive pattern of behavior? If you look at the following four basic configurations, you will sense the distinctions.

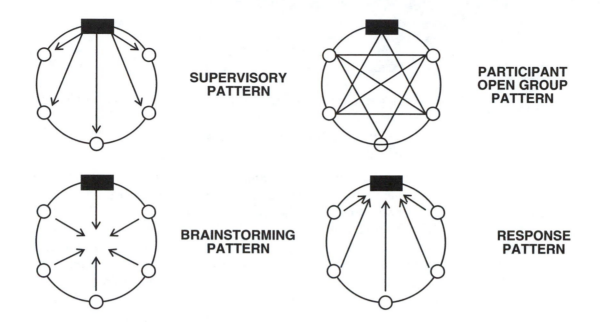

SUPERVISORY PATTERN

PARTICIPANT OPEN GROUP PATTERN

BRAINSTORMING PATTERN

RESPONSE PATTERN

If the arrows indicate the primary direction of communication, the "supervisory" pattern is obvious enough. The boss is in control. The "participation pattern" is probably a lot more interesting for a committee member, since inputs are respected and encouraged and the democratic process is more evident.

"Brainstorming" has a unique role in committees. It is not exactly interactive because the idea is to gather ideas and not to judge them. The point is to get as much "on the table" as possible. The arrows point to the table. Critical interaction comes later. The "response pattern" is the last configuration and may seem odd to you if you have never attended a public hearing for testimony. The USDA, the FDA, and other agencies hold public hearings to listen to public opinion. The event usually consists of dozens and dozens of people who speak their mind about the issue in question so that agencies such as the Department of Agriculture and the Food and Drug Administration can gather data for decision making.

It is valuable to be able to see how teamwork structures itself. It is useful in the sense that there are roles implicit in the structure. We do not ask for too much privilege in the authority model. It is highly supervised. We do not ask for too much leadership control in the democratic model. We seek help in the event of anarchy. Given the opportunity to create a team or improve a group effort of some kind, our awareness of the social structure should help the community of the group.

The Balance of Trust

A task committee does not just get a job done. There is a route to be taken if the job is to be *well* done. The social maintenance and task functions serve as the path of least resistance. It is not easy to gather five or eight or twelve people to work on some task. They must get along at the very least.

In a committee setting there is a way to go about building community. Let's look at *social maintenance* first. **The prevailing spirit of a successful team can be measured with one yardstick: trust.** If there is a shared trust, then the team will perform at its very best. At the outset of this book I noted that trust is the key to good business. In that sense trust is the bond between employee and client. Trust is also the key to the many successful shared activities that employees undertake inside a company environment. The trust between employees, and the trust between employees and supervisors, is a critical element in the construction of community. In a committee, trust is encouraged if there is good faith among members and no fear or distrust. Openness and an open-minded group encourage trust.

The opposite of an open group is a *closed group,* "closed" in that the doors of communication are closed. A number of causes are thought to be the problems, all of which lead to a destructive result: distrust. **If there is distrust, a committee is going to lack community and spirit, and achievement.** We can identify specific situations that can cause problems, particularly in the workplace.

- **Ranking**—If the roles, and particularly the powers of the members, are worn like stripes on their sleeves, there we will be a decline in openness.

- **Rules**—If there is some kind of governance *behind* the scenes that regulates the team, there may be problems. As we observed earlier, committees often reflect company structure.

- **Communication barriers**—If communication is restricted between members, there will definitely be problems. Ranking or ruling can cause this to happen.

- **Hidden agendas**—If everybody is out to protect "number one," members soon lose faith in the group. If the team has cliques, they are equally destructive. Members in either case serve only the purposes of their allegiances and not the purposes of the group.

Now compare these conditions with those of an *open group.* The open group means open communication. The communication is honest because there is trust. The trust is developed as a result of a number of conditions that can be cultivated in committees of all kinds.

- **Flexible relationships**—A first priority is to be sure any managers on the committee maintain a relaxed and trusting relationship among the members. To put it bluntly, assigning a manager to a committee does not make that member a leader. This is a frequent error that is made daily in business and industry. The safest role for any manager is to trust the group members. The result will usually be reciprocation: trust will be granted to the manager, not as a leader but as a group member!

- **Self-control**—The group that is open and trusting is usually going to be a group that establishes its own ground rules. If the members determine how they plan to go about their chores, this democratic style is more likely to encourage openness. Any policing is done from within, not from the company.

- **Shared tasks**—The members should not *assign* tasks. Instead the members should *agree* to tasks.

- **Shared interests**—Group openness will be encouraged if the members, as we noted earlier, are there because of sincere interest and shared interest. This condition will create trust from the outset.

- **No secrets**—Limiting information or limiting inputs is a dangerous practice. Facts should be openly accepted. Facts are needed for the task function of the group. Also, the feelings of the members should be openly expressed and accepted. Feelings are very much a part of the social maintenance function within the community of the group.

In sum, the best group is the group that builds community with trust. Openness is the path to trust and it *can* be encouraged. A group that is spontaneous and tolerant and appreciative can encourage openness, which is a path to trust.

"Openness" must not be misunderstood. Openness is not emotional ventilation. The idea is not to lose control by being a "blaster" or otherwise abusing people. We *must* take care in our comments. We *must* behave appropriately. There is a social fabric that develops from teamwork, and honesty is handled thoughtfully to keep the team healthy and positive. The negative results of ventilation are obvious. Unfortunately, many a coworker confuses ventilating with openness. These behavior patterns are not to be confused. The same hand can make a fist or a handshake. A member must not be a threat to other members.

A good participant is interactive, but the role is a positive one. Members must maintain objectivity and work toward resolving conflicts rather than creating them.

Sample 7.B

MEMO

To: Computer modification/Conversion Committee

Ely 2311 B	Pat 3014C
Paul 4029F	Bruce 7022 D
Mark 3030C	Joe 5223F

From: Don Bryce
Subject: Agenda and Planning
Date: November 3, 200X

A meeting has been calendared at the most convenient hour I could find for all of us: Thursday at 2:00. If this time works well, let's keep it for subsequent meetings. I have reserved the third floor annex committee room near Pat's office.

Here is a proposal for the overall agenda. I suspect we will need to meet twice a month for the time being. Look over the game plan and see if you think it is sensible/practical. Let's make the game plan our sole agenda for the first meeting. Once we structure, we can take it from there.

Preliminary Plan of Action:

1) Needs analysis

2) Upgrade option

3) New systems options

    A. IBM

    B. MAC

    C. Vendor systems

4) Funding models

    A. $100,000

    B. $300,000

    C. $450,000

5) Possible solutions?

6) Most viable option for recommendation.

This will get us started. See you Thursday. Do you think one meeting per item will be necessary?

Addressing the Task

Committee dedication is one issue. The other half of the challenge to achieving successful committee conduct is the control and development of the actual *task* that is the purpose of the group. Community is half the strategy; the other half is organization. A good team is intent upon a well-organized effort. Essentially, we need to look at the charge and simply ask, How should we tackle the problem? Any committee that will make recommendations, establish policy, or otherwise perform decision-making tasks must be prepared to look at the logical order of events that have to happen. There is an effective game plan that is the very common sequence of activities used by many committees.

Build the basic guideline that controls or shapes the mission of the group by designing an agenda—either for a single meeting or as an overview of tasks that must be handled during the course of many meetings. An agenda is an outline of the activities that have to be undertaken. There are five critical phases to include, whether you have one hour or fifty. The agenda in Sample 7.B is one of the endless variations of this plan of action.

---

1) State the problem:

   What is the charge?

   What is the question?

   (Questions often concern policy)

2) Decide on an investigation strategy.

3) Analyze the problem.

   (What is the problem behind the question?)

4) Suggest the possible solutions and analyze them.

5) Select the best solution.

The first phase of committee activity is used to define the *problem* or task. The job from the outset is to identify the charge of the committee *exactly*. If there is any error in this critical phase, the labors of the group may be entirely wasted. This phase may be quite difficult and time consuming if there is any question or argument about the issue. A committee cannot resolve a problem unless the members first agree that there is one. Courts of law are particularly sensitive to whether a violation of a legal code really *is* a violation. Similarly, a committee must examine its task closely. If the committee is a decision-making group, the task is a problem that must be identified. **The problem must also be examined and understood before solutions are pursued.**

As a tactic for any group, the members must "get it in writing," as the saying goes. A committee must be sure that a thorough discussion identifies the problems or tasks of the group. The best way to achieve this goal is to develop it as a written document, whether it is only one sentence that asks a question or an entire page of "ifs" and "ands" that define a complex issue with many contingencies. Often a simple question—"Should a southbound off ramp be constructed at 85th Street to relieve traffic congestion?"—will serve the purpose of a committee.

The second critical phase is for the team to decide on *strategy*. This involves the time-frame issue, the commitments of the members, the nature of their task and similar concerns. As a group they need to get out their calendars and agree to the strategy that will serve the purpose of getting the job done properly—preferably before quitting time. The number of meetings and their frequency and length should be established early on even if the effort is only a guess. As a tip, I would always suggest increasing the number, and never the length, of the meetings, and avoid any time frame that is not part of the nine-to-five scheme of things.

The number of meetings can best be judged by looking at the four critical phases of the task function. The first job is to *analyze the problem*. This may take one or more encounters. Second, the team needs to *suggest solutions*. The research or information gathering becomes a problem here if the issue is a complex one—the construction of a bridge, let's say. Then, after all the material is gathered and discussed, the members must weigh the results to study all the possible solutions. From this analysis they *determine the best possible solution* or outcome that the committee can recommend.

An agenda may very well be needed to outline *each* of these critical phases of committee work. The agenda is a helpful organizing tool in that the paper planning offers guidelines for the people who have agreed to follow it as committee members. This agreement, based on *participation,* gets a committee off to a firm start in a spirit of shared authority and effort.

Summary

- Group decisions are likely to be superior to individual decisions, but arriving at the decision can be a slow process.

- Group decisions are desirable because they represent group agreement. The values of the group members are respected elements of the decision.

- Group decisions also are desirable because they represent group contributions. The contributions of many people help analyze, refine, and perfect decisions with more thoroughness than would be available to an individual decision maker.

- An open atmosphere with a focus on problem solving is the ideal decision-making environment. The committee members must be trusting, honest, and cooperative.

- Any type of task group should be composed of dedicated members who can work together in the interest of a common goal.

- The team effort will depend on several conditions:
    - the available time
    - the extent of the work involved
    - adequate participation of members.

- Committee outcomes are usually of four types:
    - *Consensus* or total agreement is the ideal goal but is not easily achieved.
    - *Compromise* is an agreement that has been adapted to the needs of all parties. It is probably the most common and most practical approach to making decisions.
    - *Forced decisions* respond to timelines or outside forces.
    - *Indecision* means that there is no useful outcome.

- The committee is a community of its members. Their business is their task, but the task function is only half of the interaction. As a community, they live out a social task of maintaining the community. The *social maintenance* function is as important as the *task function*.

- Committee structure is either authoritarian (a supervisor is in control), democratic (the members control by voting), or unstructured (casual).

- "Open groups" share in task and social maintenance functions and are the desirable model because

  ✔ member relationships are flexible;

  ✔ control is internal and democratic;

  ✔ work tasks and social interests are shared;

  ✔ there are no secrets because the group is open to all ideas.

- An orderly agenda helps the open group conduct its activities. The agenda should follow a predictable plan of action:

  ✔ The group must state the problem.

  ✔ The group must decide how to investigate the problem.

  ✔ The group must analyze the problem.

  ✔ The group must then offer solutions and analyze them.

  ✔ The decision-making challenge is the final step; the group must select the best possible solution.

Activities Chapter 7

During the course of several weeks you will complete the following exercises:

1. A writing project in the form of a memo to your task committee.

2. A performance project as a participant in a task group.

3. An activity analysis (2 forms) that evaluates your team and your role.

4. An observation form that evaluates another task committee's efforts.

These four tasks are explained below.

As you read Chapters 7 and 8 you will be asked to participate in a task-group committee. The committee will meet in the classroom on several occasions to decide on a subject and identify the problems to be investigated and resolved.

1) *Your first major task will be research. You are to write a committee briefing in the form of a memo to your committee (500 words or more, typed) Provide a copy for each team member several days before the final videotaped meeting. Each member of the group should act as a resource concerning the subject under discussion (or some part of it).*

2) *Your second major task is to participate in committee meetings and especially the one that will be videotaped. Your team will be observed by one other task team during the final meeting.*

   You will be actively involved in short preliminary meetings to select a topic and define an agenda for the video session. The task-oriented activities are these:

   - *How or who will explore the problem?*

   - *Is there a strategy for the meeting? (Wide open? Tight agenda? Short reports?)*

   - *How will you limit the analysis to one hour?*

   - *Who will offer solutions?*

   - *How will the team judge the decision?*

3) *Subsequent to the video session, your team will meet a final time to enjoy the playback and discuss the tape. Your third major task will be to complete a group analysis of your task group and a self-analysis of your role (see the appendix, pp. 403–406).*

4) *There will be an additional opportunity to observe another task group in its effort to learn the skills of committee management. The fourth project consists of your observations concerning the group you observe (see the appendix, pp. 401–402).*

Note: This videotape activity is not a panel presentation addressed to observers. You and your team are meeting as a task group to resolve an issue.

Group Roles and Leadership

Work in Progress

8. Clients as Leaders

As I have mentioned previously, a significant part of my job lies in being able to maneuver my company successfully around the psychological obstacle courses the clients sometimes present us. At Tempo, this came in the form of the two owners, their drastically differing approaches to leadership, and the resulting lack of companywide communication.

Tim Simmons, as you may remember, likes to "tinker" and had many ideas about how the installation should be done. Over time, it became clear that in addition, he liked to call the shots at Tempo and make most of the company decisions, large and small. He busied himself learning about basic networking equipment and by our next meeting had a full agenda for the installation! While I certainly do not mind ideas or suggestions from my clients, I have found that the "steamroller" approach Tim was using was annoying and ultimately counterproductive for Tempo. Tim was regarding the LAN installation totally from his point of view as president and he was not taking into account the needs of the many other users.

The vice president, Sharon, on the other hand, is really interested only in her areas of the business—design and production. Outside of those two areas, she is happy to let Tim make all other business decisions, especially when they concern anything technical. She made it clear to me that as long as the network installation did not shut down production or interfere with her design software, she did not have any opinions about it.

This leadership style probably worked adequately when Tempo was a two-person operation, but now with 40 employees, the communication of basic needs and information throughout the company just was not happening. Tim's ideas about the LAN setup were exclusively from his perspective. It had not occurred to him to find out what needs his employees might have for the network. Knowing that I would need that information to design an efficient network, I informally polled the employees myself.

The front-office employees concerns were quite different from those of the president. They felt hot and cramped in their work areas and wanted to make sure that no new heat-generating computer gear would be squeezed into their workspace. They wanted a separate computer room/wiring closet to be created to relieve crowding. I also realized that while the boss wanted all the newest bells and whistles, his workers wanted a simpler, less intimidating system that they could learn quickly. Finally, one worker pointed out that the wiring in the office area was inadequate for the existing equipment in any case, and he worried that adding to it without upgrading the electrical circuit would make the computers "crash more than they already do!"

Now that I had gotten a sampling from the employees, I had a more accurate "big picture" view of Tempo as a corporate entity and was now ready to begin a preliminary design in preparation for a formal sales presentation. The problem was that the employees had a better grasp of the system needs than the company president. Since he alone would be responsible for accepting the final bid, I realized that there was a challenge here.

J. Q. C.

The Roles of the Group Members

In the last chapter we looked at the basic theory and functions of task groups in business and industry. Let's now assume that your chance to work on an important committee arrived in your e-mail this morning. Maybe you even have been asked to lead a committee. Now what? To this point we have been analyzing what makes a committee tick. It is a somewhat different matter to turn all our discussion into practical applications. So far, all we have is a sense of what should happen if the committee works well. The only specifics we have examined concern the agendas to get the task activities organized. Who should be on the committee? Who should lead it? What should you do if you are a leader? What should you do if you are a member? Let's look closely at the issues so that you have some guidelines for committee work in a company setting.

It is a fairly noticeable fact in the workplace that we assert personality traits in our work commitment. An employee is not simply a biomedical electronics tech, for example. Amy is a very bubbly, cheerful employee who always has an upbeat spirit. Her company role will take on the character of her personality in the minds of those around her, including coworkers and clients. Similarly, Jack, who is an appliance repairman for a major city utility, is a gruff old character who has seen one stove too many. He is as well known as Amy but for other reasons. Amy is remembered fondly. Jack is not. Restaurants call City Light and ask for a service call and remember to add "Don't send Jack." **The simple fact is that we play out who we are in the process of playing out what we do for a living.**

When a committee takes shape, there is a good chance that the members—our fellow employees—will live out their usual rhythms on or off of the committee. You are not likely to appoint the laziest fellow in the plant and expect to see him emerge as a powerhouse in committee. Because employees are fairly predictable, they are reasonably easy to examine as potential committee members. All we really need to identify are the actual roles that usually develop in committee work so that we can somewhat predict who will live out these roles in the name of an effective task group.

We already have divided the committee activities into *task functions* and *social maintenance functions*. One is as important as the other, although the maintenance activities are never seen or heard of outside the group meetings. Let's go a step further and subdivide the task functions into about five roles. We can do the same for the social maintenance functions.

Every committee will have members who have certain strengths. Any number of these strengths are oriented toward decision making and getting the job done. These are the *task-oriented* people. We can give them names and look at what they do.

- **The resource**—Anyone who provides information acts as a resource. From facts to cited opinions to quoted testimonials, these members are fundamental to the success of the committee because they keep the group well informed.

- **The probe**—Of course, someone has to ask the questions. I sit on one standing committee that has met once a month for years. The best probe in that group is Terry. She often asks questions. She is very objective, speaks very softly, and makes the group think very cautiously about its ideas. Her role never changes.

- **The idea person**—Some group members contribute solutions and not facts. They tend to look at alternatives. They are good listeners and follow the issue under discussion. Their openness and willingness are their strong points.

- **The analyzer**—Then someone has to pull it all together. The hope is that there is a committee member who excels as a problem solver. It is a trait of people who cannot resist organization and organizing.

- **The timekeeper**—Some supervisors enjoy two- and three-hour meetings—or longer. After one hour members will often begin to leave by the side door. There is no one to stick to the agenda, to move the agenda, to demand that all the work be done or tabled by quitting time. A group needs a sense of urgency, of movement. Someone *must* shape the group and keep it from getting bogged down.

If you have had the opportunity to perform committee work, you realize that people live out these roles on any committee. **The point is to see that the right talent is placed on the committee to get the job done.** Who is well informed? Who cares enough to ask serious questions? Who is skilled at developing proposals?

The task function of a group is never a surprise to anyone, and the major players are easy to observe. The social maintenance function—building community—is the job that most people haven't thought about. As a result, many people seem to think that they have no skill for identifying potential members who could help build community for a group. **The truth is that social maintenance roles stand out even more than task roles because employees openly exhibit behavior, whereas the rational task of problem solving may be more of an intellectual process that you do not see at the watercooler.** In other words, good logic may be less obvious than a smile or a compliment or a word of encouragement.

There are several social maintenance roles that usually develop in groups.

- **The listener**—People who listen with enthusiasm are people who are respected, probably because they are showing *their* respect in turn. The good listener is very flattering; it is a great feeling to be taken seriously. The good listener builds trust and does not waste committee time.

- **The humorist**—Chuck is the wit at his company. He spontaneously puts people in roars of laughter at meetings and is so good at this role that he is often a guest speaker at gatherings elsewhere in his facility. Comedy is relief—if it is handled correctly. By the way, another employee in Chuck's department tells formula jokes—"Hey, did you hear the one about . . ."—and *never* has cracked a bit of wit at a meeting! Wit is creative; jokes, on the other hand, do not serve the purpose of spontaneous humor.

- **The referee**—Some group members feel very uneasy if there is contention or conflict. They will often jump in to be peacemakers. If the cohesion of the group is threatened, they react. There are at least three whistles these referees will blow in a situation that they sense is not working properly:

    Peacemaking is important and any emotional conflict calls for a referee to restore harmony.

    Squelching also calls for the referee. If members cannot get a word in edgewise, the talker must be restrained. Every member should be encouraged to take part. The big talkers should be politely or bluntly regulated. Balance is important, or the goals of compromise, fair play, and agreeable outcomes will be jeopardized. "John, I think we should hear from Dale and Mitch . . ." will usually do the job.

    Compromise is the other whistle call. This is peacemaking also, not between emotional differences but rather, between the *rational* differences of proposals.

One group wants to spend $5 million on development. The other demands $10 million. It does not take much math to see the final recommendation once the referee sets to work on the decision-making task.

Because these social maintenance roles are very constructive ones on a committee, it is important to seek out members who will contribute these roles. Usually the personality type is a force behind the contribution in the social maintenance functions, so it is easy to identify the right people among our coworkers.

In contrast, the bad apple in the barrel easily damages committees. Again there are usually signals to use as guidelines. A brief survey of the *committee types to avoid* may regrettably bring to mind people you know who are best left out of any group with which you want to work.

- **The monopolizer**—On a scale of extremes the person who talks the most *may* be monopolizing the group effort. If this is happening, the member must be squelched—with skill, of course. Do not confuse this type of behavior with healthy enthusiasm. Members sometimes worry about monopolizing, and they become very self-conscious. If they are concerned, they are not likely to be monopolizing. The role of being a good listener is the first task. If they listen *and* talk, the balance should solve the problem for people who think they contribute too much.

- **The withdrawer**—The person in the back of the room by the door is a different issue. A member should contribute or depart. In industry the problem is that employees may be forced onto committees. Suppose a supervisor puts Wally on the safety advisory board—because he takes more sick leave for job injuries than any other three employees! Punishment is the *last* possible reason for being on a committee. Membership is not always in our control. All we can do is encourage involvement, or eliminate the member if possible.

- **The blocker**—If someone seems to put up roadblocks, they are not helpful members. Bill was a friend of mine, but he was a wet rag on any committee because of his hound-dog pessimism. No matter what anyone would propose, Bill would insist that "They will never let you do that!" Fortunately, he was so predictable, he was comical.

- **The competitors**—Competitors are too aggressive on behalf of their own personalities. Self-importance is a problem at times, especially since a committee is a team. Any coach knows that the prima donna is his worst team player—especially if the same person is his best athlete.

- **The lobbyist**—If some people are competitive because *they* are number one, others are competitive because their *proposals* are number one. These are members with an agenda. They are unlikely to follow and accept the path of the group.

- **The storyteller**—War stories waste an amazing amount of committee time. We are all in the trenches and have earned all the badges. Telling stories is seldom problem solving. Up to a point, it is an element of openness and trust, but storytelling must be handled firmly, if kindly.

- **The aggressive personality**—Really aggressive behavior—mean gossip, name-calling, threats, yelling—occurs from time to time. The instances sometime involve substance abuse or psychological problems or both. Very few orderly personalities are going to risk friendships, working relationships, or job security. Owen virtually tried to lead a revolt at his plant and had the "Don't Tread on Me" flag over his office door. He was a union man who went off the deep end. Ken, another employee, was less political. He simply hated just about everyone he worked with except the person with whom he happened to be speaking. All his coworkers saw through him. Committee work with these men was interesting to say the least. These personality patterns can be handled. A healthy aggressive reaction is one solution. The door is another. Being sure to isolate the type is the best bet; keep them at bay and off the team.

- **The joker**—The joker is not the humorist. Wit is one thing. Joking—particularly at other people's expense—is another issue. The distinction is *not* subtle. Humor that is critical of other people is risky business. Television personalities who specialize in this sort of humor are called "put-down artists." The expression states the problem clearly. Basically, we have fun *with* people, or we make fun *of* people. The first is humor. The latter is the work of the joker. The joker is a negative force to be dealt with. If you *never* laugh at a put-down, you will help control the problem, but jokers should be avoided on a committee.

In sum, in most companies there is a tendency for committees to be composed of people who are *informed* about an issue or who are *concerned* about it. This is, however, a *very* limited perception of the mechanics of the committee process. At times we are restricted, and informed members will be the best we can hope for. **If prospects are sufficiently large, it would be more effective to compose the committee so that both task and social maintenance roles are carefully chosen.** The "resource" is only one of a number of valuable team member roles that should be in place. A committee that is highly informed can remain ineffective if there are ten resources but no other task and social maintenance functions represented on the committee.

The Role of the Leader

Since you will probably work with committees of various sorts and sizes, you need an understanding of the tasks of a good leader. If a committee performs poorly it is quite possibly a leadership problem. Learning to identify a good leader will point up an inferior one. Given the opportunity, you may find yourself in a leadership position. Our discussion may be of particular value in that event. There is often a conflict on committees because there can be more than one leader, and this problem is quite common in the workplace. Let's begin with the issue of dual leadership because it illustrates a conflict in corporate environments.

What is a leader? This question preoccupied many minds in the late eighteenth and nine-teenth centuries. At the time of Napoleon, the treatises on the nature of leadership usually focused on the image of the hero. It is probably not coincidental that the debate occurred at the historical moment when monarchy was in decline and democracy was on the rise. You have heard the saying "a hero is born, not made." Is a hero born with godlike strengths, as some philosophers thought? Or is a great leader a well-cultivated and well-bred child of class and society? Is leadership a matter of personality or charisma? Or is leadership a skill? *Charisma* is the word that was most often used to describe John F. Kennedy's worldwide popularity in the 1960s. Charisma also explains military leaders such as Patton and Britain's Montgomery. But how do we explain Bradley or Eisenhower? *Skilled* leadership would probably be the answer. Our military academies are designed to cultivate highly trained tacticians, not charismatic heroes.

If both personality *and* skill create leadership, we should probably allow for both possibil-ities, which is probably appropriate. But is there more to it? What if a leader has both the gift of charisma *and* the gift of statesmanship, but nobody claps? In other words, the leader is not a leader without the *permission* to lead. **A leader must be *seen* as a leader in order to have a *role* as leader.** If a leader is not *perceived* as one, there is no power, and power is the ultimate measure of leadership success. What makes a Juan Perón (Argentina), a Ma-hatma Gandhi (India), or a Martin Luther King is the perception of the figure as a leader. This perception grants the authority—the power—to the leader, who only then can lead.

This may seem more like a major issue for historians, but it strikes home in any company. Corporate and business leadership is unique in that most figures who would represent leadership—from the shop floor foreman to the company vice presidents—are appointed. That is to say, they are *not* elected, at least not by the employees. Recall the pyramids of power. The power of a company usually delegates itself downward. Managerships are ap-pointments. There is some use of voting at times, but the voting is among the upper levels of the company. This capitalistic practice, which is absolutely normal in corporations from coast to coast, coexists inside a democratic nation-state where, to the contrary, most offi-cers of city and state are elected from the bottom of the pyramid. We, as democratic citi-zens, vote upward. **The *appointed* leader is delegated his chores from above. The *perceived* leader is delegated his chores from below.**

You will often see this conflict in a business environment, and nowhere will it be more glaring than in committee situations where two leaders may be quite visible. The ap-pointed leader will be the company manager or anyone who has been assigned to direct the committee. The perceived leader will be the person who takes control of much of the group activity because he or she is *seen* as a leader. Some authorities refer to the per-ceived leaders as "emergent" leaders because they will do just that: emerge. They are in the right place at the right time. It may be that certain leaders are leaders only of the his-torical moment. However, it is also likely that the person is admired outside the context of a committee structure. You and your coworkers probably know who the perceived leaders are even without a committee setting.

In other words, every company has *appointed* leaders, who are granted the respect of their *authority*. But every company has *perceived* leaders who are granted the *respect* of authority by the employees around them. As you can guess, the best company leaders will be those who are appointed leaders but who *also* are, in the eyes of the employees, *perceived* as leaders.

The AA Manufacturing Company had a big burly bear of a man who was a manager and who got along famously with just about everyone. When a new CEO came on board she passed over the popular manager when promotions were due. Little wonder. Popularity is a threatening force to be reckoned with. It is an ideal condition for the workplace, but little trusted by the bulk of the appointed managers who, through no fault of their own, are leaders of another type. Everyone was surprised that Stuart did not get the promotion. They failed to realize that his popularity made him unpopular with the few supervisors above him who were in a position to grant the promotion.

> *I was willing to wait my turn, but after I didn't get the promotion I realized that I might as well look elsewhere. Actually, I didn't see the social aspect of it until people started to object to what happened. I really thought my resume looked pretty strong, and I didn't think popularity could work against me!*

Can you be a leader? Of course you can be a leader. We all are at times. Circumstances create the need for our roles. I, for example, have led a number of groups. Far more often I have simply been a member. I have even been the member by the back door who says nothing. The prospects are clearly going to depend on the situations. You might say that, in a committee of ten, your likelihood of having to be a leader is only one in ten. In general, we are more often members of a team.

Leadership Practices

It was once fashionable to see leadership in terms of attributes. If you ever see an old 1920s or 1930s version of the *Boy Scout Manual* you will see the attributes of leadership—and good citizenship—celebrated throughout the book. Baden-Powell, who founded the movement in Britain at the turn of the twentieth century, was greatly concerned about the appropriate attributes for leadership and citizenship. In the United States the Boy Scout movement took a more outdoors and environmental orientation, but the texts to this day reflect the values of historical leadership characteristics. On a more practical note, because perceived leaders have a special strength gained from the respect of coworkers, we should look more at the group needs than at character building. **Leadership is a craft; there are practices you can utilize to keep a committee well oiled and on its way.**

A leader should, first of all, be the man or woman of the hour. We all have our moments of glory, and often the role of the leader is *situational*. We suit the moment. This situation—the best person for the job—is the first condition. When voting on a leader in any formal way, the "best" should be the person who can handle the *moment,* at least for committees. If *you* are going to consider the role of leadership, ask yourself if you fit the moment. Too many leaders are easy volunteers—both the appointed and the perceived types—and they may not fit the circumstance. A specific committee may call for specific knowledge, for example, which demands a very narrow band of members and a leader who knows the issue.

The second condition of leadership is *membership*. The leader must be a willing member. Leaders who see themselves as elite may not do a very good job at leading. Their status as appointed leaders will get in the way. **The better vision of a leader's role on a committee is a desire to be a team player.** The leader-member can still coach, but it is far better to coach from inside the team rather than from the outside. Committee members rarely need "managing" anyway. They need direction, and so it is rather easy for a committee leader to be seen as a member, since the committee leader rarely has to assert any authority other than to refer occasionally to agendas and deadlines.

Third, the leadership style has to be *appropriate* for a committee. Recall that committees evolve out of various group systems that operate either in an authoritarian, a democratic, or an anarchistic fashion. Circumstances will determine whether the leader creates, controls, or is a product of these systems. It depends on which theory of leadership we accept and what actually happens in our work environment. In any case, one point is clear: the *wrong* type of leader will create problems in a group that rejects the leadership style. For example a democratic leader who wants to encourage alternatives and options and policy among the group membership will not be a big hit in the U.S. Marine Corps. Similarly, an authoritarian will not fit democratic groups very well. A drill sergeant will be less than popular in a community club association or in an environmental group or at Rotary.

An *authoritarian leader* determines policy for a committee. The role is valuable but primarily in a highly supervised environment. In contrast, the *democratic leader* provides guidance and has faith in the members of the committee, who are there to construct policy by unanimity. They vote policy. As to the anarchy model, any leader who waffles—either out of laziness, or fear, or indifference, or inexperience—will leave a group wandering, and wondering what to do next. The result is usually quite unsatisfactory if a group has outcomes to provide. If a group exists to have fun, the *laissez-faire leader* is appropriate because the only agenda is having a good time; the members of many leisure groups meet because they enjoy each other's company or enjoy some sort of shared activities.

It is not always clear which type of leader is the best. The leader must fit the committee, but the reality of the workplace is far too complicated for an easy generalization. Be aware that there is a noticeable difference of opinion in the workplace. If you speak to the captains of industry, they will defend authority. If you speak to communications specialists, they will defend the democratic style of leadership and the democratic system of committee practices. In other words, managers want to manage; voters want to vote. Perceived leaders, the best option, probably tend to be democratic ones. If a shop foreman has less respect than the union rep who works beside him, the condition may be a response to the issue of democratic style.

The most desirable leadership attributes encourage team motivation, and the leader who *guides* decision making probably has the edge. If a manager determines policy, employees will do the job; however, if they *help* the manager determine the policy, their role not only shifts but there is an attitude change as well. They *enjoy* doing the job. For example, think of how little incentive we have when the job is hard and the pay is meager. As a manager of the men and women who are in her work environment, a manager may not be able to offer pay as an incentive, so she offers the alternative: involvement. The employees work as a team and they work toward goals *they* help determine. She can lead the way and watch the clock and the calendar. And of course she can be a resource as well as a leader in these groups. The employees assume initiative as a result.

On a somewhat more practical note there are a few more conditions to leadership skills.

1) If possible, the leader must lead the members of the group in terms of their expectations, not in terms of the hopes of the leader.

2) The leader must have knowledge of the issue in question.

3) The leader must have good communication skills, and committee skills in particular, especially a clear understanding of the task functions and the social maintenance functions; know the job and know the people.

4) The leader must be willing to spend extra time on the activities of the committee—more so, at least, than any other member.

5) The leader must interact freely and avoid taking sides or joining any clique in the committee.

6) The leader may have to take some risks. Group behavior can be assertive. Leadership is not for the timid.

7) Leadership is for the patient! Patience is a must.

An effective leader recognizes that the role involves a broad spectrum of activities that run from housekeeping to managing discussions to nurturing the group community. In practical terms there are tips that will help if the task is up to *you:*

1) **Establish the climate.** Arrange for the room, adjust the seating, decide on whether there will be coffee, and so on. Seating is very important. The shapes of tables will affect a group. A long narrow table with a leader off in the distance is not the democratic style. In the same sense, a clutter of folding chairs is not the authoritarian style.

Sample 8.A

Engineering/Electronics Division

October 15, 200X

Dear Committee Member:

A meeting has been scheduled for the CAD for Industrial (Electromechanical) Technical Advisory Committee.

Thursday, November 4, 200X
3:00–4:30 p.m.
Location: The Commons at Franklin

AGENDA

1. Welcome and introductions

2. Review and approval of minutes from September 22, 200X

3. Old business
    a. Tech - status report: Alex Riley and Skip Hudson
    b. Placement report: Fred Edel
    d. Drafting: Fred Edel

4. New business
    a. DRD committee . . . group discussion, with commentary by Nancy Smith
    b. Charter
    c. Limits of responsibility
    d. Limits of authority
    e. Programs affected

5. Next meeting date/time/location

6. Next meeting agenda
    Review

Please confirm your attendance by 10/28/200X by calling Evelyn Walker at 123-0000. Your continued support is very important to the success of this program. We look forward to seeing you again.

Sincerely,

Lynda MacDonald
Electronics & Engineering Technologies

Encl.
LM/jmt

2) **Plan the agenda.** Design the first agenda and apologize for it. All later agendas should be designed with the committee's assistance. Notice the agenda in Sample 8.A.

3) **Open the discussion.** Explain how this situation came about.

4) **Control the interaction of the members.** The primary maintenance function of a leader is to "balance" communication. In any group there will be assertive members and every possible level of involvement, down to Peterson who is always asleep. The leader needs to pass the microphone, so to speak, with skill.

5) **Control the flow of discussion** as well as the inputs. Not only do the members need to strike a balance, the agenda must move forward. One useful device is to allow a specific time frame—fifteen minutes or an hour—for each issue of concern. Stop the discussion to move one. Blame the clock. Another meeting can reschedule discussions.

6) **Do not appoint subcommittees** if you can avoid them, and do not use them as an avoidance mechanism to keep a group tidy. Subcommittees are often used to try to put a gorilla in diapers or to carry out other thankless complex tasks. Do the job or do not do the job, but do not put it in committee.

7) Instead, **brainstorm** when there are barriers. Brainstorming is often practiced incorrectly. Committees think they are brainstorming if they struggle for solutions or options or alternatives. True, this is the idea, but the *original* concept is better yet. There is to be little discussion of the solutions or alternatives, and *absolutely no criticism!* Also, someone has to write fast and as precisely as possible to document the suggestions, which are discussed in another phase of the meeting or at another time.

8) **Summarize** the findings and discussion from time to time during a meeting. A summary of this type is an extended paraphrase of sorts and will keep the group focused on the issue and guided by the agenda. Misunderstood ideas will also be evident.

9) **Maintain control** by asking all members for their ideas. Help distribute the discussion and help move the discussion. Use your imaginary coach's whistle if you need to. More than one meeting of the suit-and-tie sort has been brought back to order with a roaring two-fingered whistle—and a lot of laughter of course.

This agenda follows the usual order of business: approval of minutes, old business, new business, future plans.

Sample 8.B

Techtron

February 14, 200X

Dear Committee Member:

A meeting has been scheduled for the Industrial Power & Control Committee:

> Tuesday, March 1, 200X
>
> 2:00 p.m.
>
> Office Staff Lounge IB2430C

AGENDA

1. Welcome and introductions.

2. Approval of minutes from June 9, 200X.

3. Old business

4. New business

    a. Program update

    b. Review of proposed changes

    c. Proposal for increasing output

    d. JATC announcements

    e. Program participation by advisory members

    f. Date, time, location of next meeting

Please confirm your attendance by February 24, 200X by calling Susan Schmidt at 531-0000. For your convenience, a temporary parking permit is enclosed for your use. We look forward to seeing you again.

Sincerely,

John Heyden
Technologies Support Services

Enc: Parking permits
JH/ss

Committee Protocol

Meetings run the gamut from very casual to very formal. Most of the larger corporations will use the full spectrum. Formal or informal, the typical committee uses a written agenda to hold a meeting. A number of common practices help expedite the work of any group, particularly a task group. The first order of events is the agenda. A meeting should not be a surprise; announce it in advance. It should not be a mystery either, so route an agenda as a memo to the members. Simply put the usual memo details at the top of the agenda:

TO: Members of the committee

FROM: Leader's name

SUBJECT: Purpose and date for meeting

DATE: Of the *memo*, not the meeting

The agenda should be fairly clear to the members. Use an outline format but do not be too skimpy with the wording, because it can be next to impossible to read someone else's topic outline. Notice Sample 8.B. Earlier we discussed the basic five-point strategy of task completion, which also makes a good agenda format:

> **PROBLEM**
> **STRATEGY**
> **ANALYSIS**
> **SOLUTIONS**
> **DECISIONS**

There may be more that has to be considered, and often meetings have to reflect previous and future meetings. As a result you may use a format such as this:

> *Business:*
> 1. **Action items**
> 2.
> 3. **Discussion topics**
> 4.
> 5. **New items**
> 6. **Other**
> 7. **Adjourn 9:00 (or "Good of the Order")**

Notice the focus on new business. Many meetings look at old business as little more than loose ends; the action items are usually the new agenda.

If you call a meeting, announce the end of the meeting, and try to stick to it. If you break the promise, fewer members will return the next time. A leader wants to be known as a good time manager, particularly when managing other people's time.

I made a few suggestions for opening a committee meeting if the committee has not met before (see pp. 157–159). For subsequent meetings the usual routine is to more or less follow the agenda provided on the meeting announcement and proceed in some manner similar to the following:

- **Call the meeting to order.** Of course, you do not have to put it this way.

- **Vote to approve the minutes.** Minutes are the written highlights of the previous meeting. Few people will have read them, and few people are focused on the new meeting in the first five minutes, so usually the team simply votes approval. Peterson, of course, has not fallen asleep yet, so he may complain that such and such did not happen at the last meeting. If the minutes are challenged, the issue can be removed to "new business" so that the discussion is reopened. Usually the minutes are quickly approved and are handled with a vote. The minutes of a meeting appear in Sample 8.C(1) and (2). Notice that the minutes usually expand on the agenda.

- **Call for new business.** Any items that members did not have placed on the agenda before the meeting can now be identified. These new items are last-minute considerations, and so they get the last minutes of the meeting—if time allows.

- **Announcements.** Some meetings are little more than announcements. At other times there are none. Do not invite much opinion for information items. They are not action items.

- **Move to the primary agenda items.** For the rest of the meeting you must balance task and social maintenance functions.

- **Watch the clock** and give everyone credit for allowing you to end on time. Adjourn. If anyone blocks adjournment, try to reschedule the issue of concern.

It helps to know a few basic practices of what are called "parliamentary procedures" or "rules of order." Rules of order they are indeed. They are very handy, but you do not have to be able to say "The vote is on the motion that the report of the bylaws be made a special order of business." It is very hard to even follow such comments much less speak them:

> It is so moved that the amendment be thus amended before the last motion is substituted for the original motion.

Drawing a blank? Anybody would. Basically, you need to know how to use only a half dozen rules of order that are the real organizers and workhorses of any meeting. In particular, when you are ready to vote, these basic devices are handy tools.

Sample 8.C(1)

MINUTES OF THE MEETING

Committee name: LAN-Network Specialist Committee

Place: Conference Room IB2430C

Date: April 5, 200X

Time: 4 p.m.

Present:

Tom Baines
Patricia Smith
Joe Keri
Dave Mulligan, Workforce Training Director

John Hought
Katherine Stoddard
Dennis Fitzgerald
Anne Long
Pat Gibbons

Meeting Proceedings

Anne started the meeting by requesting that the committee address the #5 item on the agenda, titled "Roles and responsibilities." Tom informed the committee that the Ethernet side of the laboratory was up and running. The Tokenet is now being installed. Paul is having problems with the hard drives and the SCSI (scuzzy) connections. Patricia wants to install a financial software program. Tom emphasized that the installation at this time was being done to facilitate use and that permanent installation would be accomplished later.

Tom continued to explain how the permanent installation will be done through Facilities. Facilities will be involved in the spring and summer. Dennis will probably be the cite manager for the raceway installation. Tom wants to be part of the raceway installation.

MOTION/APPROVED:(1) To hire someone to make final drawings for the raceway. (2) Have final drawings ready by April 18, 200X (10 days from April 6, 200X). (3) Check with Budd about Brad.

Anne then moved the discussion to item #3, "Short-term needs."

Tables: Patricia received the desk and table she needed from John. The table was 8 feet long. She also asked for an additional table.

Chairs: Katherine stated that the new lab could have the chairs in 3405.

ACTION: Get chairs from 3405 into TB.

LCD unit: This unit is here and securely stored. Tom Baines has access to it. He mentioned that it was very expensive and needed to be kept under lock and key until it was secured in TB. June had just prepared the requisition for Dave's signature to order the special overhead projector that accompanies this unit. This projector needs stronger lighting to illuminate the work. It costs about $5000 plus $625.00 for the projector.

Sample 8.C(2)

MINUTES Page 2

LOCK-DOWN CABLE: Tom mentioned that this needs to be ordered for the overhead projector and LCD panel.

SOFTWARE REQUESTS

A system to order software for the TB lab is needed. It was suggested that a memo be generated requesting software and routed to each user for their initials. If it is approved by all parties, it will then be given to Anne for her approval, and ordered. All orders for software and computer equipment go through Tom's office so he will automatically see them.

Who will install the software and monitor it? Tom was concerned about trashing the computers with too much software. This led to a discussion about a backup system for the computers. Backup tape equipment was ordered for this purpose.

BUDGET

Dave asked how the budget was being handled. He was told by the committee that they were given orders to spend it. Over $122,000 was designated for equipment.

Dave felt it was important for the committee to get a handle on how much was in the budget at this time and how much more would be available in the future so the committee could prioritize and plan their needs for equipment, etc.

It was suggested that Anne get the LAN budget for next year, rather than have Dave as a go-between. Dave indicated he would encourage this. Anne was more concerned at this time with the model of this program.

CD ROM—Patricia will view a copy of a training program on CD-ROM.

RESPONSIBILITIES DEFINED

John stated it was assumed he would have responsibility for the lab.

SECURITY

Once NOVELL goes down what will happen? Tom stated that if it goes down, it will take several hours to come back on line again. John concurred.

Anne suggested that John write up guidelines on security. She was also encouraged to talk with and work with the technicians on this.

ACTION: Guidelines for security issues in TB should be written up.

NEXT MEETING: 12 noon on 4/20/200X

A Few Rules of Order

"Do we have a motion?"

You may or may not want to use the precise expressions that are identified here. Some committees do; others do not. What matters, when someone asks for a motion, is that any proposal must be articulated—and written down—so that precise wording is the end product.

To Second a Motion

The idea of a "second" has one important value. If no one seconds a motion, the motion dies. And of course if no one other than one member is interested, it *should* die. The use of seconds stops crank proposals because there must be a second interested party to continue discussion. Some one must say "I second the motion" to show interest.

The Call for Discussion

The proposal is now open for comment. A formal committee that plays by the rules is only at this moment ready for discussion. The truth is that most committees talk an issue to death and *then* develop the proposal or motion. By then the only "discussion" left will concern the way a motion is stated. Your call for discussion is most likely to be of the second type.

Friendly Amendments

Members may offer to change a word or entire sections of a proposal. Vote on each suggestion and add or subtract these amendments with the committee's approval.

The Call for the Question

The friendliest question of all after a long meeting is the question, should we take a vote? Usually the question is not asked until the answer is an obvious yes. Of course Peterson may say no, but the committee spirit will decide whether to go on with the discussion or vote. Then call the vote:

All in Favor?

Nine in favor.

In a corporate environment, the minutes of a meeting are far more substantial than the meeting announcement and agenda. The minutes represent the outcomes.

Sample 8.D

Personnel Training Office
North Building 2304-A

SYSTEMS & PROCEDURES—TECHNICAL ADVISORY COMMITTEES

Meeting Setups:

1. Set date & time, with input from Board Chair, member(s), & Ron Frazer.
2. Reserve room:
   • Board Room—Ginny (761)
   • Dining Room - Lynda McNalley (145)
   • Technology Center—John (377)
   • Staff Lounge or Conference Room
3. Review minutes from previous meeting for actions and/or resolutions to be addressed at this meeting; set agenda accordingly.
4. Check with Board Chair, and Ron Frazer for additions and changes to agenda.
5. Finalize meeting announcement (use masterdt.doc on BDM disk; save on YYYmaZZZ).
6. Get parking permits from security to be included in announcements.
7. Send out announcements to:
   Committee members (use List Processing file BDM for labels)
   Ruth (378)
   Ivan (245M)
   Janet (245B)
   Members/Internal
   BDM binder
   Correspondence file.
8. Check into coffee and refreshments.
9. Copy membership list, and use it to keep track of RSVPs. Call members approx. 1 week before meeting to confirm attendance.

Meeting:

1. Arrange for someone to take minutes.
2. Print copies of agenda and previous meeting minutes to be approved.
3. Pass around sign-in sheet, with column for name, organization represented, and phone number.

Minutes Preparation:

1. Type up rough draft of minutes within 24 hours of meeting, if possible. (Use masterdt.doc on BDM disk; save on YYYmnZZZ.doc)
2. Note clearly on minutes any actions or resolutions to be followed up on.
3. Send rough draft to Ron Frazer and member(s) in attendance for revision and additions.
4. Finalize minutes.
5. Prepare any other correspondence regarding actions or resolutions to be sent with minutes.
6. Send minutes to:
   • Ruth (378)
   • Ivan (245M)
   • Janet (245B)
   • Members/Internal
   • BDM binder
   • Correspondence file.

All Opposed?

Poor McCarthy. He votes down anything.

Abstentions?

One. Peterson is dozing again.

Do be sure that a committee allows for abstentions. Members may want the *right* to be neutral on issues, and they usually want the vote in the records.

How you decide to vote approvals in policy is complicated. There are many formulas:

---

1) **The simple majority of those in attendance.**

2) **A two-thirds majority of those in attendance.**

3) **The unanimous vote of those in attendance.**

4) **The same conditions can be applied to *all* members of the committee, which means that absent members count!**

Sample 8.D is a set of company procedures that explain company regulations for calling, holding, and documenting meetings. This is actually an *abbreviated* version, but it will give you an idea of the complexity of details involved in dealing with a community of people.

The Team Player

Because the likelihood of your being the leader of a group is statistically small at any given time (or else we would all be chiefs), then we should also discuss the role of the members of the group. In a way, this entire text is about being a member—of a company as well as of any of its committees. There is no simple point to be made. Your entire contribution to a company will shape and condition your role or roles in committee work.

If you were on a committee with me I would want to be able to count on you in at least five specific ways. **The first two attributes I would look for concern your awareness of how a group works and how a leader helps or hinders a group's success.** In the last two chapters we have tried to focus on these problems. If you understand the distinctions between task activities and maintenance behavior you will be a better member. You will understand what to do in a group. Also, if you understand the working relationship between a leader and a team, you will be an asset to the group because you can help strengthen the efficiency of the leader. Remember, the real power of a leader depends on the group. Leaders can be only as effective as groups encourage them to be.

The remaining three conditions are obvious from our discussion of other committee matters. A good member is knowledgeable. **You help to the extent to which you *know*.** A good member also can be judged by attitude. **You help to the extent to which you *care*.** Finally, a good member can be judged by involvement. **You help to the extent to which you *share* the importance of the group task.**

In very practical terms, how can you possibly achieve all five goals? I would simply ask you to do one thing: get involved. How? My answer will seem odd: take notes during every meeting. Do not keep the minutes, although this task would keep you on your toes. The leader needs ideas, ideas that get lost if you do not jot them down. Get in the practice of passing judgments. Write down what people say if you wish, but more importantly, write down what *you* think about what they say. Take the usual objective notes at a meeting, but take personal notes and keep *involved* so that you will be creative and reactive. Put your initials by the best ideas and offer them if you think they may be valuable. Taking notes is a very successful method of putting yourself to work, with your own initiative, in a committee situation. It works. You will be a very desirable member on any committee—on all five counts.

It is interesting to note that a committee is simply a miniature culture of the workaday world. At the outset of this book I compared IBM with a primitive society. The point was that any corporation is a model of a society. It *is* a society. Similarly, a committee is a society. It even has norms and rules. It certainly has expectations. In a sense, then, the committee is a large enough environment for us to talk about the pyramids of power and the interactions of coworkers, but in a small enough community where we can see many of the complexities on a small scale—a scale of, let's say, you and a half dozen coworkers and a supervisor. This small community allows us to clearly see desirable and undesirable interactions with clarity.

Our discussion of committee member traits and leadership traits can be combined. They stand side by side as characteristics of interaction that allow us to work side by side. On the small scale of our committee, we realize that managers should reflect leadership traits, and employees should reflect team-spirit traits. This is a profile of behavior that works well.

GOOD BEHAVIORS TO DEVELOP

Managers/Leaders	_Employees/Participants_
Flexible	Team players
Encouraging	Take directions
Trustworthy	Personable
Knowledgeable	Thorough
Open	Take initiative
Responsive	React positively to problems

In either capacity—manager/leader or employee/participant—the valuable behavior we appreciate is quite obvious. Similarly, the undesirable behaviors stack up just as neatly to define taboos we must guard against if we want to get the job done, either as a supervisor or as the employee.

POOR BEHAVIORS TO AVOID

Managers/Leaders	_Employees/Participants_
Bossy or weak	Lazy
Avoid responsibility	Tardy
Uninformed	Defensive
Unavailable	Complain
Take advantage (of people)	Rebellious
Exploitative	Dependent (needs a nursemaid)
Never respond to needs	Inflexible

The committee is a smaller version of the larger reality of a company. The right style for a manager and right style for an employee are mutually effective in both environments. In other words, the management style of a company generates the leadership style that structures its committees. Good committee leaders and good committee participants reflect strengths that are desired by both the group and the larger unit, the company.

Patterns of Interaction

Being an attentive committee member depends on skills, and one of those skills is awareness. Several additional useful ideas may be of value in committee environments. The first of these concepts is what is called a *sociogram*. A sociogram is a way to *draw* what is happening in the goings on of a committee. Meetings are difficult to track. You will usually have a sense of how people plan to vote, so to speak, but there may be more to look at. What people say is simply the obvious measure. There are others.

We have touched upon the matter of seating. The geography of a space is readable if you look carefully. You want to note *where* people are. You might also note *who* people talk to. In a committee situation people do not usually talk to everyone. They may respond to certain people. At times, they may ignore certain members; they may whisper to still others. *How often* people speak is also an obvious consideration. I have had discussions tabulated to count how many times each member of a committee spoke. This is an interesting quantitative analysis—particularly of leadership. A good leader does not usually monopolize by speaking the most (the *duration* of comments) or, as my tabulations would indicate, the most often (the *incidence* of comments).

The sociogram is a drawing that can plot some of this material in an informal way. Draw a sketch of the table and the chairs. Put initials by each chair. After the meeting is well under way, draw lines between speakers. Add arrows if the comments seem to be going in one direction. Increase the number of lines if the frequency is intense. This drawing is a unique way to plot the interactions of an otherwise very four-dimensional event. The results can be useful. A sample sociogram of a meeting follows.

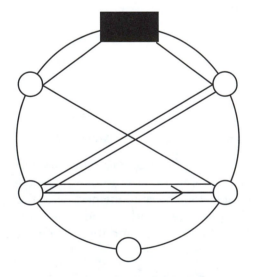

THE SOCIOGRAM

THE SOCIOGRAM

The sociogram is an easy way to identify interaction patterns. One line may represent, for example, five comments between two people. You can add arrows if the direction of the comments seem to be consistent. The results often reveal political allegiances, personal friendships, likes and dislikes concerning issues, and other points of interest. The agenda is likely to change the social interaction pattern but not the circuits and networks, which change infrequently.

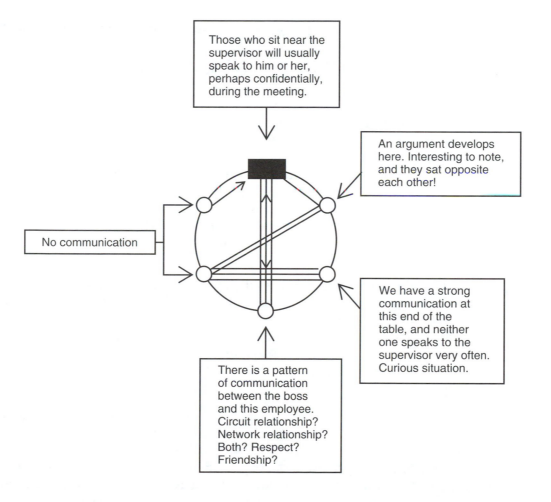

Those who sit near the supervisor will usually speak to him or her, perhaps confidentially, during the meeting.

An argument develops here. Interesting to note, and they sat opposite each other!

No communication

We have a strong communication at this end of the table, and neither one speaks to the supervisor very often. Curious situation.

There is a pattern of communication between the boss and this employee. Circuit relationship? Network relationship? Both? Respect? Friendship?

Groupthink

There is an enormous body of literature about good group management. Although there is plenty of bad group management, the example of bad management is nothing to aspire to and so very little literature is focused on the risks of poor group behavior. Everyone, not unwisely of course, wants to be concerned with doing the job correctly. Because poor group decision-making leads to poor judgments, we should look at the problem so we have a handle on what *not* to do.

The worst of the worst in a committee is what has been called "groupthink." Groupthink is a kind of cheerleading that gets out of hand. **If a group suspends its critical sense of an issue, particularly because of ingroup pressures, there will be trouble. There will be groupthink.**

In the business world groupthink is not uncommon, but since the fittest survive the economy, few businesses will go far if the CEO and the board have their heads in the sand as a result of groupthink. Cheerleading will stop once a company is in the red. In the political sphere, the risk of failure is just as great but the time it takes is not clearly spelled out in economic terms. The time it takes to be "found out" can be disastrous and costly. Irving Janis, an authority on the concept of groupthink, devoted an entire book to examples that were strictly military or political.* He argues that the Department of the Navy created the setting for the defeat at Pearl Harbor because of groupthink. The navy chiefs of staff had a tendency to cheer their way into thinking they would not and could not be attacked. They then rejected the growing evidence on the western horizon. They went to the officers' club rather than listen to top secret warnings from Washington. Groupthink, according to Janis, played a major role in the disaster.

Among other historical examples, he examines the Bay of Pigs event at length. The Bay of Pigs was a U.S.-sponsored "invasion" of Cuba that resulted in a humiliating defeat. This embarrassment to President John Kennedy was another groupthink situation. Kennedy and his younger brother Robert, and their close friends, formed a tightly knit group of decision makers. Kennedy's charisma and the exuberance of the group resulted in a cheerleading environment rather quickly. They could do no wrong. They sponsored the badly bungled invasion of Cuba.

Today's weekend paint-ball warriors would have done it better. Why the disaster? Because of poor judgment, no critical analysis, and no outside information. Granted, Hoover and the FBI did not like or help Kennedy, but other secret information on Cuban strength was available from the CIA. It was ignored. In fact, anybody in Florida could have explained the obvious dangers. Fidel Castro was as popular in Cuba as Kennedy was stateside. Only a handful of Miami Cubans could pretend otherwise, and that was exactly the group Kennedy sent to liberate Cuba. The result is history. The cause was groupthink.

* This discussion is indebted to Irving Janis, **Victims of Groupthink**, *New York: Houghton Mifflin, 1972.*

The 1998 national elections resulted in the worst Republican defeat in the history of the party in an election the Republicans thought would be a landslide victory. What went wrong? Groupthink. Republican leaders convinced themselves that their impeachment proceeding against President Clinton was a just cause and a clincher for the election—even though the Harris polls repeatedly showed that 70% or more of the American people found the proceedings objectionable. The results clearly demonstrated the dangers of cliques and ingroup pressure.

The evidence for groupthink in military and political spheres is significant. There is plenty of evidence in business and industry also, but the results have probably doomed most such companies to oblivion. It is for this reason that we should know about the mind-set of this phenomenon and be alert to its dangers. Specifically, information must not be censored. All options must be considered. **Critical analysis should be encouraged in all the activities of the committee.** Critical thinking—especially "criticism"—is *not* to be viewed as a negative attitude.

Summary

- Committee members tend to perform one or more predictable roles on a committee. A committee can be selected to reflect a desired composition of task roles:

  - *The resource* provides information.

  - *The probe* asks the right questions.

  - *The idea person* is there to think.

  - *The analyzer* is there to package the inputs.

  - *The timekeeper* watches the clock and keeps the members in step.

- The desired social roles are no less important:

  - *The listener* respects people and builds trust.

  - *The humorist* lightens the load.

  - *The referee* resolves conflicts.

- Undesired roles can emerge in committee sessions but can often be avoided altogether by selecting employees who do not exhibit negative traits.

the monopolizer	*the lobbyist*
the withdrawer	*the storyteller*
the blocker	*the aggressive personality*
the competitor	*the joker*

- A leader is not a leader without permission to lead. Authentic permission can come only from the group members who support a *perceived* leader.

- The *appointed* leader is delegated a leadership task by powers outside the committee.

- In a business environment there can be a conflict between the two leadership situations unless the appointed leader seeks recognition as a *perceived* leader.

- A leader is always a conditional leader who responds to a particular leadership situation.

- The leader
    - ✔ establishes the climate,
    - ✔ plans the agenda,
    - ✔ opens the discussion,
    - ✔ controls interaction and flow as needed,
    - ✔ summarizes,
    - ✔ maintains appropriate control, and
    - ✔ knows a few "rules of order" if they are needed.

- Whether leader or participant, the committee member must meet the expectations of the other members. The leader must be open and trustworthy in the eyes of the committee. The committee members must be committed team players in the eyes of the leader.

Activities Chapter 8

Present a memo to your instructor that develops a discussion of a personal experience that relates to the chapter. The memo should be 500 to 1000 words, typed. In it you can explain how you have related the text or the lecture discussions to some event you recall. This exercise will give you a better understanding of the material because you will explain incidents in terms of your perceptions of workplace structure and communication.

Select one of the following suggestions and develop an analysis that involves your current employment or a former position.

- *Discuss your roles in several activities that involved group work. Were your roles consistent or did you find that your motivations changed your roles at times? Explain.*

- *Identify a good leader and explain the reasons why you think the person is effective. This can be a foreman, a manager, a union representative, a coworker group leader, or some other employee.*

- *Explore a historical experience in which a leader was not trusted by the rank and file. Was the person kept in place by management cronies? Was the distrust made public? Explain the events.*

- *Examine a leader who was ignored. This can be someone who was assigned a leadership role or someone who did not want it. If this led to a reorganization because of group action, discuss the events.*

- *Discuss a cohesive group situation in which the response was very positive and all the members were good followers of a leader.*

- *If you have ever been involved with a decision-making process that developed into a struggle between the will of the majority and the rights of the minority, discuss the event and explain the outcomes. This topic selection, and the preceding one, might involve town councils, special-interest groups, and other organizations that are not employment related.*

- *Explain a work-related group activity in terms of roles. Use the lists of roles in the chapter and interpret the various group members in terms of the lists. Examine how the roles affected the group process and the outcomes.*

Committee Presentations

9

Work in Progress

9. Presentation Skills

I came to the point in the bidding process with Tempo when it was time to take all my data and suggestions and turn them into a presentation, with the hope that it would be one that would inform and dazzle. This is one of the "big guns" in my sales strategy because I'm good on computers, and accordingly, I take a great deal of time and care in putting a package together.

I called Simmons back a few days after my last visit to Tempo and suggested that we have a formal meeting with Sharon and the office manager and a spokesperson from the production crew. After the informal poll I took (see the previous chapter) I realized how important it was to include representatives from all the sectors of Tempo. Tim was only lukewarm about this idea but not unwilling. We set a time and date, and I got busy!

First, I compiled all the specific measurements and sketches I had taken from the Tempo building. With Visio design software, I created detailed layouts of the areas involved in the upgrade. I even included details like plants and bookcases to give the finished product a nice touch. I did two of these Visio floor plan diagrams. One showed the Tempo LAN layout as it currently exists and the other showed what my proposed changes would look like.

Next, I designed the PowerPoint slide show that I wanted to use to open the presentation. I imported the Visio diagrams onto "slides" and used some of the animation features in Power-Point to make arrows zoom over key areas. I wanted to strike a balance with the animation and not overdo the "cute factor." I knew that Tim would be primarily interested in facts and figures, while Sharon's interest could be best captured with creativity.

Feeling satisfied with the slide show, I moved on to create the last part of the presentation. These were the foam-backed display boards showing colorful graphs with cost tabulations, improved performance quotes from clients, and other basic information. This I did for Tim, knowing that this was the format most likely to catch his eye.

The big day arrived, and the presentation went off without a hitch. I could tell that Sharon and the office manager were impressed with the slides (especially the Visio material), while Tim and the production lead were more interested in the graphs on the boards. We concluded our meeting and I now had to wait and see if Tempo would ask me to submit a formal bid.

J. Q. C.

Effective Small-Group Presen- tation

For outreach, companies depend on specific employees to handle such activities as sales promotions, public hearings, civic gatherings, social functions, and similar situations that involve the public. Your role is not likely to involve these areas of concern. In this sense "public speaking," as you might generally imagine the term, is an unlikely career interest for you unless you happen, by some circumstance, to find yourself involved in such events. On the other hand, the day-to-day humdrum of employee committees and meetings is likely to be part of your world. And as you progress in the company, your potential supervisory roles or your potential technical expertise might, from time to time, put you at the head of a boardroom table, or behind a podium, or in front of a projector screen with a pointer. Are you prepared?

Could you, for example, stand before five new employees where you currently work and deliver an in-service training seminar? Think of the CPR training volunteers from the fire department. When they come to your company to offer training sessions in cardiopulmonary resuscitation, they are offering to instruct a group, typically in the range of 10 to 25 employees. Could you deliver the training session?

Are you prepared to sit before a group of ten supervisors and explain why you have selected a certain product for a bulk purchase? Think of committee meetings you have attended where your company purchasing agent or other authorities came to explain the logic of a purchase agreement. Could you do it?

If your unit of a company is blamed for an on-site accident in an office building, could you call a meeting of your employees and explain their responsibilities and rally them to do a better job? Think of the pep talks, or Monday morning musters, or emergency meetings you have had in the workplace. Someone took charge of those encounters. If you are a supervisor someday, will you know how to handle such situations?

In company settings, the committee presentation is the workaday variety of public speaking. Only a few agents of a company are expected to actually speak to a "public," meaning the population outside the company gates. Within the closed environment of a company, a great many more employees make presentations before modestly sized groups—composed of perhaps three to fifteen people, or an occasional larger group that may involve perhaps as many as fifty. The committees are typically made up of a particular sector of employees, and the group leaders are often supervisors or technical specialists. The meetings are often called in order to hear presentations. Just as commonly, a speaker is one item of a committee agenda that involves any number of other issues that will be handled at the same meeting.

It is not difficult to determine if committee presentations will be an important consideration in your future. If you are self-employed, the role of committee presentations will be limited to your supervisory tasks when meeting with your employees, and your contract discussions with clients. The clients will typically be sole proprietors or, possibly, co-owners of a business. You will encounter a committee only if you decide to try to negotiate a contract with larger business environments or with public agencies.

If you are a company employee, you may occasionally need to speak to a small group. The reason the oral presentation is a popular communication vehicle in companies is obvious enough, since most of the communication that is undertaken is spoken. Studies show that managers spend approximately 80% of their time in conversations and discussions. The bulk of their communication is rendered in conversation. Presentations appear not to be identified as a workday percentage in such studies, probably because the use of presentations will depend on company practices. Small business owners with ten or twenty employees may find they need few presentations. A hospital with 500 employees may use more presentations in one week than the smaller company will use in a year. Another reason why it is difficult to calculate a percentage for employee involvement in presentations is that many employees will never deliver a presentation. Others will find the task a weekly chore. However, observe that the positions of two of the most probable speakers for committees—supervisors and technical authorities—are likely to be ones you may hold sooner or later.

The Committee Presentation

The good news is that the committee presentation is quite unlike public speaking in several important ways. Most of this chapter will address the craft of speaking before a small, often familiar, group in an enclosed working environment. Most of what we will examine involves the very traditional craft of public speaking, but in the form of *in-house speaking*. In fact, very little literature is dedicated to this niche of public speaking. The communication textbooks tend to discuss committee meeting management at length, but the focus, as you have seen in this text, involves leader roles and member roles and group interaction. The texts then move on to public speaking, usually in the broad sense of the word. In the middle is the crossover environment where you work, where you hear people speak before committees, and where you may need to speak before a committee. What is the good news, then? It is that your speaking efforts never enter unknown territory.

Because most people are apprehensive about standing before any kind of audience, it is important to realize that, for example, you *can* sit down. It is important to understand that the committee is likely to be composed of coworkers you know and respect. You must realize that the numbers may involve five people but not five hundred. Everything will frame familiar territory. You will be inside the company; you will probably know the room; you may have had technical support from the graphics department to create visual aids with *Photoshop* or *PowerPoint*. Finally, you will be the man or woman behind the pointer because you are respected—either for your leadership or your expertise, or both. This does not sound like everybody's vision of public speaking does it? That is because it isn't. It is not a bad dream, and it is not exactly public speaking either.

Public speaking is a balancing act between spontaneous presentation and careful preparation. A presentation can be read line by line, but there are few specific applications for a presentation that is read. Presentations can also be memorized, but this tactic is more of a tool for someone who will repeat a presentation for, let's say, hundreds of public appearances. Presentations can also be impromptu, but improvising calls for a quick-witted talent, and improvisation takes daring—and involves risk. That leaves one practical way to handle a presentation: it must be prepared and practiced. **Because the other types of speaking are more difficult to handle, the presentation that is prepared with care but spoken with a natural and relaxed manner is the preferred technique.**

Notice that there is a hint of theatrical deception in a well-prepared presentation that is then delivered with a spontaneous style. Part of the skill of the delivery is *appearance*. A speaker wants to appear relaxed even though he or she may be nervous. The speaker wants to appear casual even though the presentation is carefully designed and rehearsed. There are even touches of the craft of acting in the behavior of a speaker. When you speak to a friend the scale of your behavior is one-on-one. When you hear a public speaker who seems to be a little tedious it is possible that you are seeing someone who is behaving one-on-one but speaking one-on-ten. **It takes a little practice to develop a slightly larger sense of behavior for larger numbers of people.**

Distance is the force at work that creates resistance to the transfer of energy if you have an audience. It is basic electronics: it take power to push power. If you sit at a table with a committee of ten, only the person to your immediate left and the one to your immediate right are within one-on-one distance. If you speak one-on-one, your powers diminish as you reach out to the other eight. Because distance is an inherent consideration in a company presentation, large or small, a little practice will result in a little more voicing intensity, perhaps slightly larger facial movements, and a little more sweep in gesturing. This is not hamming it up. The changes are subtle, but the person in the back row will be thankful, and it may be the CEO.

The eyes do the hardest work in this little company theater. You use your eyes one-on-one for most speaking situations, and you must guard against that habit in a larger speaking environment. Speakers are inclined to favor certain members of the audience for eye contact—perhaps because the members are within one-on-one distance or, if there is open discussion, perhaps because the speaker is in direct conversation with a member of the audience. The eyes have a great deal more work to do in this setting because you are there to address the group and *not* several individual members. The eyes must watch them all. **You must listen with your eyes and understand what people speak with their behavior. You must also *involve* everyone by using eye contact.** The eyes can overcome the distance that separates you from those you want to involve in your presentation—because involvement is the key to speaking success.

The time you allow for a presentation will vary from a brief five minutes to a sustained talk that will last twenty minutes or longer. If you are one item on an agenda, the brevity will be appreciated. If you are the focus of the gathering, the timetable is in your hands, since listeners have probably provided an hour for the activity out of a common habit everyone has in the way they schedule appointments.

Committee presentations are generally brief. In the case of committees, the clock starts ticking the minute the meeting gets underway. The value of the people at the meeting is more than the head count. If your company accountant is among them, her perception will be quite different from the perceptions of others. For her, the fifteen people represent a mathematical value expressed as

number of people x length of meeting x hourly value

Here we have a unique element that distinguishes the committee presentation. The expression "time is money" is an important consideration and should be part of your strategy for preparing a talk. You must not waste the time of the audience. More importantly, you do not want audience members to arrive at such a thought. In this respect it is critical to time your activity wisely. If you come up short on the time allocated, the problem is not critical if the presentation is thorough and you do the job. At the other end, if you take too long, the reactions may begin to show: yawning, fidgeting, clock watching, and other signs of impatience. Pity the poor soul who, for example, delays a seminar group at a large conference when all the members know they were due in the dining room twenty minutes ago for an all-expenses-paid premium lunch at a four-star hotel.

Preparation is the key to timing; in order to keep to your projected time frame, you must prepare with care, because speakers who misjudge their forums are in trouble. An inverse reaction sets to work: a speaker will keep an audience on overtime because the speaker thinks every minute of the information is important. However, the reaction of the listeners can defeat any value the overtime might have had; **for every minute the speaker runs beyond the expectations of an audience, the speaker runs the risk that the audience will begin to respond negatively—which can defeat the *entire* presentation.** Needless to say, grinding to a halt is no solution either if important points and a summary end up omitted. Preparation counts.

Basic Presentation Goals

An effective presentation means a great deal more than one that is *heard*. The challenge is to meet desirable levels of audience response.

Was it understood?

Will it be remembered?

Will it be put to use?

Four simple watchwords can help you build and sustain an effective presentation that has audience appeal:

CLARITY

VIVIDNESS

EMPHASIS

WORD CHOICE

Each of these measures can be used to judge your overall effectiveness at what you have achieved with an audience.

Clarity and Vividness

The challenge is to present the subject with clarity and to discuss it in such a way that you enhance the listeners' understanding of the material. This would seem to be obvious enough, but presentations in the workplace are often made because of new products, new events, new concepts, or new realities of any sort that call for explanation. You are not likely to make a presentation that covers familiar ground. The idea is to mark off *new* territory to help the audience understand the unfamiliar. This is a central focus that is typical of in-house speaking in company settings. There are two challenges to committee presentations of this sort.

First, there is a learning curve involved in both informational and persuasive presentations. Since the intent of the speaking effort is often supposed to explain new technology, or new policy or some other unfamiliar material, the audience depends on the speaker as the source of a learning experience. There is a second challenge to introducing anything new: resistance to change. **Over a period of time you will notice that an enormous percentage of committee presentations underwrite change.** As a result, the presentation that is designed to present information can be controversial even though the information—perhaps a description of new apparatus—seems straightforward.

Emphasis

You, as the speaker, are in complete control of the importance you intend to give to key ideas. The emphasis is easy to control and easy to project, but project you must. One reason college teachers write on the blackboard or write on overhead projections is because the emphasis creates a predictable response. Watch your classmates and notice how they respond to an instructor's discussion. Notice that the greatest number will take notes when an idea is presented in writing. This may or may not reflect any deep-seated learning style, but what is clear is that the visual experience *repeats* the spoken experience. In other words, the repetition creates emphasis. **Because an audience depends on the speaker to identify what is important, the speaker has to signal the listeners.** Several practice are helpful:

> • **Maintain proportion.**
>
> • **Use emphatic behavior.**
>
> • **Use repetition.**
>
> • **Highlight transitions and lists.**

These are practical suggestions that can help make a presentation a success with an audience.

1. *Maintain proportion.* If four main ideas in a presentation are of equal value, you must make that clear to the audience. The most practical way to convince them of the mutual importance of four ideas is to treat them with equal regard—and more or less equal discussion time. If you run out of time during a fifteen minute presentation in which you covered three of the four ideas, you cannot simply say

> *Of course, it goes without saying that additional overtime would help fourth quarter earnings.*

The listeners are going to respond only to the proportions you present them.

2. *Use emphatic behavior.* If you want to make a point, give it a little oomph. Punch it with a wave, a pause, a turn of your voice, a tap on the table. Emphasis of this sort is natural to most people because everyone colors dialog with various devices that are characteristic of their manners of speaking. Voice coloring is a strong indicator, and listeners will pay attention to such cues. By the same token, a monotone is going to create problems. A monotone fails to project the speaker's judgments so that a listener can hear the speaker's position. In addition, the monotone discourages effective listening because the interest level of the audience will start to decline.

3. *Use repetition.* Repeat key points. If you highlight the key points in the introduction, that is the first mention of the main ideas. The body of the presentation will again identify each central idea in the course of discussion. Repeat the key ideas in the conclusion.

4. *Highlight transitions and lists.* Remember that the audience has no idea where you plan to take them in the presentation. Even if they know the subject, the subject does not explain *your* perception of it. Listeners need to be told what is important, so you must emphasize key ideas, as I noted above. You must also organize those ideas for them. Transitions will allow the listeners to move with you as you travel through the discussion. Signal movements in the discussion so that the listeners move right along. Your job is not very different from that of a tour bus guide explaining the sights to tourists.

> **After** the collapse, the engine overheated.
>
> **Subsequently,** we discovered that the Nolan Company also had a model on the market.
>
> Even **before** Benetti arrived, the committee was ready for the change in our legal staff.

Transitions will shift a listener's attention as you move along. The audience must, after all, move with you.

In addition, use words that *list* because they *organize* and signal *priorities* if the list moves from most important to least important. The following four sample comments are unrelated, but they are organized in sequence. The power of organization is apparent in the way you will tend to relate the four sentences.

> *The lion's share of this money needs to go to research. That is the **top** priority.*

*A **second** issue, equally important, is the problem of intermittent failure in the drives.*

***Third,** we need new uniforms.*

***Last,** some part of the facility should be earmarked for lockers and a decontamination room.*

Word Choice

The final touch that will guarantee your success with an audience is language sensitivity. Technical specialists need to make every effort to be sure they are communicating, and the single most important component is vocabulary. Because technical fields develop highly precise terminology, the likelihood of confusion is extremely high. As you now realize, the audience is, as usual, the issue. The best approach is to be prepared. Anticipation of the audience is the best preparation for vocabulary choice. You can also have definitions prepared that are either part of your presentation or are set aside for possible use.

You can anticipate the vocabulary needs of the audience, and in a company setting, it is probably not going to be too difficult to determine. You can get on the phone with your supervisor or walk over to Building Six and meet a few members of the committee. If you were asked to speak by your supervisor, your supervisor will have a sense of what the audience makeup calls for because he or she has judged your presentation to be a necessary event for a specific group. As another option, if you speak to a few of the committee members you can find out the degree of familiarity the committee will have with the subject, and the language that relates to it.

Anticipating the situation is one tactic. On-site response is another. You can and must respond to the listeners *during* the presentation. From glazed eyes to shaking heads, to questions, audiences communicate any problems they are having. Be sensitive to their needs. Select vocabulary with care.

On a related note, humility is a winning attitude for a speaker to have, and sympathetic allegiances with the audience will not go unnoticed. If a speaker parades his or her expertise and ignores the audience perceptions, there will be a problem. There are enough employee training programs in today's business world to limit this dilemma through training, but the problem will never go away entirely. Professional people can sound terribly bright and terribly important and there is an occasional temptation to come forward with a good bit of overpowering knowledge effectively wrapped in very sophisticated vocabulary. **Favor plain speaking unless you are confident that you are among people in the know.** If a committee is composed of engineers, you have an excellent opportunity to use technical terminology so that it can do what the language is intended to do, which is to create technically precise comments. If, however, a group is not equipped to deal with the complexities, try to be prepared. This situation will slow you down, it will affect your vocabulary, and you will have to explain terms. To achieve these ends you must watch the clock and either limit the subject matter or lengthen the presentation.

Committee Presentation Basics

There is no reason not to be well prepared for a committee presentation if you are a reader of the *Wordworks* series, because you are familiar with the tool of choice for the proper preparation of a committee presentation: an outline. Developing a speech is as simple as developing a properly constructed outline. In one way or another, every volume of the *Wordworks* series discusses the outline as an organizational tool. I will review the basic concept briefly, but the central concern in this case has to do with the fundamental patterns that are characteristic of the two basic speaking presentations you are likely to make: *a presentation that informs* and *a presentation that persuades.*

Then, too, your particular needs as a technical specialist enter the picture. There are specific patterns you can use to logically organize various technical presentations you might make. You might, for example, be called in to identify ways to cut costs for commuting time for a company that has engineers rushing back and forth from the office to a job site. In this case you would follow a pattern from *problem to solution.* You might be given a quarterly performance report that explains additional expectations that are being added to your job description. You are to meet with the executive board on Friday at 1:00 to explain your response. In this case you have a problem to solve but you do not have the opportunity to define it in your terms. You have a set of criteria or objectives and you might respond with a *solution to given needs.* Perhaps you are given a timeline to submit a proposal and you are not sure of the available funding. Present the talk by *prioritizing* so that you move through the list of suggestions from head to tail. The committee will welcome the logic and decide what they can afford by beginning at the top of your list of priorities.

The outline can be used for either the instructive presentation or the persuasive presentation and it can easily meet your needs regardless of your specialty. There is, however, another reason for using the outline. Since I noted earlier that reading a speech or memorizing a speech are high-risk options, delivering a presentation from an outline is *the* standard option and the strategy of choice. You have seen speakers use note cards, of course, but the note cards usually contain only pieces of an outline. The outline is the central axis that can give you continuity and control, and it keeps your timing in line. It is the *script.* When you marvel over the speaking skills of television anchors and other speakers in the "talking head" category, you will often notice that the eyes are slightly off camera because they are using a TelePrompter. What's on the TelePrompter that keeps the golden voices from ever uttering an "uh"? An outline, of course. The outline is a set of cues. And it is worth noting that if professional speakers use cues from time to time, the occasional speaker should use them *all* the time.

The Order of Events

Your speaking agenda will involve a number of basic considerations.

We will examine a few of these considerations in detail.

- **Know why you were asked to speak.**
- **Anticipate the audience.**
- **Anticipate your image.**
- **Know your time frame.**
- **Identify the purpose of the talk.**
- **Isolate the objective.**
- **State the keynote idea.**
- **Identify subordinate ideas.**
- **Limit the scope to the available time.**
- **Outline the delivery.**
- **Construct a brief summary.**

Situation Analysis

A presentation, as I mean it here, is a highly specific type of public speaking. A presentation in the context of this discussion is an in-house work-related talk that is given before fellow employees or supervisors. It will fulfill a requested or required task with highly specific parameters. You might hear this from your supervisor:

> I got a call from Watts over in production; I think he is in building D. He says a group is flying in from the Memphis office on Tuesday. They need to see the new assembly that Testing approved in January. I've reserved the conference room next to the boardroom. Can you meet them there at 11:00 and show them what we have? There is a catered lunch in it for your efforts.

This is not the sort of situation in which you need to think of something to talk about, or frantically hunt for an opening punch line. This is business as usual.

In addition, a group that attends a presentation in a company environment is selected. The link between the subject matter and the audience is quite defined. In the preceding sample, for example, the audience represents a "need-to-know" group. Another common audience includes decision makers with the power to act. Two traditional speeches—the information speech and the persuasive speech—are often used to address these audiences because of the way in which the subject matter links the presentation and the audience.

Identify the Mission

If you have to speak because you are asked to do so, you are responding to given objectives. This pinpoints the intent, but you need to think out the situation.

Who asked you to speak?

Why were you asked?

What do you need to present?

Who will be in the audience?

What are the audience expectations?

What outcomes are expected?

Any text on general-purpose public speaking will identify most of these concerns, but pay close attention to the first two and the relations these two unique concerns have with regard to the other interests. The remaining questions define an avenue to the desired outcomes. The last question is the actual goal that is defined by the request for your presentation. What is unique is that you are given a mission to accomplish. You need to examine the *situation* because the context of a company presentation is narrower than the usual speaker-subject-audience focus of public speaking.

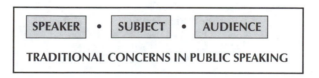

SPEAKER • **SUBJECT** • **AUDIENCE**

TRADITIONAL CONCERNS IN PUBLIC SPEAKING

The driving force behind the company presentation is a specific situation in which there are usually management objectives behind the scenes.

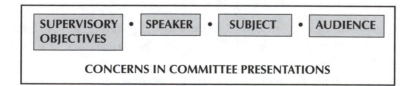

SUPERVISORY OBJECTIVES • **SPEAKER** • **SUBJECT** • **AUDIENCE**

CONCERNS IN COMMITTEE PRESENTATIONS

In general, you will be asked either to present information or to convince a group to accept an action or make some kind of commitment. The latter is usually not your task, which is fortunate because the audience is a special challenge if the presentation is persuasive.

Identify the Audience

The next task is to examine audience expectation. This may not be critical if you are a shop steward and you are meeting the union stewards for a contract compliance analysis. Many committees are composed of birds of a feather, and the speaker is often "one of them." Just as commonly, you should take measure of the audience with whom you will be dealing. Because your status as an employee will be that of a technical specialist, you will usually want to think out predictable considerations about people who will be hearing what you have to say:

1) **What is their *knowledge* of the topic?**

2) **What is their *interest* in the topic?**

Since the presentation will usually inform or persuade, we can generalize the degree to which these two considerations will be critical.

REACTIONS TO SPEAKER'S *INFORMATION*	AUDIENCE RESPONSE AREAS	REACTIONS TO SPEAKER'S *CALL TO ACTION*
CRITICAL _____	KNOWLEDGE and UNDERSTANDING are IMPORTANT	
	INTERESTS and ATTITUDES are IMPORTANT	CRITICAL _____

AUDIENCE RESPONSE GOALS

In a delivery that is informational, the audience response area depends on knowledge and understanding. In a delivery that is persuasive, the audience response often depends on the interests and attitudes of the members of the group.

If you develop a presentation to provide *information*, the likelihood of judgmental reaction might be very low. If you develop a *persuasive* presentation, you are, in essence, asking your audience to commit. This intent is another world entirely. You might be asking for support or votes. You might be asking for a grievance settlement on behalf of a union member. You might be asking for a million dollars in risk capital for a high- (or low-) risk R&D program. Obviously, knowledge matters in every case, but the involvement of group interests and attitudes will be critical if you intend to persuade, probably more so than in a situation where a presentation passes as an information item on an agenda.

Of course, these sample situations are relative. For example, I attended a meeting with a guest speaker who dropped in to explain a $10 increase in monthly parking fees. It was a single agenda item among many—except that he was lucky to leave in one piece. It was merely information in one sense, but to everyone present it was a 100% increase in parking fees, which had cost only $10 a month in the first place.

The fewer boundaries there are between the speaker and the audience, the better the likelihood of a successful presentation. In company environments the space boundaries are often no greater than a large table or a room full of chairs. The serious boundaries are differences between the speaker and the listeners. Professional respect can eliminate many boundaries that listeners might define, and this is why the way in which a speaker handles himself or herself is extremely important. The remaining challenge is to keep boundaries from being marked *during* the presentation. **Because your message—indeed, any message—is going to be interpreted, you must take care to avoid creating boundaries.** You must make a strong effort to get listeners to understand your presentation, but to the extent to which it is possible, you want listeners to understand the material in the *way* in which you wanted it to be understood. This is a difficult challenge and it depends on careful consideration of the group.

Identify the Topic

Regarding the actual presentation, the first concern is the subject. Since you are not picking something interesting to tell a Rotary luncheon, this matter will probably fall into place without much attention or concern. Company meetings are usually responses to perceived issues, and any purpose you would fulfill with a presentation will be highly defined. All that matters is that you understand the parameters in a *precise* way. You do not want to talk about what does not matter, and you do not want to omit some considerations that should matter. If you have to tell the plant-safety committee how they should meet the new OSHA regulations for shop safety, you do not need to spend ten minutes explaining the last group of regulations that were presented to the company three years ago.

State Your Purpose for Speaking

Write down the purpose. This is a serious challenge, since the word *purpose* can mean a great many things. Interpret the word to mean the answers to two specific questions:

What do you want to do?

What do you want to say?

Sample 9.A

PCB Basics:
An Outline

PURPOSE: My presentation will describe the different steps it takes for a wave-soldered PCB to become a final product.

OBJECTIVE: I want the committee to understand the procedures and timelines for the task.

A. The traces on the bare PCB boards are used for electrical connections.

    1. PCBs may come with traces on one side only or traces on both sides, or they may be multilayered, consisting of individual sandwiched circuit boards.

    2. The boards have holes in them for the parts.

    3. Some boards have silkscreening and some boards do not.

B. The boards are connected in sheets by small tabs. Depending on the size of the boards, these sheets come in different quantities that make flow-soldering easier.

C. All parts are prepped so that they will fit into the boards, and the parts are cut and bent to fit the holes in the boards.

D. Loading parts into a PCB is similar to putting a puzzle together.

    1. Loading can be done by machine or by hand.

    2. ICs are loaded first, and then the smaller parts are loaded.

    3. The polarity of parts such as diodes, transistors, capacitors, and ICs must be observed when installing them.

E. The loaded boards are then flow-soldered, which involves a process by which the boards are slowly run over a wave of solder.

F. After the flow-solder process, the parts are checked to find any problems and the solder is touched up where needed.

G. The boards go either to Testing or to Inspection.

    1. The boards are tested to see if they will work properly.

    2. The boards that do not work are reworked after troubleshooting, and they are then sent back to retest.

H. Finally, the boards are routed to Inspection to be checked for workmanship.

State the Audience Objective

Here we have another general term that can mean a great many things, and it can also be confused with purpose. Again, see the *objective* as an answer to a very specific question:

What is the desired audience response?

As you can see, the two words *purpose* and *objective* have two distinctly different points of reference: the *presentation* and the *reaction* to it. In-house speaking, like public speaking, pivots on two goals:

What you want to *say.*

How you expect the audience to *react.*

Once you have a strong sense of the preceding conditions, you understand your goal and you are ready to draft the outline.

Draft the Outline

Develop this document by using full sentences throughout. The sentences are important in that they will express the complete logical intent of the presentation.

> *First,* place the purpose and objective at the top of the outline.
>
> *Second,* structure the speech point for point as it will be presented.
>
> *Third,* use the traditional indentation method for subheadings to divide key points and to identify the subordinate points in a distinct fashion.

You are probably familiar with the most popular outline pattern:

I. _____

  A. _____

    1. _____

      a. _____

◀ *Here is a practical end result of outline organization. This is the plan for a brief committee presentation.*

You are not likely to go beyond these four steps. If you use only two categories to build the outline, you can conveniently number the primary ideas and identify the subordinate features (usually the evidence for each major point) with bullets.

1. _____

   • _____

   • _____

   • _____

   • _____

2. _____

   • _____

   • _____

   • _____

   • _____

Outlines are used to show two distinct types of progressions. The *subordination* of ideas is one progression. The indented steps signal the subordination.

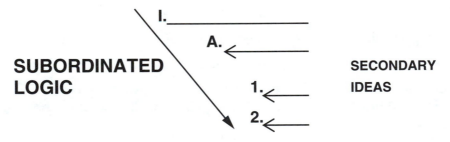

SUBORDINATED LOGIC

I._____

   A. ←

      1. ←

      2. ←

SECONDARY IDEAS

The other progression moves from central point to central point to *coordinate* ideas of equal value, which must be clearly identified for the listening audience.

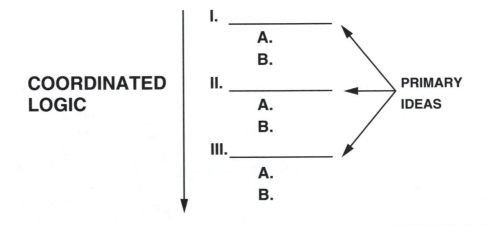

COORDINATED LOGIC

I. _____

   A.

   B.

II. _____

   A.

   B.

III. _____

   A.

   B.

PRIMARY IDEAS

Essentially, *subordinated* ideas are signaled by indentation (the diagonal arrow in the figure) and *coordinated* ideas are signaled by an aligned column (the vertical arrow in the figure).

As a script for a speaker, the two progressions have somewhat different objectives. The *audience* must have a clear understanding of the point-for-point or vertical logic of the central ideas. The *speaker,* in order to explain the central ideas, must have a keen understanding of the subordinated evidence or supporting logic that builds each important idea. The result is a simple orchestration of events for both speaker and listener.

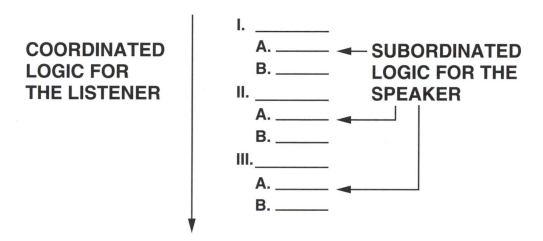

The outline is your script; it can be as brief or as thorough as you wish. How long is long? I have not seen them run longer than three pages of single-spaced text for technical talks of varying length that ran as long as thirty minutes. A single-spaced text is not, however, a convenient medium in any case. Double-space the text so you can reference it quickly. The three-page outline becomes a six-page outline if it is double-spaced. You might be perfectly comfortable with one page. If you decide to use the outline as your presentation script, you might print it in all caps using a sans serif font such as Arial or Helvetica. See if it looks more readable in 14-pt bold. If the lights will be dim (because of a slide presentation, for example) perhaps 16-pt size will help. I understand that the Orator font was designed for public speaking, with the idea that it was larger and easier to read. Obviously, any kind of fine print could present you with a problem.

With a computer you can do a surprising amount of script writing with your outline. Look at the following outline fragment.

III. The heating system will use 14″ diameter exposed ducting.

   A. The facility will have no suspended ceilings to cover the ductwork.

   B. Duct heating will gain added efficiency from radiant heating that is unobstructed.

   C. Fully exposed installation is the most cost effective method.

   D. Installation costs, maintenance costs, and fuel costs are low.

Suppose I design the outline as a script so that I can quickly see the outline as a series of cues. I will put the key idea in CAPS. I must emphasize each key idea in the outline. I will replace the Roman numerals with a transition word to signal my movement down the list of key ideas, so I will type the word **THIRD** in 18-point bold underlined. Then I will insert bold italics cues on the right to tell me when to show the audience the duct plan for the ceiling and the bar graph of costs. The result, ready for the podium, looks something like this:

THIRD **THE HEATING SYSTEM WILL USE 14″ DIAMETER EXPOSED DUCTING.**

   A. The facility will have *no* suspended ceilings to cover the ductwork.

   (Show duct plans)

   B. Duct heating will gain efficiency from radiant heating that is unobstructed.

   C. Fully exposed installation is the most cost effective method.

   D. Installation costs, maintenance costs, and fuel costs are low.

   (Use chart)

The listener must understand the primary ideas *and* the movement from one primary idea to another. *List indicators* or *point-for-point transitions* are important cues for listeners:

One. The evidence shows . . .

Two. Our objectives point toward a . . .

Three. We can achieve . . .

First, the contractor began . . .

The second problem was that . . .

Then finally the third crisis occurred.

Using these transitions in the outline will remind you to speak them.

Although most uses of the outline will probably not involve questions, questions make useful transitions in a public speaking situation. If you are going to develop an informational presentation, the key supporting ideas can be stated as questions, which you then answer.

I)   In what circumstances do you earn overtime?

II)   How many overtime rates are there?

III)   Is there a ceiling to overtime?

IV)   Are all standard deductions taken from overtime?

Once the outline is complete, gather any visual aids or samples or similar material. Decide where each item fits into the sequence of your presentation. Add these *points of display* to the outline at the appropriate locations. It will be helpful if these cues are highly visible. I suggest printing them in color or highlighting each *display cue.* Do not use a sentence for each display cue. The sentence outline is used to structure the spoken text. The display cue is only a reminder to show the audience a display, and it need not consist of more than a word or two.

Once you have reviewed and revised the outline, you can set it aside in favor of either a *topic* outline or 3″ × 5″ cards. The topic outline might prove to be the best tool for you. During the presentation, place it on top of the sentence outline and use it for your cues. It is a faster reference and you will not have to pause to consult it. If, on the other hand, you forget what the topic outline is indicating, briefly consult the sentence outline underneath it. If you feel a little disorganized when you speak, use a 21/2″ × 21/2″ Post-it to mark your place and move it down the page as you cover points.

You can use a pen or pencil to mark off the topics of discussion if you do not mind the audience watching the repeated gesture.

The 3″ × 5″ card tradition is probably more popular with college students than with corporate employees. Recipe cards will do the job, but I only occasionally see a speaker use them. The idea is to have the cues in small groups of flip cards. What goes on the 3″ × 5″ note cards? I would suggest that they need little more than the material in the outline, and, possibly, items that you find yourself forgetting, since your notes are reminders (see Sample 9.B).

Cards have one distinct advantage if there is no podium or table. Standing before an audience with 81/2″ × 11″ sheets of paper held in two hands will block a speaker's natural animation. The 3″ × 5″ cards are small and can be held in one hand. Both hands remain free. The preference is yours, but try to anticipate the environment. If the lights are out, there could be trouble. Besides, you cannot always be sure of what will happen. During a very momentous occasion—a presidential inaugural ceremony—the great poet Robert Frost was supposed to read a poem he had printed on a sheet of paper that he struggled with because of, of all things, a roaring wind.

The 3″ × 5″ card option, like the topic outline, can be used *with* the sentence outline as a backup. Place the sentence outline nearby. Use the cards as your cues. If you get lost, check you sentence outline—and do not be embarrassed. You have seen every one of your college instructors "lose it" more than once. Typically, they digress to explain something and then forget where they were. I have gotten so comfortable with the classroom presentation environment that I ask the class, "Where was I before I got off on this other issue?" You may know a committee well enough to ask the same question at times, but at other times getting lost does not look good and the sentence outline is there to get you back on track without any questions.

Be certain to number the cards boldly on the upper right or left. Print the notes boldly. Do *not* use the reverse of the cards; this will cause confusion. If you are using visual aids or any sort of display model or other items, indicate the point at which each item is to be presented during your presentation. Use a colored pen to indicate the item on your cue cards.

We have now reviewed the basics. A presentation depends on the basics. An *effective* presentation depends on a good deal more. Consider the visual aids indicated in the preceding paragraph, for example. What visual aids are appropriate? What graphics are inappropriate? How can they be designed for the most effective communication for an audience in a committee situation? Should you show a model of a product? Should you demonstrate it? Should you use handouts? There is a great deal more to think about if you are to come to terms with an effectively managed presentation. I will examine each area of significance and explain the lore. Even the obvious sometimes is not so obvious. For example, I attend meetings and regularly watch handouts flutter around a circle of committee members, either distracting attentive listeners who pass them on or taking the attention of other members who are looking for something to read. The table becomes progressively more cluttered, as does the focus of attention of the members. Yes, handouts are just that: literature to hand out, but you want to learn to use them as your closure when you no longer want the direct attention of the group.

Sample 9.B

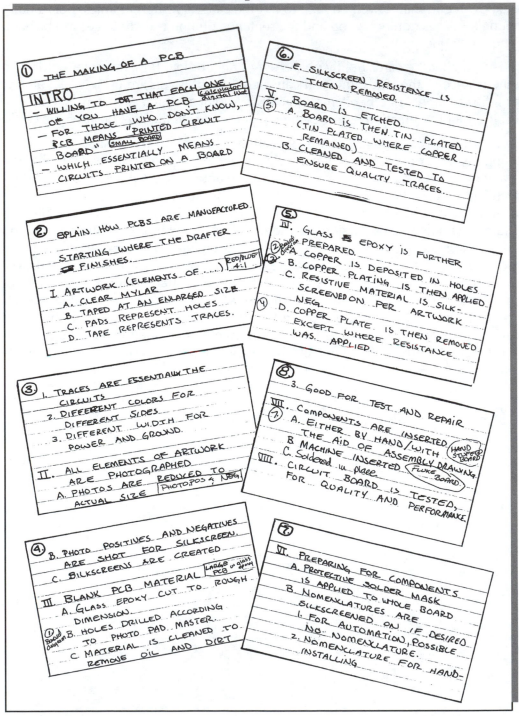

① THE MAKING OF A PCB

INTRO
- WILLING TO BET THAT EACH ONE [calculator/digital watch] OF YOU HAVE A PCB
- FOR THOSE WHO DON'T KNOW, PCB MEANS "PRINTED CIRCUIT BOARD" [SMALL BOARD]
- WHICH ESSENTIALLY MEANS CIRCUITS PRINTED ON A BOARD

② EXPLAIN HOW PCBS ARE MANUFACTURED STARTING WHERE THE DRAFTER FINISHES. [RED/BLUE 4:1]

I. ARTWORK (ELEMENTS OF ....)
   A. CLEAR MYLAR
   B. TAPED AT AN ENLARGED SIZE
   C. PADS REPRESENT HOLES
   D. TAPE REPRESENTS TRACES.

③
   1. TRACES ARE ESSENTIALLY THE CIRCUITS
   2. DIFFERENT COLORS FOR DIFFERENT SIDES
   3. DIFFERENT WIDTH FOR POWER AND GROUND.

II. ALL ELEMENTS OF ARTWORK ARE PHOTOGRAPHED
   A. PHOTOS ARE REDUCED TO ACTUAL SIZE [PHOTO:POS & NEG]

④
   B. PHOTO POSITIVES AND NEGATIVES ARE SHOT FOR SILKSCREEN.
   C. SILKSCREENS ARE CREATED

III. BLANK PCB MATERIAL [LARGE PCB w GLASS EPOXY]
   A. GLASS EPOXY CUT TO ROUGH DIMENSION.
   B. HOLES DRILLED ACCORDING TO PHOTO PAD MASTER. [Board diagram]
   C. MATERIAL IS CLEANED TO REMOVE OIL AND DIRT

⑥
   E. SILKSCREEN RESISTANCE IS THEN REMOVED

V. BOARD IS ETCHED [⑤]
   A. BOARD IS THEN TIN PLATED (TIN PLATED WHERE COPPER REMAINED)
   B. CLEANED AND TESTED TO ENSURE QUALITY TRACES.

⑤

IV. GLASS & EPOXY IS FURTHER PREPARED. [② Board diagram]
   A. COPPER IS DEPOSITED IN HOLES [③]
   B. COPPER PLATING IS THEN APPLIED
   C. RESISTIVE MATERIAL IS SILK-SCREENED ON PER ARTWORK NEG.
   [④] D. COPPER PLATE IS THEN REMOVED EXCEPT WHERE RESISTANCE WAS APPLIED.

⑧
   3. GOOD FOR TEST AND REPAIR

VII. COMPONENTS ARE INSERTED [⑦]
   A. EITHER BY HAND/WITH THE AID OF ASSEMBLY DRAWING [HAND STUFFED BOARD]
   B. MACHINE INSERTED [FLUKE BOARD]
   C. SOLDERED in place

VIII. CIRCUIT BOARD IS TESTED, FOR QUALITY AND PERFORMANCE.

⑦

VI. PREPARING FOR COMPONENTS
   A. PROTECTIVE SOLDER MASK IS APPLIED TO WHOLE BOARD
   B. NOMENCLATURES ARE SILKSCREENED ON, IF DESIRED
      1. FOR AUTOMATION, POSSIBLE NO NOMENCLATURE.
      2. NOMENCLATURE FOR HAND-INSTALLING

The Parts of a Presentation

The Introduction

The committee presentation must have a beginning, a middle, and an end. A talk is not a discussion, and its organization is not random. There is an introduction; there is a body; there is a conclusion—often followed by a discussion period. You have the option of inviting questions during the presentation, but this practice will lengthen the delivery in an unpredictable way. In addition, until you are skilled at managing an oral presentation, you could find yourself sidetracked by questions that could damage a tightly organized presentation. Define the parts of the presentation, avoid questions if they will distract your attention, and follow the outline of the intended talk.

The conventional introductions are intended to serve two practical goals:

- **Gain attention.**

- **Preview the talk.**

A little background is appropriate at times, or you can state key points of discussion to briefly outline the plan of your talk. Again, do not belabor the key points. State key points simply and clearly, and do not try to develop ten points in ten minutes. Public speaking is very limited in the range and depth that a presentation can accomplish. The overachiever is bound to encounter problems once the listeners have lost the continuity of the talk, and the problem can start in the introduction. There are practical openings you should consider.

1. *Opening with Management Objectives* Introductions to speeches are often designed around a few traditional practices such as opening with provocative questions or a startling incident or a few statistics that will arouse interest. However, remember that an in-house speaking situation is often a response to a request that you speak. That request, from your supervisor let's say, is usually the result of some need within the company. That need defines the committee that will convene, the role you are to play, and the subject of your presentation. In this sense the opening cliché you see on television is not far from the mark:

> Ladies and gentlemen, I have asked you to be here today because. . . .

Recipe cards are an old and respected approach to handling a presentation, and they remain popular because they assist in maintaining order in a delivery.

Then the plot of the episode reveals itself. In company environments, thousands of meetings of this sort are held daily. Because you will be responding to someone's request, you can open with an immediate comment on the intended objective. This springboard opening is the classic, but do it without the cliché.

> Now that we are here, let me explain this meeting. Jim called last Thursday morning and said that the Houston office had a fire. We still don't have the assessment, but when Karen called from Houston it looked like all the record storage was involved. The fire wasn't the problem. Smoke damage and water from the sprinkler system went everywhere. They need our records-backup files and we have to figure out how to get them there. Jim has asked me to estimate a response time and a strategy.

The springboard opening is remarkably tidy and sets the presentation in motion with a stated objective—usually wrapped in an immediate historical backdrop. It has the additional value of pointing out the importance of the subject, the point of the meeting, and your relevance.

Open a presentation by explaining what you are going to talk about. Do not complicate the statement. If you are responding to a management objective, explain the history of the situation after you identify your purpose. Here are three sample openings:

> My department has had a good year and we need to spend $30,000 by July 1 in order to fulfill the use of our allocation.

> We were granted approval for hosting the Sony team from Tokyo, and I have a tentative itinerary to share with you.

> Mary Ellen asked me to explain the new state inventory tax procedures. The procedures are now going to be based on business license categories.

Any of these statements will set the stage.

2. *The Withheld Objective* Having urged you to open with the central idea, I should identify exceptions to the practice, or else the presentation could encounter difficulties. As you have probably realized in your readings throughout the text, nothing works all of the time in communications. **If you have to prove a point before you can make the point, it is not wise to open with a declaration, not even a tentative one.** The concept must first be explored and developed so that an audience can see the issue clearly. For example, while working at an area children's hospital I once attended a speech delivered in the hospital auditorium. An authority on some aspect of pediatrics was there to urge a stop to high-school driving safety programs. Of course, no one knew his position at first. He knew the idea was considered outrageous, so he spoke for half an hour about statistics that demonstrated mortality and accident rates. His proposal emerged only at the end of his speech when it seemed to be a logical conclusion.

The other case for withholding the central idea of a presentation concerns audience expectations. In many instances a forthright proposal is unquestioned; at other times it will seem blunt or judgmental.

Ladies and gentlemen, you are all fired. Let me tell you why.

A bad-news presentation is a special case. A bitter pill takes special handling. If an audience is likely to have a negative reaction to a presentation, the presentation must first rationalize the outcomes and you, as a speaker, must make the effort to communicate the logic to the audience.

Thank you for being here. It was a difficult year for AlCo. I have spent some months wrestling with this issue. Let me begin by having Gordon explain the charts you see here. Accounting may also be able to answer a few of your questions.

There is a specific reason—apart from courtesy—for withholding the central idea or proposal in the case of controversy or of bad news. If you begin with a proposal and then demonstrate its validity or effectiveness, the pattern of the logic is basically *deductive,* meaning that the opening statement is supported *after* it is proposed. This is usually the most practical approach to a presentation in a company setting because it provides the audience with the point of the discussion at the outset. This method provides a strong sense of what you are doing and why. The two exceptional situations we just observed do not fit the mold. In a case where there will be audience objections, the evidence must be explained, and the *inductive* approach is often the better option. It is the reverse tactic. First, provide all the evidence and supporting material. Otherwise the listeners have little to convince them that the proposal is a correct judgment. They must see the outcome in the making.

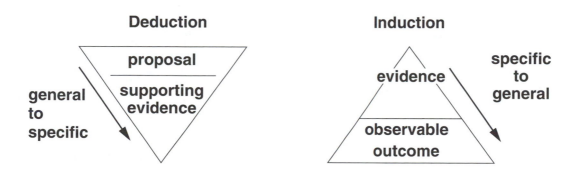

Among the other options for openings are the three I initially identified:

3. *Opening with Questions*

Can we continue hiring without a new facility? Can we maintain the efficiency of our personnel if the exisiting space limitations are frustrating the employees?

4. Opening with an Incident, Story, or Dramatic Illustration

The two new personnel officers were brought on last week. Rhonda has them in Sarah Ferguson's office, where she had maintenance install folding tables for desks. I told her, "Rhonda, you just hired two people to help you hire 200 more; where does the board plan to put the 200?" She said the first two were the immediate problem.

5. Opening with Statistics

When we were still on Baxter Avenue we had 75 employees in a building that had 6000 square feet of space. We doubled the space by moving here, but the 12,000 square feet lost any sense of spaciousness as soon as we reached 150 employees five years ago. Since that time the footage per employee has declined from 80 square feet ten years ago to the 63 square feet each of you enjoys today.

6. *Opening with Quotations* There are additional openings you can use for presentations. Notice that one of the preceding samples involved a quote. It is occasionally appropriate to open with a relevant quote. For general speeches, such quotes could come from anyone from Robert E. Lee to Hillary Clinton to Bill Cosby. For company presentations the more likely choice will be the company president or another executive.

7. *Opening with Humor* You can also open with a witty story—but never a joke. Jokes often strike educated men and women as ranging from corny to in bad taste; either way, they are hardly a peg above the knock-knock material from old vaudeville routines.

You will have very limited use for any of these openings in a committee situation. Even with a little effort to make them reasonably serious, notice the result if I apply these devices to the Houston fire. First, I will open with statistics.

We have a problem here this morning. Half of all the disk drive records are in Houston, along with 30% of the hard-copy records for the southwest units. They total over 77,000 accounts. We lost them last night. All of them.

Not bad. Let's try another one.

We have a problem here this morning. Have you ever wondered how Florida businesses survive those hurricanes? What about their account histories? What do they do if they are washed out into the ocean along with the buildings? And what if their computers are full of saltwater?

You can see that they will work if they do not seem too contrived. Judge the situation and use them if the presentation merits such openings. The *stated objective* is probably the more practical tactic, and regardless of the other approaches you might take, move on to the statement of key points as the bridge from the introduction to the heart of the matter.

Key Points and Support Evidence

The structure of your presentation will be designed to coordinate key ideas, whether they are coordinated in terms of time or space or some logical sequence. The discussion you provide for each point will be based on the evidence you have gathered. Whether informational or persuasive, the evidence will usually involve familiar material that an audience will understand. Statistics, visual aids, analyses of details, explanations of data, and other options are effective. If they are appropriate, demonstrations, models, photographs of events, and other prospects can become part of a presentation. Quotations can be used as authoritative support, but in a public situation you have the interesting option of inviting an authority to the event as an information source (not as a primary speaker). It is commonplace to see financial officers at meetings where they are consulted as needed.

Ways to Explore Key Ideas

- **Demonstrations, simulations, examples**
- **Experiences**
- **Analysis of details, technical descriptions**
- **Explanations**
- **Statistics**
- **Definitions**
- **Visual aids**
- **Guest authorities (or quotations)**
- **Participation from the audience**

For technical presentations that are designed to address a committee of selected or interested parties, the subject of the presentation will dictate the features of the presentation. Within that area—probably the area of expertise that brings you to the meeting—predictable practices are used to develop the talk. Examples and illustrations are but two of a number of typical features that are valuable either as supporting material or as direct evidence of the logic of the talk.

1. *Examples* Models will usually be the most likely example for an engineering technician making a presentation. Display models are popular and add interest to a presentation. In fact, they can be the entire focus of the presentation.

---

2. *Illustrations* For many engineering specialties, drawings substitute for models. Civil engineers or structural engineers operate on a scale that will make models, even scale models, impractical at times. Other illustrations include schematics, photographs, slides, charts, graphs, and a number of other visual possibilities.

3. *Histories* The background of a situation is often a significant focus of attention. Perhaps the evaluation of a new product is used to explain a current patent application. If earlier versions were inadequate, the superior nature of a new invention, let's say a new professional grade of camera film that is chemically stable without refrigeration, can be measured against the history of the older products.

4. *Experiences* Remember that you may be at these meetings in a supervisory capacity. Supervisors spend a good deal of time dealing with people problems as well as with technical problems. People problems are often discussed in terms of experiences. It might be up to you to present an analysis of complaints about the mail room services. Each complaint is an experience to be analyzed before suggesting a solution to the difficulties.

5. *Comparisons* If you have a product that you want to present to a meeting, one way to effectively capture attention is to compare the product to one or more of the competitors' products. This can be done with charts or products or discussion. Comparisons, of course, are not limited to products. A good many presentation topics—from theories to parking lots, from production schedules to use of facilities, from policies to personnel—can be handled in comparisons if the approach is merited.

6. *Statistics* Numbers matter. Whether a presentation is based on the statistical probabilities of failure in a stress analysis for a metal fitting, or on the number of client calls per day of the week, the numbers run the corporations—and they usually are not bottom lines. Financial statistics represent only one of many considerations. Statistical material represents a major feature of presentations because the numbers frequently are the major considerations for decision making.

7. *Definitions* Terminology takes on a special importance in technical specialties of the sort that are likely to be your concern and the concern of your audience. One reason you need to think out the nature of the audience is so that you can measure their understanding of your vocabulary. You can, of course, respond to the moment if people appear puzzled or ask questions during a presentation. However, spontaneous definitions, particularly definitions that are complex or technical, might be inadequate. If you anticipate the needs of the listeners you can prepare important definitions before you speak. In fact, the idea is to make the definitions part of the presentation. They can also be presented as visuals.

8. *Explanations* All the presentation features identified here will call for an explanation. Essentially, these elements of evidence help build either an informational presentation or a persuasive presentation. A speaker has the task, however, of explaining the significance of each source of evidence. Usually, points of evidence are also linked, which is part of the explanation process that leads to conclusions.

9. *Quotations* Quotations are popular features of public speaking, but they are less likely to appear in committee presentations except for the occasional use of *opinions*. Executive opinion, legal opinion, and technical opinion may be very important in a presentation. I do not use the word "quotation" in this context because the actual quotation may be an entire executive briefing or legal briefing that runs pages. It may be frequently mentioned, but the document may or may not be quoted. It is often the opinion expressed in the document that will matter.

10. *Participation from other parties* A committee meeting is not usually held in an auditorium and seldom even uses row seating. As a result, the role of the speaker does not isolate him or her from the audience. As a speaker you have the option of inviting the audience to be involved in your talk. You can also invite a guest authority to share the event as a resource. In both situations the participants become part of the evidence that is developed by the presentation.

Transitions

Regardless of the pattern of organization you adopt, you should use key transitional phrases to move the listeners along and to be sure that they perceive the presentation in an orderly fashion. In the case of chronological organization, many familiar transitions will readily come to mind when you speak. It might help you to cue yourself. As I explained earlier, you can highlight the desired phrasing on your outline or on your notecards if you think you will forget transitions. You could also replace the Roman numbers on the outline—I, II, III—with capitalized words and highlight the point-to-point transitions: *FIRST, SECOND, THIRD*. As you speak, focus attention on key transition words with a slight pause, some voice emphasis or a gesture:

> *afterwards, just prior to this, before, then*

In the case of spatial organization, the transition words move the eye of the reader or the mind's eye if there are no displays.

> *Notice on the far left . . .*
>
> *At the top of the scale . . .*
>
> *These dotted lines show you the supporting walls of the floor below.*
>
> *On center, we decided to position . . .*

Logical transitions are often handled by a wide variety of terms:

therefore	*however*
accordingly	*evidently*
as a result	*the outcome was*

Conclusions

Conclusions often *summarize* if a speech primarily provides information. Conclusions *identify actions* if the speech is primarily persuasive. In the second case, the audience is directly (or indirectly) called on to adopt the actions by approval or by actual performance of the action.

1. The Summary You will demonstrate better control of both the subject matter and a listener's response if you add a conclusion. It does not have to be an elegant affair, nor does it have to be drawn out. Quite the opposite, a brief conclusion takes little time, but it will allow you to control the outcome and not just the delivery of your presentation. The most convenient conclusion is the *summary.* Highlight the main points of your talk. If you think back to the suggestions for the introductions, you will recall that the response-to-desired-objectives introduction (the "springboard") was the one I suggested that you might find most useful in a company setting. Given an introduction that identifies goal objectives, the conclusion will obviously draw on those needs to sum up.

> *To close, please note that we do not have the $150,000 for covering the shortfall. As I explained, First Bank will loan us the required sum. As the chart showed you, the interest on the loan will double the shortfall costs before payoff on the borrowing. However, the note can be carried for three years, as I suggested a moment ago. I do not see a more cost effective solution.*

2. The Call for Action Some meetings are designed to encourage the listeners to act on specific recommendations. If a presentation is supposed to persuade, you will need a conclusion so that the intended desired actions are not left to chance. If there is going to be a request for action, make the request at the end of the presentation. Do not leave the decision in the hands of the audience. If you have demonstrated the need for an action, the timely opportunity for support is *now.* If at all possible, do not leave desired actions for another day. As a worst case, try to avoid the classic corporate mothballing strategy:

> *This proposal absolutely deserves attention, and I would move that we form a committee to study both the implementation possibilities and whatever is involved in the legal implications prior to voting on implementation.*

Whether well meaning or devious, the call for a subcommittee can bring a persuasive presentation to a halt. On the other hand, perhaps the subcommittee is a victory of sorts in the cautious world of company dealings.

To encourage action, a presentation can end on an *appeal.* The appeal does not have to be emotional or sentimental or in anyway related to feelings. It can be quite logical and it can call for actions for analytical or technical reasons. A mixture of sentiment and logic will obviously work as well:

We lost Mark because he burned out. We lost Larry because his wife said enough was enough. Okay, Molly left because she was pregnant, but Elroy, as all of you know, took that position at Pomeroy and Pomeroy and took a pay cut! That's four architects—good ones—in two years. I think this proposal is practical and humane. You know we can't afford to lose one more talent over in Design.

3. *The Statement of Importance* There is a third option—the *statement of importance*—that can be used with either the informational presentation or the persuasive presentation. Because the in-house presentation is usually requested, that demand was initiated by some special need, as I noted earlier. Such a situation makes it particularly easy for a speaker to draw a presentation to a close by stating the importance of the subject matter. At times this closing can be used to try to win over an audience of employees that seems hesitant. More commonly, closing with importance simply reaffirms the reason for the meeting in the first place:

> *If the Mexican market opens to Erickson, we have to enter the market at the same time. If we don't position ourselves for the market, it will be lost in two years. It is important to have Harrison set to work on this by June.*

Regardless of the nature of the summary. Do not run out of time or you will lose this important dimension of the talk. The repetition of your key ideas is often critical.

Having noted that there are three desirable conclusions for technical committee presentations, we should observe the very practical fact that short presentations—perhaps brief comments that reflect a single meeting agenda item—often *stop* rather than *conclude*.

Committee presentations can often stop comfortably without a conclusion. This will not be the case if a talk is supposed to be persuasive, but if the intent is informational, the speaker is done when the sum of the information is said. This is a *very* common practice, particularly if the presentation is an agenda item. Given the opportunity, the speaker might judge the audience's understanding of the information by asking for questions:

> *If you have any questions I would be glad to try to answer them.*

Presentation Logic

A listener must be able to follow the intentions of the speaker. It is often said that a written presentation is a challenge because a reader cannot seek clarifications if the text is in anyway confusing. Similar difficulties arise in a spoken presentation, particularly if there is no opportunity for listeners to ask questions. In addition, remember that a written document can be *reviewed;* a spoken presentation must be *remembered,* which is a far more demanding task. Your role as a speaker is to make your presentation memorable. I do not mean that your words will be unforgettable but that listeners will *understand* what you said and *remember* important aspects of the talk, which is what you should hope for. The body of the presentation carries the greatest responsibility for the task. Your demeanor matters. Your delivery matters. Your presentation, however, is the substance of the matter.

The body of the presentation can be handled in a number of ways. Fundamentally the issue is organization. The question is, how do you organize a set of circumstances in terms of central themes and how do you analyze them and present them to a public? The situation under discussion may have occurred and is now history. The situation may be occurring and is ongoing. The situation may be awaited and will occur in the future. The task is to organize the reality of any situation and make sense out of it. You have a number of alternatives you can use to build the framework of the logic of the presentation.

Criteria Approach

In the same sense that a meeting agenda is driven by company needs, company interests can drive a company presentation. At times, a presentation is not exactly in your hands. **There may be strict topics, priorities, or criteria that shape what you present if you are asked to support or express the concerns of a supervisor or a recommending board or a committee.** Such a presentation is then shaped by specific objectives or topics that state conditions of varying importance, usually by some kind of priority.

> Our first short-term goal is lower price.
>
> Our second short-term goal is fewer returns for service.
>
> Our long-range goal is stronger customer loyalty.

If you are in total control of the presentation and there are no company guidelines or supervisory recommendations, then you can rank the presentation to suit your own needs. Most uses of a prioritized approach begin with the most important consideration and move down a list to the least important.

The criteria method of organization, like the others discussed next, can be explained with a simple diagram.

CRITERIA STRUCTURE

- **Most important element**
- **Second most important**
- **Third most important**
- **Least important element**

Topical Approach

Topics can also be assigned any order of discussion without criteria. In other words, you can speak about events or concepts or policy or any subject matter by providing any order you choose. In this situation you might have the special task of meeting company expectations if the presentation will be designed to serve as a backdrop for such considerations as legal policy, legal precedents, corporate policy, or corporate precedents. For a topical presentation you might want a supervisory review of your intentions (your outline) before you make your presentation. You must be certain you are not misunderstanding the expectations of the powers that you are being asked to represent or the expectations supervisors might have of what you intend to say. Here is a simple representation of the topical concept:

TOPICAL STRUCTURE

- **First subject**
- **Second subject**
- **Third subject**
- **Fourth subject**

If you are under no constraints, topical presentations have no order except the one you assign to the parts. However, I am including here those situations where your objectives have been suggested by your supervisor.

Chronological Approach

Perhaps the easiest organizational tool is *chronology*. **It is very likely that many presentations you might develop could be best presented in terms of a timetable.** If you are an engineering supervisor and you have to establish a critical path schedule for a revision to a storm runoff system, you can establish timelines and present the material from groundbreaking to backfill. Explain the new engineering project as a statement of management objectives and proceed to define the tasks, in order of occurrence.

> City Engineering will arrive on July 1 to reroute traffic. The barricades and portable fences will go up before the end of that week. B & A Contracting will begin the digs by July 8. The conduit will be coming from Memphis to the sites during that week.

Chronology is easy to organize:

```
┌─────────────────────────────────────┐
│          CHRONOLOGICAL              │
│            STRUCTURE                │
│                                     │
│  1998 _____         │
│                                     │
│  1999 _____         │
│                                     │
│  2000 _____         │
│                                     │
│  2001 _____         │
│                                     │
│  2002 _____         │
│                                     │
└─────────────────────────────────────┘
```

Space Relationships Approach

A presentation is often organized by *space* rather than by time. If we are dealing with the interior of a computer we are dealing with components and it is appropriate to look at parts. Those parts exist as physical elements of a spatial relationship. Subassemblies, for example, can be represented graphically for an audience:

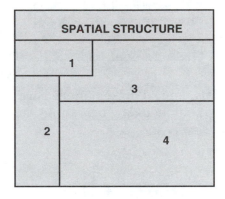

Using a product model or visual aids or both, you could explain the situation to a committee.

> *Notice the distribution of the components. The motherboard sits here. To the left is the disk drive. The fan was positioned correctly, as you can see to the right. Nonetheless, failures from overheating began to occur nine months after release. There is no geographic pattern. In fact, the biggest problem area is Boston, not a hot location like Phoenix. Now look carefully at this fan assembly. We have guessed that there are 10,000 of these—and they are all defective. They are blocking air because of the position of the components, which then overheat, as you see here, at this point.*

Logical Approaches

Both the *time* and the *space* approaches can organize with great clarity. They are vivid and they are easily understood. Another approach—*logic*—is possibly a more demanding matter. Perhaps logic is less vivid than chronology or physical makeup. If a speaker relates events, and listeners hear the story unfold, the presentation will probably be vivid. If a speaker demonstrates a product or reviews a pictorial schematic drawing, observers will see the story unfold. Logic, as any lawyer knows, is a more demanding craft—but equally critical.

Consider this case from a domestic court hearing. A husband and wife divorce. Sometime later the former wife goes to the former husband's office in an angry mood. He tries to not let her in. She pushes her way through the door and creates a scene. His secretary calls 911 and they end up in court. The former wife claims the former husband bruised her by pushing the door against her. He claims she hit him with the door. Both argue an intent to harm. As you might guess, it is difficult to examine this sort of situation very clearly. The lawyers, as in any case, argued the logic of it for two days. The first decision was that the door was not a weapon and there was no attempt by either party to use it to harm. The decision was that since they were no longer married, the former spouse had no right to force her way into his office; the *logical* issue concerned forced entry. Because both parties were pleading bodily harm, it was difficult to see the obvious.

- **Cause-Effect Reasoning** What kinds of logical analysis are typical for a presentation? *Cause-to-effect* reasoning is one. Determine the causes. Determine the effect. There are many variables. If the causes are known, the central issue is, what are the effects? If the effects are known, the probable question is, what was the cause? Perhaps both are unknown or poorly understood.

- **Problem-Solution Reasoning** Another logical approach in a similar vein is *problem-solution* analysis. Here the relationship is also "causal." As I will explain in another chapter, a presentation that proposes a contract offer for services is usually a response to a need, and it is, therefore, a solution to a problem. A presentation can dwell on problems if the concern of a meeting is based on defining them. A presentation can focus on solutions if the intent of a meeting is directed at solutions. Both subject areas—problem and solution—can also be addressed.

Conceptually, a problem-solution presentation involves two subject areas, even though they may not receive equal attention:

> **PROBLEM-SOLUTION STRUCTURE**
>
> - **problem**
> - **solution**

The challenge of a problem-solution presentation concerns both ends of the issue.

- You must thoroughly examine the significance of the problem.

- You must examine the possible solutions in order to determine the best possible solution.

Problem-solution discussions and cause-effect presentations do not have to devote equal time to what superficially appear to *be halves*. They can be quite lopsided if there are no contentions about certain conditions.

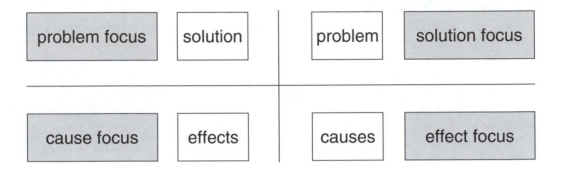

The proportions focus the interests of the listeners. Whichever concerns merit the greatest share of the presentation assume the greatest value for the listener.

Polarized Thinking

If a situation is reduced to an analysis that views the situation as a problem with solutions, be wary of the likelihood that the issues involved might become oversimplified. If a *problem* is oversimplified, it can be misunderstood. If a *solution* is oversimplified, it can be misunderstood.

Either it works or it doesn't.

There are good employees and bad employees.

They are either with us or against us.

For example, the comment "pesticides solve problems" is a risky oversimplification. Early pesticides that would not deteriorate created more problems than they solved once they entered the food chain. In truth, an insect pest is a complex problem, and a pesticide is a complex solution. Avoid oversimplification by being precise. For example, the Texas cattle fever tick is not just a bug to be sprayed. If I speak about this pest, I must be very specific. It is a tick, it carries a deadly disease, it does not harm cattle that acquire resistance to it, it is localized in the Rio Grande border district, and so on. Since the pest is a complex problem, the solution must be viewed precisely. To be specific, a pesticide is a solution in this case, *but* it must be a specific one for bovine ticks, and it is called an *acaricide*. Precision allows me to avoid the idea that all pests are bad or that all bug sprays are good.

The signal of polarized thinking is black-and-white logic. Look a little deeper into the issue that is being polarized to see if the evidence that is being used is causing the over-simplification. In other words, we can fool ourselves into seeing complex issues in simple ways, but also the understanding we have of the issue may be oversimplified initially. Why? Because the evidence can be inadequate, questionable, or wrong. Very likely the absence of enough data will be the problem. In medical research, control studies with human subjects begin modestly. Perhaps a two-year study is conducted on twenty people. If risks are low, a medical school may try 100 people. If results are positive, perhaps 5000 people are accepted as volunteers for the next step. However, even at this point, the data can still lead to oversimplifications based on limited evidence. The data at the first two levels, and even at the third, may be inadequate.

Persuasion

In a persuasive presentation there are behavioral objectives that affect the organization of a presentation because the delivery is intended to stimulate or inspire a response. There is every intention of compelling people to act on what they hear. Persuasion involves several types of modifications in the audience. One matter concerns the way listeners *rationally* understand a need to which they should respond. Another matter concerns the way they *emotionally* respond to the speaker and the subject. The idea is that the emotional response will support the rational decision so that there is no conflict. The goal is to have the listener feel attracted to the speaker's requests in a favorable way. The only way that support will be convincing to the listener is if heart and mind join in the perceived response.

In a persuasive presentation for a committee, there is not a great deal of politicking in the delivery. The presentation should be solidly grounded in information and should be as informative as the informational delivery. The difference is the role of the information in the two presentations. *Information is the goal in a need-to-know situation. However, in a persuasive presentation, information is being used as proof.* In a need-to-act situation the information is intended to demonstrate the value of action and to motivate listeners to take that action.

As a single event, a speech is only mildly persuasive if we allow for the fact that listeners are long in arriving at their positions. Those beliefs and values represent life experiences and not twenty minutes. Persuasive presentations are difficult if you want to *change* listener attitudes but somewhat more convincing if you seek to *modify* attitudes. Prospects are brightest if you intend to arouse a commitment where there is already a *supportive* attitude among listeners.

- You can seek to **change** the attitudes of an audience.
- You can seek to **modify** the attitudes of an audience.
- You can seek to **commit** supportive attitudes of an audience.

These challenges reflect different intensities of audience resistance, and the first of these three challenges is the most difficult.

For a persuasive presentation the basic steps involved in the delivery are straightforward.

> - **Gain attention.**
> - **Explain the need.**
> - **Resolve the need.**
> - **Identify the action needed.**
> - **Ask for commitment to the action.**

Our discussion frequently refers to a desired action as though there were only one. In truth, there will more likely be a number of desired actions. The summary of a persuasive speech may have to outline a series of action steps.

Be alert to the occasion when facts are not particularly persuasive. The audience, any audience, will always interpret and evaluate. The challenge is to control, shape, mold, direct, channel, or otherwise take an audience to a desired end.

Facts can trigger emotional reactions. Be aware of that possibility if an audience is sensitive about matters under discussion. The truth of the facts does not mean there will not be a reaction to them!

If there is any possibility that an audience might be either resistant or skeptical about a presentation, be sure to approach the talk with tact and logical clarity. Persuasive presentations are particularly challenging, and bad news is a particularly bitter pill, as we observed earlier. At times, persuasion is linked to some serious problem. The tensions can mount during sensitive periods such as a serious economic downturn or a threatened product recall or a pending contract deadline with a major union affiliate. If your role is supervisory and you must make presentations, be sure that you use a tactful manner and be intensely logical. By *tactful* I mean that you should be suggestive and not provocative. By *logical* I mean that you should speak in facts. **Guard against making any assumptions that can be taken as truths because this shifts your position, and it might be challenged.**

If you do presume that some event is true or will occur, acknowledge the fact that your perception is not a fact.

> *Usually the Minneapolis plant can match demand.*
>
> *If history is an indication, this bid is probably undervalued by half.*

When you do state opinions, be loud and clear about the fact that your perspective is your own.

> *Personally, I see a settlement within the week.*

> *I think we should hire Johnson.*

> *We can, it would seem to me, contract with Texas Instruments to produce the chip at a better price.*

> *I would suppose that fourth quarter earnings were perfectly normal if we look at other industry earnings.*

Word selection is important in a persuasive presentation. It is important to select the appropriate vocabulary so that there are no negative reactions to the words themselves. Rather than signal an objective and shared meaning, even a simple word will often function as a label for a personal understanding of the word. If you say "chip" I probably share your objective meaning, but if you say "overtime," you probably have identified a label for various attitudes toward overtime or various experiences with overtime. In this case, the understanding of the word becomes an interpretation. Although there is no way to remove *interpretations* from words, you can usually stay in the arena of objectively shared meanings. You are a technical specialist. You can discuss the chip, talk about its advantages and disadvantages and otherwise describe it, and the listeners will respect your views even if they challenge your criteria. In a technical presentation, you should be able to limit most value judgments that an audience would make based on your terminology. Obviously, you cannot call the chip a "piece of junk" or a "cheap import" or a "lousy product" because this will invite listener reactions, which may not be supportive. The tactful approach will maintain a proper balance.

If you are in a supervisory capacity, the issues become more complex and you need to be sensitive to the I-them relationship that is inherently a part of labor-management realities. In this case you must be tactful, but you must also be alert to audience values. Remember that words are interpreted. Some words have objective meanings some of the time. At other times the same words have highly subjective meanings: *downsizing, market corrections, closure.*

The corporate world is highly dynamic. It changes constantly in response to market forces. In a word, survival means response to competition. Nonetheless, in the day to day humdrum, employees prefer to see their world as a stable one. There is a conflict here that is difficult to balance for both employee and employer. Committee meetings introduce new ideas, and new ideas bring about changes that are often resisted. The potential conflict is obvious in the way each employee will interpret the seemingly harmless word *new.*

Summary

- The committee presentation is a frequently used method of communication in company settings. These presentations involve a delivery that usually ranges from five to twenty minutes. The presentation may be a minor agenda item of a meeting, but it can also be the focus of the entire meeting.

- The committee presentation is unique in that your supervisor might desire a hand in determining the purpose and the plan of your presentation.

- Your basic goals for delivering a committee presentation are based on the needs of the listeners:

  - ✔ Speak with clarity and vividness

  - ✔ Emphasize important points

  - ✔ Use a vocabulary that will be understood by the audience.

- Analyze the audience profile by considering the group's level of *knowledge* and *interest* in your subject. Also consider their *attitude* toward what you plan to say.

- Develop a *full-sentence* outline to use as a script for a short presentation.

- Use point-for-point transitions to clearly organize your central ideas during the delivery.

- Develop a *topic* outline to follow during the presentation.

- If you use the popular 3″ × 5″ cards that are often used by speakers, keep a full-sentence outline nearby as a backup just in case you become confused for any reason.

- On the 3″ × 5″ cards and on the outline, indicate the points at which you intend to use visual aids.

- Open the presentation by explaining the reason for your being there to speak and by highlighting your intentions.

- Develop the talk in specifics: experiences, technical descriptions, explanations, statistics, authoritative sources, and visual aids.

- Summarize the key points and clearly identify any call for support or action you expect from the audience.

- Timing is extremely important for a successful presentation. Avoid running beyond your allocated time because you will risk a negative audience response, and you may not have time for the summary.

- The usual structures that speakers use to organize a committee presentation include organization by prioritizing, by topic, by chronology, by spatial organization, or by logical organization.

- Logical organization for a presentation is frequently intended to discuss either causes and effects or problems and solutions.

- A persuasive presentation usually will demonstrate a *need* first; the call for *support* or *action* follows and becomes the second focus of the presentation.

Activities Chapter 9

During the course of several weeks you will be asked to complete the following exercises.

1. A writing project consisting of a "script outline" for a committee presentation.

2. A performance project that will be a five-to-ten minute committee presentation to your class.

3. a critical evaluation that will be a self-analysis of your presentation.

4. An observation form that evaluates another speaker. (You may be asked to complete several of these forms.)

These four tasks are explained below.

As you read Chapters 9 and 10 you will be asked to prepare a brief presentation for the class. The presentation will concern your technological focus and will show the class how you would deliver a discussion before a committee. In the presentation you should explain the subject, examine how you would organize the presentation, and show the class the various visual aids you developed. Explain how you plan to use the aids. Discuss the audience you might encounter. Sum up. Close by asking if there are questions.

---

1. To prepare to speak to the class, first develop an outline. Type it and submit it to your instructor as an exercise.

2. Develop the visual aids. Use the aids as the central theme of your discussion. You need to explain how and why they are important. Carefully think out the size, the type, and the content of the aids.

Using the outline and the aids, you will complete the performance task when you deliver the brief presentation to the class.

3. You will then have the opportunity to watch the videotaped presentation with a small discussion group consisting of those members of the class who made presentations on the day of your presentation. Your third major task is to complete a self-analysis of your presentation after you see the playback (see appendix, pp. 411–412).

4. You will have the opportunity to see a number of presentations. You will be asked to complete one or more observation forms (see the appendix, pp. 409–410). Submit the completed form to the instructor as the fourth exercise.

Note: If you are a pre-major consider the following options.

1) *If you wish, you can replace the technical subject with a topic of personal interest. Discuss this option with your instructor.*

2) *You might also take a current sporting or news event that you can reconstruct with photographs from magazines or newspapers. Gather as many useful photographs as possible and develop a presentation that explains what happened. Explain why it happened from your perspective.*

Audiences and Visual Aids

Work in Progress

10. Visualizing Technology

At this point, I need to step away from my story and discuss the power of using visual aids in helping potential clients understand the technology you are offering them. Investing in technology (like a LAN upgrade) is a fairly abstract proposition for most customers and it demystifies things if they can actually see what they are buying and see how it works. For example, trying to explain the benefits of full-duplex cabling versus half-duplex can be explained easily on a graph but becomes lethally dull and hopelessly confusing by word of mouth alone! If you can enlighten your audience without patronizing them, they are more likely to be open to your real message, which is, "hire me or buy from me."

My other presentation goal is to help my clients understand why they need what I am offering. This last point may sound obvious, but you would be surprised at how vague a concept it can be in many a customer's mind. A well-crafted presentation not only will show clients what the technology can do for the company (perhaps save time and money), but a presentation can also get them excited about what it can do for them personally. The trick is to turn an abstract concept into something real. In my case, it can be as simple as having the audience imagine how nice it would be to have no wait time on downloads. The particulars don't matter as much as making sure that your potential customers can see and really understand that this technology will be helpful to them.

I use visual aids as the cornerstone of any presentation I do. I like to use a mixture of foam-backed display boards and a short PowerPoint presentation to introduce the key points I want to make. I create floor plans with Visio software showing before and after shots (like the weight loss ads!), and these are some of the best tools I have. I usually find a table where I can prop up the displays, but I always have my own small easel where I can place the boards beside me in easy view of the clients. I find it effective to put the "bottom line" information on the boards, so if attention wanders during the rest of the "show," the main messages are almost unavoidable. Usually, the statistics on the boards speak loudly enough for themselves. I also carry product specifications and brochures that explain (visualize) important features and help account for certain costs.

I use the notes and information I have gathered about the client and the client's needs and slant my presentation accordingly. In other words, my prior contacts with the company will let me know the level of technicality I should use as well as the tone I should take. Some businesses will be formal and no-nonsense, while others will respond to humor and snazzy graphics. Some clients turn the event into a bag lunch; others are deadly formal. You never quite know.

Finally, I remember to stop talking and give the customers plenty of time to think of questions and to ask them.

J. Q. C.

Audience Analysis

Audiences are not easy to put into categories, but you can see patterns in most company settings. Audiences in company environments seldom gather in large random groups unless there are special events: a new building is dedicated or a new CEO is brought in. The likelihood that you will be called on to address such events is remote. You will be fortunate enough to deal with small groups of people who have been selected to hear what you have to say. Their commitment to you and your talk will depend on the dedication or spirit they bring to the meeting. You can influence the audience by knowing something about their interests before you arrive. As you speak to a group of people, you can respond to their needs if you are alert to their reactions to you. There is a spirit of dialog in committee presentation, although no one may be speaking but you.

10

At times a committee is highly responsive, perhaps because they are volunteers—for the annual fund drive, for example. At times the group is responsive because they are in some sort of allegiance with you—as in the case of a union meeting. Other groups may be called together because of a need-to-know situation, and they are likely to be interested, but they may not see themselves as people who are there to swing into action. Audiences can, as I have noted, also be resistant. I use the example of employees who might be resistant to change at times, but you might also encounter hesitancy from executives if, for example, you propose dramatic spending increases. Executives may have an abiding respect for changes, but perhaps not changes that are costly.

If at all possible, you want to match the purpose of your presentation with the expectations you hope to receive from the audience. In public speaking this challenge can be difficult, but in the in-house speaking situations of a company environment, the groups you will encounter are likely to be small and the group members often familiar. The significance of the typical committee audience in a company is that there is little guesswork in establishing the link between the purpose of your presentation and audience expectation.

Knowledge and Interest Analysis

Anticipation of audience need is essentially an effort to anticipate the listeners' *knowledge, thoughts,* and *feelings* with regard to your subject matter. In an effort to judge their knowledge, you need to determine their understanding of your subject. In company settings you should be able to predict the answers to these questions as long as the committee audience is a familiar one.

The anticipated audience is a point of reference for the entire presentation. The presentation is constructed for them, and the way to correctly plan what you will say is to anticipate what their response will be. If an audience is receptive, the response doubtlessly coincides with the speaker's intentions. If an audience is not receptive, the response will not coincide with what the speaker desired. For information-oriented presentations, the audience will usually be receptive. Persuasive presentations are more likely to encounter resistance because company politics can be a problem.

Audiences run the gamut from those that will have high expectations of you to those that will walk in with negative assumptions. The vast majority will wait and see. If there is going to be a prevailing mood, it helps to know it. For a workaday presentation, the audience will file in and file out, and moods may not matter significantly beyond the Monday mood, the midweek mood, and the Friday mood; you are familiar with these time frames, and a presentation is sometimes positioned at one end of the week or the other. Let's hope not.

You will usually receive positive reinforcement from some members of a small audience. Interest or enthusiasm is easily projected and easily seen. At other times, interest levels might be low. The audience can be viewed collectively or individually. They can also be segmented by various allegiances, and some subgroups might be more valuable to identify than others. Some members might be more knowledgeable than others; some members might be more powerful than others.

If your audience is small, you have the opportunity to respect the individuality of each member. View each member as a unique situation in communication. As you become familiar with giving committee presentations you will notice the uniqueness among the members of any committee. You will often observe these responses, for example:

Pamela understands entirely and asks no questions.

Mitch understands poorly but asks no questions.

Luis understands poorly and asks many questions.

Alex understands entirely and is supportive and helps you explain clarifications to Luis.

Christina understands and asks what-is-in-it-for-me questions.

Tracy understands and asks economic questions.

To watch an audience, watch those who are watching you so that you constructively use your eye contact. Do not linger on anyone for longer than a few seconds and do not allow yourself to dwell on one or two people during the delivery. Move from face to face if the group is small. This approach to the audience allows you to see what their ongoing responses are as you speak.

As long as you are addressing a particular sector of the company—executives, legal officers, accounting, production, technical specialists—you can make a know-and-need-to-know evaluation without difficulty. **You design the presentation based on assumptions about the knowledge and interests of the listeners.** You cannot, for example, provide a *general* overview of a product design for an interested group of technical specialists. They will want the *specifics*. The same presentation, placed before executives, will fall on deaf ears. They will not know the meaning of digital gates, for example. They need the general presentation that the technical specialists will reject.

As we observed in the last chapter, what the audience knows and what the audience needs to know represent the key interests you will have as a speaker who is going to give a technical presentation. Beware, however, of a third concern:

What do they know?

What do they need to know?

How do they *feel* about the subject?

Every speaker has a perspective from which he or she views the subject matter of a presentation. Every listener does also. In a one-on-one discussion, a speaker has substantially more understanding of a listener's position because there is interplay: discussion defines the perspectives. **In speaking before a committee, you can only guess at listeners' perspectives. It helps to anticipate them, since they may not coincide with yours.**

Consider the example of changes in company policy, equipment, practices, production standards, production output, and similar matters. By training, executives are taught that today is the day to be improved upon. Tomorrow should be better than today, and if the executives help make tomorrow a new and improved tomorrow they are doing their job. Managers define their successes by changes. Holding the line is seldom a goal. As a result, if a manager makes a presentation in the name of improved output, for example, he or she hears the proposal from a manager's perspective: change is good. The audience, however, may hear a different message, particularly if they are not managers. Why more change? What for? We did that two years ago. Why doesn't he leave well enough alone? These organizational roles are played out rather frequently, and not always with the appropriate analysis of the reactions (if not the needs) of others in a company setting.

The Demographic Audience Analysis

There are other ways to analyze an audience that are not based on the *knowledge* and *interest* approach. It is occasionally useful to look at a group in demographic terms, which may include social status, economic status, occupation, age, gender, and other conditions. We will examine a few of these conditions so that you understand the ways in which such matters can be important. For example, each employee group you encounter during committee meetings can be viewed in terms of social status and economic status.

There are two issues to consider, since we are looking inside the company gates. First, there is the social matrix in the larger sense. Everyone in your company has a social status or economic status relative to the national culture. Although we have some grasp of that fact in our dealings with each other, it is a big picture. The little picture is easier to understand. Within a company there is a microcosm of the larger culture, and it is easier to look at a group of assemblers, or sales agents, or research and development engineers in terms of their status in the company. What to make of this knowledge is the question. You may be able to gauge attitudes toward an issue or perhaps you can foresee group expectations. For example, since we observed that managerial groups measure effectiveness by changes for the good, they are optimistic about the future. They will respond in a predictable way to the bright prospects of suggestions for a better tomorrow. Workers may not share that optimism, and they may look at the future from an entirely different perspective.

The age of committee members can definitely affect response. Younger employees will be open to change, for example. They are more excited about technological breakthroughs. Older employees have years of experience and are wise to effective and ineffective policy decisions. They can also be a little shopworn and can be prone to thinking in such terms as "oh no, not that again."

The politics of gender is another demographic issue. Managers, at least until recent years, would portray themselves as tough and virile hard-ball competitors. Meetings would be devoted to assertiveness training and involve references to football and so on. When increasing numbers of women entered management roles, the old approach did not communicate and had to be eliminated. Similarly, corporate sales seminars used to have a fairly rough edge. The customer was more of a victim than a client, and sales tactics were aggressive. There was a lot of "take control" strategizing. Women somewhat rejected all these ideas, and selling has had to be seriously redefined in recent years. In other words, all the presentations were failing to bridge the gender gap. Demographic considerations are relevant considerations for a speaker.

The list of demographic considerations is open-ended because unique forces can be present. If your company has a branch in Phoenix and you are based in Chicago, then geography becomes a consideration. Of course, you are very likely to know the makeup of a committee in very precise terms—by familiar names. This opportunity gives you a unique perception that public speakers would envy: familiarity. In-house speakers face for less guesswork than public speakers. On the other hand, no speaker can take an audience for granted. Whether you have five minutes on an agenda or complete control of a ninety-minute meeting, you are there to share ideas. **You must anticipate the audience and its perceptions, or the perceptions of subgroups in the audience, or possibly everybody's perceptions if the group is small and the faces are well known.**

Attitude and Value Analysis

A more complex approach to understanding an audience would be an attempt to map the group's *attitudes* and *values*. This is an underlying dimension of the entire discussion here, but the task of constructing an interpretation of this sort of mental territory is complex. In addition, this approach to audience analysis can involve an attempt to understand the *beliefs* of a group, which can be another difficult task. The following practical definitions for these terms will allow you to see the distinctions among the three concepts.

> An **attitude** is an ongoing reaction to transient conditions.
>
> A **value** is a behavioral guideline for a person's actions.
>
> A **belief** is a trust that something is so, that some point is true.

Even with these extremely simplistic definitions, you can see that this approach to understanding an audience is more intuitive than analytical. To explain your own attitudes or values or beliefs would be a substantial challenge. Obviously you could not explain the attitudes or values or beliefs of others any more easily than you could explain your own. Nonetheless, we can and do try to understand these dimensions of the people around us, and we can and do respect the attitudes, values, and beliefs of others. It is in this sense that our understanding is "intuitive." To say that an attitude-and-value analysis is complex is not, however to say that it is mysterious or too involved with the inner workings of the mind. **People readily and sharply display attitudes and values and beliefs.**

The focus that concerns us here is the committee setting, which is a courteous if not a formal environment. Emotional display is not the norm. Rather, in this setting, understanding the values of listeners is a matter of dialog and polite discussion. You will see the values that committee members attach to concepts and to words. In the examples of *new information* or *needed changes* that we have examined, you *can* see attitudes and values in the positions people take regarding new information or changes that affect them. There is no mystery. As I noted, there may be an assumption among administrative personnel that the word *change* means *change for the good.* Other employees will not share that definition; for many of them change may mean change for the good but it can also mean *inconvenience* or *insecurity,* or other responses that are, in fact, the opposite of the perceptions of others. These responses project attitudes, values, and beliefs and help a speaker understand an audience.

If you give some thought to your audience as you prepare a committee presentation, you will be set to go. During the delivery, you will know if your interpretation of their perceptions is accurate because small committees are interactive. The listeners are quietly talking to you during the presentation even if they do not speak. When you prepare a presentation, the image of the audience seems like the dark depths of a huge auditorium. Then you get there and find your friends Jean and Bob, and there is Sharon from down the hall. Your boss is eating his sandwich while he listens, and only one employee is someone you do not know: the new guy from accounting. This is the reality of a *real* audience in a committee setting. Harriett shows up to take the minutes just as you begin to speak. You have known her for seven years. In this situation every speaker learns that listeners will talk to you with nonverbal patterns of behavior. They will show interest, inquisitiveness, hesitation, doubt, passiveness, excitement, annoyance, indifference, and other attitudes, and you will find you can respond to those attitudes to either reinforce them or redirect them. To ignore these signals it to invite difficulties.

Real Distance and Psychological Distance

The back row may pose the greatest challenge to a speaker because of a natural tendency for the most interested listeners to position themselves near the speaker. There is some sort of predictability curve of audience expectation at work here. The most enthusiastic and expectant come forward—and are on time. The least enthusiastic file to the back and are often late, perhaps because they want the house to fill and the lights to go down.

The relationships illustrated in the following diagram suggest the problem. Distance matters, and *psychological distance* matters also. If the back row now holds the hecklers, does it still hold true that you should give each and all your attention? Yes. Committees are not randomly composed. They are, even when loosely composed, employee sectors of your workplace. They are not, of course, hecklers at all, but people *will* often tend to show their allegiances in their seating preferences, and this gives you an understanding of them at times. If you have carefully analyzed the audience, you will sense any repositioning *you* must do during your talk. One reason for never memorizing your presentation is that you want to be able to react to what you see. If your audience profile identified four key players among ten executive who will attend, the question is, where are they? If they come forward, this can affect the presentation plan. If they all sit at the other end of the table, the signals may indicate a greater challenge.

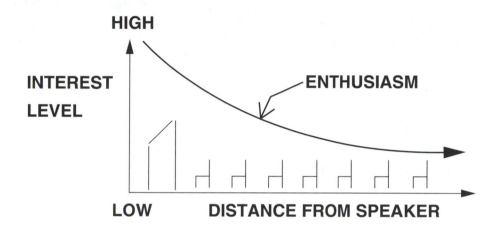

**Psychological Distance Can Be
Measured in Real Distance**

Although the curious phenomenon of seating preference is one I see all the time, it is important to realize that most meetings transpire with no need for any such awareness on the part of a speaker. People are at a meeting because the are *expected* to be there, which is a unique feature of a committee presentation. The meetings are, as often as not, composed of informational presentations, mixed with an occasional persuasive presentation. The audiences in a workplace are far less inclined to be wary than to be weary. If you are stuck with a Friday afternoon presentation at 2:30, the weary audience is psychologically distant no matter where they sit. In this everyday setting, the challenge is to build interest among all listeners, and a speaker's perception of the audience is helpful. Perhaps the greatest challenge are employees who are "committeed out"; for them, another meeting is just more humdrum. At an information meeting, the key players are not critical. The typical day-to-day meeting can present a speaker with a very simple issue: the psychological distance caused by fatigue and boredom.

Questions from the Audience

Asking questions or accepting questions is a handy way to gauge your success at making listeners understand your key points. *Audience participation* can be invited *during* a presentation, particularly if the talk is informational. If listeners can ask questions, the manner of the presentation becomes relaxed and has more of the atmosphere of a discussion. **Listeners can be very helpful, but the speaker must try to keep to the goals of the delivery as well.** Allowing the audience to participate during the presentation can also be distracting to both the speaker and the listeners.

A *question-and-answer session* has the same value for the speaker: the questions can be a clear measure of the listeners' understanding of the speaker's material. The advantage of a question-and-answer period is that it follows the presentation and does not interrupt it. The question-and-answer period does, however, present a challenge at times. Well-focused, relevant questions are not always forthcoming. In a question-and-answer period, be patient with the person who volunteers a question. People will ramble and become confused. At times there is no question. Often the volunteer is posing an observation and not a question—and these nonquestions can be long winded at times. Your patience gives you additional time to understand and evaluate the comment and offer a response. You must study the question to see what the focus is.

When a nonquestion is delivered, you could ask if the person had a question, or you might prefer to move on, instead. It is easier to thank the person for the observation and move on.

> *Thank you. That is an interesting possibility.*

> *I appreciate your concerns.*

Request that listeners raise their hands to ask questions. This puts you in control of the situation so that one or two people cannot monopolize the questions. If you request raised hands, you can avoid inviting a second question from people who are talkative or in any way annoying. They will be polite enough to not insist because of the protocol that is socially expected in the raised-hand format. If they are interruptive, their credibility will be defeated and they know that.

Also, do not hesitate to observe that a question seems unrelated to the issue, and ask for a related question. People will stray, and the questions should concern the presentation. If you are contradicted or challenged, explain that your presentation was a *sampling*. You cannot use all the evidence all the time. It is hoped that you made the case. Offer to help the questioners locate additional resources.

Questioners may try to put you on the spot. React confidently and do not try to put *them* on the spot. You cannot sidestep questions—although you can certainly say you do not know the answer.

> *I would like to know the answer to that myself. I'm glad you mentioned it.*

Questioners will often try to redefine the issues, and this greatly complicates question periods. Comments of this sort are directed at your evidence or the premises (the key supporting points) of your argument. You cannot readily repeat the evidence or supply additional evidence in a few brief responses.

Do not become heated during or after a presentation. You must not attack the information of others, or you will be seen as attacking their integrity. Speak cautiously.

> *I don't have that evidence in my findings.*

End a question period on a solid answer. In other words, do not end because time is up. End with a positive image of yourself.

> *Right! My point exactly! Well, I see by the clock that . . . .*

Question sessions can be aggressive at times, but this problem is not likely to occur in an information presentation of a technical nature. For a need-to-know group, the questions are usually impersonal and objective. Questions become provocative or aggressive when meetings have some sort of thrust that involves emotional triggers, which could be set off by the speaker, the subject, the audience, or a combination of these. These events are common but probably will not involve your speaking role. If you anticipate any controversy, put yourself in control of the question session—or omit it.

The Use of Visual Aids

A very common need in technical presentations is visual support to make a discussion vivid. I have mentioned the use of samples and products to provide demonstrations or simulations for an audience. The actual circumstances will vary, and often a hands-on approach is not a feasible approach to demonstration. In that case, all the other visual aid options can be considered. I will discuss each of the various tools that are available and you can consider them in relation to your needs. They serve many useful purposes.

Visuals

- **help technical listeners understand technical material;**

- **help observers retain information;**

- **signal important points;**

- **organize audience understanding;**

- **add variety to the presentation;**

- **can have a high impact.**

Be sure that your visual aids are constructed on a scale that can be viewed and understood by the audience. The aid is only as effective as its visibility allows it to be. If possible, construct the aids to match the location for the presentation. Calculate the maximum distance from which a viewer will look at the aids. Stand that distance from them and see what you think.

When using graphs, point to what you want an audience to see. Pointing is more effective than describing, since the audience sees the image but may not see the point of interest. Also avoid apologizing for your graphics:

> *I'm no artist.*
>
> *I didn't have much time to build these.*
>
> *If you could see this you would notice . . .*
>
> *Sorry but this slide is out of focus.*

Be conscientious. Plan ahead and throw out the slide that is out of focus!

Visual Aids

A large variety of visual aids are available to support your presentation.

Chalkboards and Ink-Marker Boards

The conventional blackboard remains the visual support of choice for educators. Although a few companies use them, many meeting rooms in company settings have white inkboards instead. Because boardrooms are often designed as elegant settings, there may be no boards at all. Anticipate the setting; you cannot guess about the boards if you will need one. The use of a chalk- or inkboard is one you know well, but if you have never been the person holding the chalk, you may not realize that there are tricks to the use of this visual aid. You have three options. The first option is to leave the board blank until it is needed, and use the board as you speak. As a second option you can place all the information on the board before you begin to speak, which you cannot do if you are an agenda item at a meeting that will involve other uses of the board. The third option is to write some, but not all, of the information on the board before you begin.

As a rule of thumb, judge how much time is involved if you will stop a presentation to write on the board. It is difficult to write as you speak, and a pause is usually necessary. The problem is that if entire sentences or entire sets of calculations or several entire schematics are placed on the board, there can be awkward and inefficient pauses in the presentation. If long interruptions are expected, have the complicated visuals on the board before you begin or have them prepared in some other way. **Use the board to *reinforce* ideas or to *express* material such as calculations.** Be careful not to talk to the board as you write or as you explain the material on the board. This is a common problem that speakers have with many visual aids. The visual is for your audience. Draw attention to the aids but do not face the aids. Face the audience.

In the case of all the visual aids, use color if you want to add a little additional interest, but also use color to clarify perceptions of the material. Inkboards usually have three or four colors of markers you can use.

Flipcharts

Everybody develops a style if they speak with any regularity, and the use of visual aids becomes part of that style. I, for example, have developed a tendency to use overhead transparencies with regularity. Dave is an administrator who works in another building here, and he is well known for going back and forth to one meeting or another with an easel and a large 2′ × 3′ tablet that he uses in all his presentations. The advantages of the flipchart are that the same visuals can be used repeatedly (they are not erased), and the speaker does not rely on boards. Since Dave often speaks in executive boardrooms, he cannot count on chalkboards.

From a public-speaking point of view, there is another value to using flipcharts. **It is popularly thought that visual aids should come when needed and disappear when not needed.** The idea is that an aid explains or clarifies and adds emphasis to a point. If the chalkboard fills up with material, the momentary effectiveness may diminish. Dave's visuals are a sequence of colorful graphs and boldly worded highlights, but because they are flipped out of the way one by one, only one visual is used to draw audience attention at one time.

Posterboards

Art supply stores carry large sheets of 2′ × 3′ card stock, posterboards, and foam core. In fact, foam core is sold in much larger sizes as well. Because these materials are semirigid or rigid, they are used to make more professional looking visual aids than the sketches that usually fill the paper tablets that are used for flipcharts. If you decide on this option you might see if your company has a graphics department that could lend a little time to you efforts to develop attractive visual supports. One particular advantage to rigid material is that you can mount drawings and photographs onto the boards along with words, phrases, math, schematics, and so on. Posterboards can also be very colorful since a few brightly colored pen markers can have a striking effect.

Feltboards were, at one time, a variation on the posterboard and had the advantage of needing no glue. As a result, the boards could be used repeatedly, but I seldom see them except in very large bifold panels for trade shows and the like.

Handouts

Endless numbers of meetings are held in every company, and at most of those meetings materials are distributed, usually beginning to the left or right of the person in charge of the agenda. Of course, many of those meetings are discussions, and the handouts are often the point of discussion. For a speaker engaged in a presentation, the tendency to proceed in the same manner is commonplace. Out go the photos or the drawings or the production brochures and other items. At communications seminars you will hear a strange contradiction. Aspiring speakers are told to *have* the handouts but they are told *not* to hand them around. Discuss the handout, show the handout, but do not distribute it until the close of the meeting. The idea here is that, like a cluttered blackboard, a handout provides distraction. And the handout might be very attractively designed to draw attention. The handout becomes competition! Be alert to the problem if you intend to use handouts.

Digital Imaging

Two very popular tools have appeared in recent years: *Photoshop* and *PowerPoint*. *Photoshop* can be used to create very elegant transparencies, and *PowerPoint* is popularly used in a similar fashion except that the *PowerPoint* software creates digital images that are projected on a screen. These programs are used in color, project at any size desired and, for the time being, even generate interest because of the technology involved. These two approaches to visuals are expensive and time consuming, and the results must be measured against their utility. If a presentation is going to be made repeatedly, then labor-intensive visuals—including digital tools, videos and motion pictures—are an option. In a company situation, the use of these aids will depend on factors that you and your supervisors will have to judge.

If you have direct access to *Photoshop* and *PowerPoint,* the question is whether you have the time to develop the digitally assisted graphic supports. Factor in the number of times you will use the material. Also factor in the relevance of the materials. For example, I once observed an upper-level manager in a computer center. He came to build one transparency for a presentation. He spent several hours doing it and used a number of sheets of film and took a good deal of time from one of the technicians. The transparency was used once. Some companies are aware of these difficulties and have personnel to handle special graphic needs. Where I work we now have a *PowerPoint* person who builds presentations for staff members.

One other point about projections is that they are often poorly designed conceptually. Visually the projections may be stunning, but conceptually they can end up playing a very minor role. Perhaps the problem is partly due to the popular idea that one key idea should be projected at one time. This approach is often adopted in the design of digital images, overhead projections, and slides. What I did not mention about the transparency that the manager took two hours to construct was that it contained *one* word. For a professional audience, the one-key-concept approach to projections may be misleading and inappropriate. In education seminars I have attended, key-idea projections seemed to make no meaningful impression on the audience if they were overly simple. The technique can be too simplistic to have value.

Key phrases may prove to be more useful than *key words* when you are dealing with a professional audience. To be certain your projections are neither too simple nor too complex, let's construct a sample. Suppose an executive is delivering a presentation to a group of managers. He has chosen to discuss *excellence*. We see him in his office looking at a beautiful transparency designed by his media services group. In the middle of the transparency is one large word:

EXCELLENCE

Suppose he asks us what we think. What should we say? You might flatter him and marvel at what *Photoshop* can do. I might ask him what his speech is about. When he says "excellence," I begin to see trouble on the horizon. Not only is the transparency underpowered because it contains only one word, it is the *only* transparency. Furthermore, the word on the transparency has been used and abused as a buzzword in management training seminars. You then ask him what he means by "excellence," and he explains that managers who excel make a company excel and that he will discuss how to achieve that end. Now, without any more fiction, let's look at what he needs to do. His projection

$$\boxed{\textbf{EXCELLENCE}}$$

will not mean much to thirty professional MBAs who are familiar with the concept. He needs to focus:

$$\boxed{\textbf{EXCELLENCE MATTERS AT ELCO}}$$

The statement is still far too obvious to bother with, so he needs to identify his key ideas:

> **EXCELLENCE MATTERS AT ELCO**
> **The power player—**
> ✔ **knows the corporate milestones**
> ✔ **aspires to production benchmarks**
> ✔ **achieves his or her management objectives**

Once the transparency takes on meaning, the audience will devote attention to it. Avoid the obvious, be cautious of the one-idea concept, and avoid slogans.

The projection concerning excellence now needs follow-up, and at least three or four more transparencies will help. One projection might define the term "power player." Three more could explore or develop the strategies identified on the leading projection. So, for example, a follow-up transparency might explain the management objectives that are the final item on the lead transparency:

> **Achieve your management objectives:**
>
> ✔ **Use employee reviews regularly.**
>
> ✔ **Follow a six-month game plan.**
>
> ✔ **Win every week and know why.**

Visual aids are tools. If they are not significant, an audience will ignore them. If you see a transparency with one or two words on it, there is probably a problem. It will sustain very little audience attention.

Overhead Projectors

The preceding comments apply to transparencies of all sorts, but the conventional transparency that is quickly made on a copy machine is somewhat more of a daily convenience than those that require digital electronics. Any sheet of paper can be made into a transparency, and the transparencies can be made in color also, although this added touch will usually involve a trip to a copy shop. Office copy machines will copy a page of text on transparency film, or you can write directly on the film with transparency markers (for example Sanford's Vis-a-Vis pens). If you will use the transparencies more than once, use a permanent marker. The temporary version of the markers will smear when touched.

If you write on the transparency film as you talk, you may want to use transparency roll film (long rolls rather than single sheets) and temporary markers so that you can wash the film for the next use, even though you intend to repeat your presentation. If you use transparency sheets, they can be layered one by one. Overlays are a unique use of transparencies. One transparency, perhaps in green, can be placed on top of another one that is handled in another color. The result is interesting and adds variety to the method of presentation.

Full pages of text can be projected on a transparency, but the image must be very large for the material to be read. It is best to highlight what you want observers to see if you use full pages of text. Although a specific projected image is not supposed to be left on the screen after you move on, you might leave the image on the screen for a while if you are using something on the order of a schematic or an architectural drawing. Always glance at a new transparency to make sure it is straight on the screen. **Remember not to talk into the screen, and do no stand in front of the projection.**

In the case of all electronic devices that you might use, try to scout out the committee room and check for power outlets and projector screens and so on. Also, you might want to use a pointer. Laser pointers are now affordable, although a common pencil will do the job. Number the transparencies and organize the sequence with care. If you keep a sheet of white paper between them, you will be able to see each one clearly before you put it on the projector.

The *opaque projector* is occasionally used for meetings, but the image quality leaves a lot to be desired. The advantage of the opaque projector is that it projects an image taken from a paper text such as a book or magazine or brochure. There is no plastic transparency. However, because light does not travel through the product, the image is a reflected image and lacks brilliance.

Slide Presentations

If an engineer is going to develop a presentation, slides may be the preferred medium. The other projections I have examined lack the striking photographic image of a slide. A 35-mm slide will project enormous images with clarity. The projection can be 8′ × 8′ or larger, which is very large for a meeting room environment. If the subject of a presentation involves the design of a building, the phases of a bridge under construction, the effects of a drug on cancer cells, or any situation that is visual in a photographic way, then slide presentations are ideal. Of the various types of single-image screen projections, the film slide is probably the most involved in time and materials and costs. In addition, a 35-mm camera is required.

Each slide should be labeled and numbered on the top edge that is visible in the slide tray so that you can see the wording just in case there is some difficulty during the presentation. Run through the collection to make sure the slides are all right-side-up and forward. Remember, slides can go in to a projector in eight ways! There are four sides, and only one is up. And they can be reversed and inserted backward also. To make matters worse, up is not up; it is down. I keep thousands of slides in carousels because of my activities and I still get confused from time to time. A *preview* is your only guarantee that the carousel or tray is correctly prepared. Be sure to lock the slides into position.

Do not limit yourself to photographs if you use a slide presentation. If you plan to use overhead projections for key ideas, and if you have several charts on posterboard, you may be looking at an awkward mix of media. If you have a media department or graphics department, you may be able to transfer the overhead material to slides and have the posterboards photographed also.

The projector screen can be straight ahead of viewers if the projector is not an annoyance. Overhead projectors usually sit near the screen, and slide projectors sit at the opposite end of the room if you have a zoom lens.

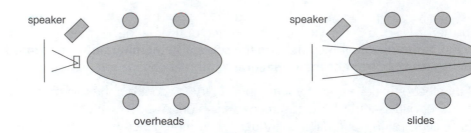

overheads   slides

If the projector is in the way, it can be projected diagonally as long as the screen does not make viewing uncomfortable for the audience.

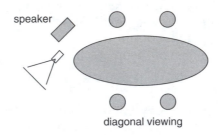

diagonal viewing

If you use a slide projector, avoid marching to the machine to change slides. Have an assistant or use a remote. Do not call the slides a "slide show." Take care to properly secure cords that might trip viewers. The well-prepared speaker might carry tape (or even carpet runners) to cover cords.

Videos and Films

A final option involves video footage and motion picture photography. This is costly for your applications and may not be convenient. For a college lecturer there is little difficulty in finding material among thousands of educational videos and films. Whether the lecture concerns preschool teachers or DNA molecules, videos and films are available. In a company setting your presentations will be far too specific to be accompanied by a video or film unless it is made for the job. At times perhaps a video can be a quick and affordable way to examine such issues as an assembly plant, a production line, or a construction site. Film rental services, however, will be of little value. You or someone in your company will have to produce the video (film is substantially more expensive). In the absence of videos and films, your presentation is likely to be stronger in any case. A film is a powerful medium and it can upstage you as the speaker.

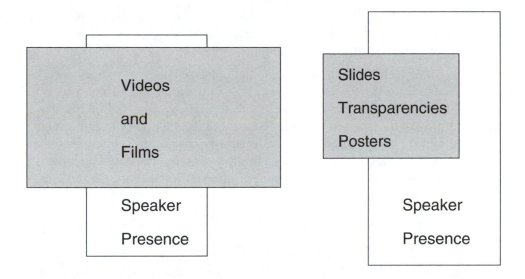

VISUALS DIVIDE AUDIENCE ATTENTION AND MUST

BE HANDLED WITH CARE

Lighting

Lighting in the committee room is an important consideration when you use any of the visual aids. In a sense lighting is a visual aid that assists all the others. If the room is bright, the inkboard and the posterboard displays are highly visible, but projections will be weak. If the lights are out, you will have brilliantly lighted electronic aids, and your posterboard displays will vanish in the darkness—and so will you. And it does not help that fan-cooled projectors are humming while you speak. A well-planned environment is not always available Communications consultants will suggest that you to leave the lights on, but that will seriously limit the quality of any electronic projections, especially if the room is overilluminated. You need a room with a dimmer. Better yet, you need a room with proper lighting features so that one switch will turn off all the lights in the front (near the screen). Another switch turns out the room lights but leaves a few on for the listeners' convenience so that they can take notes and see the speaker. Farfetched? Not at all. These are common features of meeting rooms. I have enjoyed these features for many years, because I schedule rooms where I know I will get the correct lighting. It helps.

Graphs

The visual aids identified in the preceding section are *media* that can be used to present all manner of visual materials. The *subject matter* of these visual aids will depend on the focus of your presentation and cannot be addressed in this discussion. One exception is graphs, which are popularly used in many subject areas. They can be presented in any of the visual-aid media, whether they are placed on a blackboard, constructed on posters, or designed for transparencies, or digital projection.

Visual aids are usually perceived as supplements that assist a speech that could be better presented with visual reinforcement. This is almost an understatement in technological areas. If a technical specialist makes a presentation that involves his or her expertise, it would be something of a marvel to achieve the ends of the talk *without* aids. **Technologies are often more clearly represented by visual media—schematics, engineering drawings, chemical reactions, higher math, business math—than by the English language.** This basic reality means that a technical presentation almost always enjoys the support of visual materials, including graphs.

Three variations of graphs are very helpful in a speaking situation. First, observe that *tables* are not popularly used in public speaking. Tables of data are detailed and may not project well in either overhead or slide format. Most reasonably sized enlargements of a table, of perhaps ten vertical and ten horizontal columns, will be difficult to read for most people seated around a large conference table. *Graphs,* however, make clear and simple statements—if they are used with clear and simple outcomes in mind. The simplest of the three is the *pie graph.* The pie graph is used to show *portions* of a whole.

As long as the slices are large, the meaningful point of a pie graph is usually unmistakable. I explained early on in *Workplace Communications* that technicians, engineers, and supervisors spend an enormous amount of time communicating—usually in conversations.

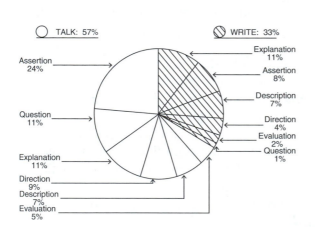

TECHNICAL OCCUPATIONS

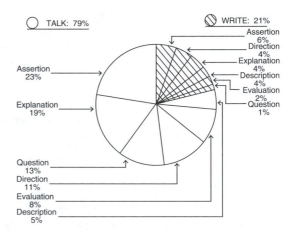

TRADE AND INDUSTRIAL OCCUPATIONS

THE KINDS OF INFORMATION EMPLOYEES SEND IN COMMUNICATION*

From a report by the Washington State Commission for Vocational Education that was prepared for the U.S. Department of Health, Education and Welfare (1976).

There are pie graphs I could use to support the statements. I could quote percentages, but my resources used pie graphs that could quite easily be converted into transparencies or posters. The two preceding tables concern technical and industrial occupations.

Here is a similar analysis of the amount of time engineers devote to spoken communication while they are on the job.

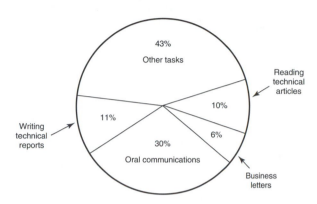

AVERAGE TIME SPENT ON COMMUNICATION TASKS BY ENGINEERS*

*From Richard Arthur, The Engineer's Guide to Better Communication, Glenview, Ill.: Scott Foresman, 1984.

The *bar graph* is a vivid tool as well. If *comparisons* are important aspects of a presentation, it might be appropriate to use a bar graph. The following example suggests the relative costs for the adoption of three different vehicle leasing contracts.

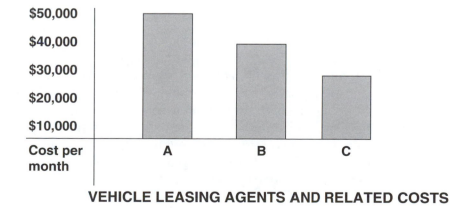

VEHICLE LEASING AGENTS AND RELATED COSTS

If you have more than one bar graph, begin with the most important. Prioritize the presentation, and show only one graph at a time. Once each has been examined, they can be viewed as a group if desired. Since the cost differential of the lease agreement is large, you might need to look at the other graphs and then compare the contracts. Perhaps you need a graph concerning service and warranty, a graph concerning buy-back provisions, and so on. These can be discussed one by one and then viewed as a group.

The *line graph* is distinctly different from the pie graph (of parts) and the bar graph (of comparables). The line graph represents time and shows movement. The movement is perceived as a trend. If your company enjoyed and suffered the boom-and-bust year of 1998, a line graph of the wave-and-trough effect might vividly illustrate the problem.

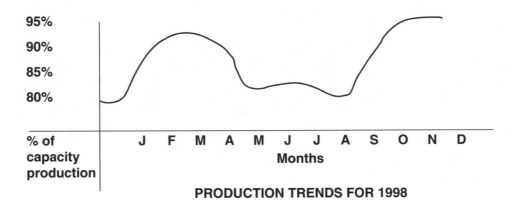

PRODUCTION TRENDS FOR 1998

Because a great many company meetings involve production, costs, and similar matters, all the graphs are common fair. If your role is supervisory, you might have considerable opportunities to use graphs. If your presentations are quite technical, other visual aids are the more likely candidates to support a talk.

Diagrams

Each technical specialty favors certain types of graphic components that support the discipline. In electronics the schematic is central. In architecture the structural drawing is critical, and so on. If you anticipate speaking before a group that is unfamiliar with the technology you represent, be prepared for the problem. The usual graphics you might use for a professional group are likely to do very little communicating to a group that is not equipped to *read* the visuals. I cannot read a schematic from supply to ground, but I can comprehend the overall idea if you provide me with a *block diagram*. The block diagram is one of two elementary visuals that can help a technician. I have used block diagrams throughout the *Wordworks* series. Here for example is a block diagram of the communication loop on p. 61.

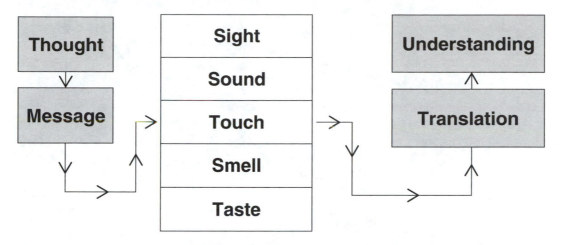

THE LOOP:
THE BASIC STAGES OF COMMUNICATION

Another useful device is the *pictorial diagram*. Very complicated processes can be diagrammed. For example, the efficiency of a physical facility is often studied to determine the most efficient flow of production activities. These concepts involve both space and time but they can be rendered in surprisingly vivid pictorial diagrams that can explain a manufacturing process in a step-by-step graphic. The graphic is designed in such a way that each step is visualized in some understandable image. A bale of cotton might represent "receiving" at the top of a drawing and a truck might represent "delivery" at the bottom of the drawing. In between, the manufacturing process for a cotton fabric can be symbolized with a steaming vat, rollers, and presses and so on. The following example is the production diagram for making wine.

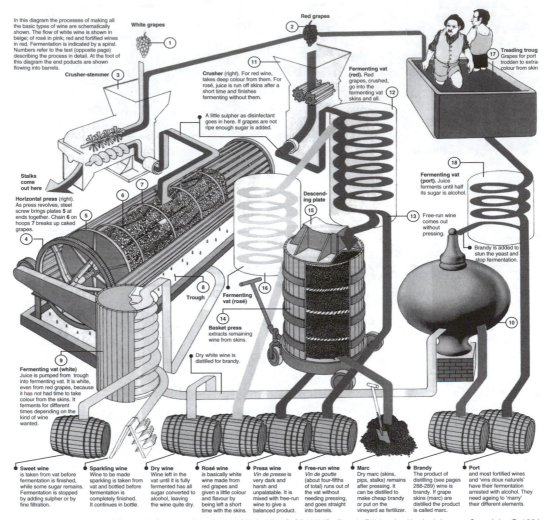

In this diagram the processes of making all the basic types of wine are schematically shown. The flow of white wine is shown in beige; of rosé in pink; red and fortified wines in red. Fermentation is indicated by a spiral. Numbers refer to the text (opposite page) describing the process in detail. At the foot of this diagram the end products are shown flowing into barrels.

White grapes ①

Red grapes ②

Crusher-stemmer ③

Crusher (right). For red wine, takes deep colour from them. For rosé, juice is run off skins after a short time and finishes fermenting without them. ⑪

Fermenting vat (red). Red grapes, crushed, go into the fermenting vat skins and all. ⑫

Treading trough ⑰ Grapes for port trodden to extract colour from skin

A little sulpher as disinfectant goes in here. If grapes are not ripe enough sugar is added.

Stalks come out here

Horizontal press (right). As press revolves, steel screw brings plates **5** at ends together. Chain **6** on hoops **7** breaks up caked grapes. ④ ⑤ ⑥ ⑦

Descending plate ⑮

Fermenting vat (port). Juice ferments until half its sugar is alcohol. ⑱

Free-run wine comes out without pressing. ⑬

Brandy is added to stun the yeast and stop fermentation.

Trough ⑧

Fermenting vat (rosé) ⑯

Basket press extracts remaining wine from skins. ⑭

Dry white wine is distilled for brandy.

⑩

Fermenting vat (white) ⑨ Juice is pumped from trough into fermenting vat. It is white, even from red grapes, because it has not had time to take colour from the skins. It ferments for different times depending on the kind of wine wanted.

● **Sweet wine** is taken from vat before fermentation is finished, while some sugar remains. Fermentation is stopped by adding sulpher or by fine filtration.

● **Sparkling wine** Wine to be made sparkling is taken from vat and bottled before fermentation is completely finished. It continues in bottle.

● **Dry wine** Wine left in the vat until it is fully fermented has all sugar converted to alcohol, leaving the wine quite dry.

● **Rosé wine** is basically white wine made from red grapes and given a little colour and flavour by being left a short time with the skins.

● **Press wine** *Vin de presse* is very dark and harsh and unpalatable. It is mixed with free-run wine to give a balanced product.

● **Free-run wine** *Vin de goutte* (about four-fifths of total) runs out of the vat without needing pressing, and goes straight into barrels.

● **Marc** Dry marc (skins, pips, stalks) remains after pressing. It can be distilled to make cheap brandy or put on the vineyard as fertilizer.

● **Brandy** The product of distilling (see pages 288-289) wine is brandy. If grape skins (marc) are distilled the product is called marc.

● **Port** and most fortified wines and 'vins doux naturels' have their fermentation arrested with alcohol. They need ageing to 'marry' their different elements.

How to Distract Your Jitters

With respect to the issue of stage fright, technical presentations may help solve your problems. First, let me be quite honest. I am entirely unconvinced by the comments in public-speaking literature and the suggestions that are offered by communications specialists. There is no research to support the idea that imagining everyone in the audience is sitting there in their underwear is going to improve your speaking. (Honest—this idea frequently shows up in the literature.) Nor is there any evidence that taking deep breaths is going to do anything for you. The "prime yourself–psych yourself" approaches are no better. People who fear public speaking—and that evidently involves most people—will not easily overcome the problem in most cases. The cures lack validity, but the cause is pure science. If your adrenaline* kicks in and your beta waves are going off the chart, the guys' underwear in the front row is not going to alter a *flight or fight* response when you cannot do either.

Fortunately, committees are going to be your environment, the members will often be familiar and friendly, and your presentations will be brief and specialized. My discovery was that, since I need lots of visual aids for my presentations, I can distract myself with my props. I get busy setting up everything. I focus on readiness and organization. During presentations, I feel that the aids are taking the heat off me. The audience is often looking at the visuals. I focus primarily on the audience and feel a little more at ease knowing that I am not the exclusive focus of their attention. I also use the visual aids to remind me of what to say, since they feature highlights as well as details of my talk. The visuals replaced my 3″ × 5″ cards long ago. Well-organized transparencies make a well-organized outline for you to follow.

I have come to see visuals as a third dimension in a talk. A talk is a two-way relationship between a speaker and the audience. If you use visual aids you create a three-way relationship among yourself, your aids, and the audience. I can address a group without visuals, but it is noticeably easier if I have them. I am more relaxed. I keep busy organizing and explaining. The audience has something else to do other than look at me. There is a third presence.

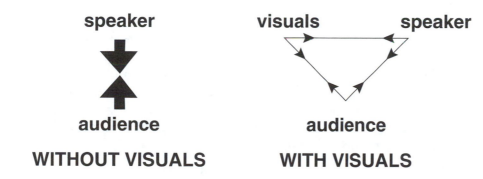

How does this calm my butterflies? I think the visuals (or simulations or demonstrations or whatever you use) become a *partner* to the presentation. Instead of a two-way effort that connects speaker and listener, the presentation becomes a three-way effort because the visuals become a supplement to the speaker's efforts.

Consider using visual aids as a way to alter the direct confrontation with an audience. Allow yourself to step aside. Allow visuals to become key players in your presentation. Look at yourself as the *spoken* half of the presentation. Look at your aids as the *visual* half

Adrenaline is epinephrine, a very powerful hormone trigger.

of the presentation. The aids take on stage presence and appeal to different levels of comprehension for an audience. That stage presence is comforting, because it can be extremely helpful for the speaker who will defer to the visuals and discuss them. **The visuals become your partner in the presentation.**

Having heard this idea about *triangular speaking,* observe that it will work only for speakers who genuinely use and need some assortment of visuals. In my daily dealings I use hundreds of transparencies. I *need* them. Math instructors go through yards and yards of transparency film roll, as do electronics, chemistry and other instructors. Any technical specialist can use the concept of triangular speaking. Your presentations are very likely to involve visual aids, and they should have a calming and confidence-building effect. You can even sit down! Many overhead projector users sit beside their projectors. They are a team.

Two final tips concerning the jitters: Do not drink coffee or tea the day of the presentation. Save these caffeinated beverages for your reward after the meeting, and if you use a pen or pencil to point at a detail on a transparency, realize that the projected image is ten times larger. Hit the glass platen firmly with the tip so it does not show that you are a little shaky. Otherwise, the projection will amplify your movements ten times for the audience.

Delivering the Product

Communications skills are important for today's engineers and engineering technicians. Because the technical knowledge of technicians is often the keystone of company success, they are central to interactions within the company. Technologies were always complex, but now there are many more technical specialty areas, and they are often obscure or little understood by an audience any larger than the specialists themselves. In addition, technology defines a new and better world; you represent *change,* as you have seen. As a result, public reaction to technology can be mixed. Your expertise cannot only be *little understood,* it can also be *misunderstood.*

If your specialty concerns computers, you doubtlessly view the skills you have as some small contribution to the betterment of society. Engineering technicians *use* science; they put science to work for a better world. Perhaps we seldom stop to look at technology in this way, but some technical fields—medical research and development, for example—have a very sharp focus on human need. If you were a service warranty tech for an MRI manufacturer, you would be viewed with trust by staff and patient alike. They are willing to make the leap of faith to adapt and use a new technology. But this is not always the case. If you are a computer specialist, you will still encounter millions of people who are resistant to or who perhaps fear the technology. In sum, when the public does not understand, the public will often assume an *attitude* toward what they do not understand.

You must adjust any presentation as you deliver it so that your listeners understand the information and accept you in relation to it. You must be sensitive to the rise and fall of audience levels of understanding and interest and reactions. You must be able to adapt accordingly. If, for example, a concept you explain leaves the audience in confusion, you can go into reverse. You can halt, back up, and attempt to clarify the issue. There is a risk in plunging forward if important points are not explained. A slow and steady speed is often important. The sensitivity of the speaker is important. Speaking is a commitment to the listeners.

Credibility

Because you are probably going to speak about matters that are related to your technical expertise, there will be immediate respect from the audience regarding your status as a professional. If a memo preceded the meeting, perhaps it explained your background, if only in a sentence or two. With this as an initial impression, the audience will look at you favorably. Then you must fit the role and the image to look the part. Finally, as you speak, the impression the listeners have of you must be reinforced. The reason for making every effort to understand the audience is to build *trust*. How? Be alert to your listeners' behavior. As I explained earlier, they will project their perceptions with nods, looks, smiles, and other signals of moods and attitudes. If you respond to them, the presentation will serve a subtly different purpose. **The *typical* presentation *tells*. The *effective* presentation *shares*. Sharing your ideas means that you communicate with *and* respond to the listeners.** A presentation that is shared will build trust. Do not overlook the fact that your visual aids are another measure of your credibility. You need to look as professional as possible, and the preparation of your visuals deserves time and effort so that they too look reasonably professional, or at least competent and interesting.

Variety

You have experienced a monotone presentation at some meeting or another. What is it about a monotone that does not work? After all, no one at the table speaks with breathless excitement, so why is the monotone of a speaker always boring? The answer is "variety." The monotone lacks movement. It lacks change. Sad to say, if a speaker uses a monotone, the same speaker is also likely to brace one hand on either side of a podium if he or she is going to deliver a presentation. This habit locks the body in a rigid position, and again there is no variety. There is no color. The problem is a simple one: without variety a speaker can lose the attention of a listener. Listeners do not focus on the message alone. *They focus on the medium also.* The medium needs life. It needs enthusiasm. **The message does not live by words alone. Keep the body free to move. Keep the voicing natural and varied.**

Reading a presentation is usually a worst-case scenario for monotone speaking. Manuscripts violate a popularly accepted standard in public speaking: the conversational style. If a presentation is read, the language is stilted, and the delivery is not even vaguely akin to conversation. In addition, a paper tends to be read rapidly, faster than the pace of normal speaking. Speak naturally. Move naturally.

On the other hand, do not think of a speech as a performance or as acting. Do not contrive movements or voice patterns. The results will appear artificial. It is easy to temper any tendency to exaggeration. Recall how politicians speak at public rallies. Perhaps the national Democratic and Republican conventions are examples that are easy to recall. The style of the politicians is usually peculiar. It is a *manner* of speaking and it is very artificial. It is, of course, often humorously imitated simply by emphasizing every word in a sentence that can possibly be emphasized.

> *I WILL,* and *I INTEND,* to make *EVERY* effort, to *TROUNCE* any and *ALL* opposition
> to this *GREAT* legislation! And *I PROMISE* to do so. So *THERE.*

Notice that the cadence is a droning cliché that you would never use in speaking. You should make no artificial effort to speak to a group other than to be sure that the person in the back gets the message. That person may be ten feet from you or thirty and that distance is the only challenge. Eye contact, a slight increase in voice volume and voice characteristics, and some natural gestures should do the job.

Be aware of the tendency to rush through your presentation. Speakers who have not had much experience do not properly pace their presentations, perhaps out of some self-consciousness about pausing. *The pause is important.* You can catch your breath when you pause, and both you and your audience then have a moment to think. Besides, a pause can mark a transition, and it can be used for emphasis.

Conversational speaking runs at approximately 160 to 180 words per minute. You need not tape record yourself to monitor your speed, but note that there is a conventional standard pace. Be careful not to rush because you then will *appear* rushed because you are violating the normally expected pace to which your listeners are accustomed.

Make a conscious effort to speak every word properly. Do not delete final consonant sounds.

doin'	**is**	**doing**
havin'	**is**	**having**
perfomin'	**is**	**performing**

Also avoid the "ums" and "ahs" if possible.

Be wary of the occasional "tick" in behavior that can set in. Any repetitive behavior can be distracting or comical. These tendencies run from saying "Now, then . . ." ten times to tapping a pencil on the podium, to frequent clearing of the throat, to pulling on an earlobe or constantly looking over the narrow frames of reading glasses. These quirks are as annoying as the "ums" and "ahs" we use.

Be sensitive to the gender politics of pronouns and use the plurals—*they* and *them.* You can also use plural nouns that do not refer to *workmen* and other such terms: *security, clerical workers.* Remember that the polite *he or she* expression that you would use in writing will usually sound overly formal and awkward in a speaking situation.

Practice the presentation before you deliver it. Practice will encourage confidence, practice will allow you to master the use of visuals, and practice will build a stronger and more orderly delivery. Practice will also determine a very important consideration: how long is the talk?

Dress

As you realize from reading earlier chapters, your image is "speaking" before you say a word. Your audience will be "sizing you up" from the moment they see you until you leave the committee meeting. You will not have five to twenty people *staring* at you, but you are dealing with many people when your have an audience, and it is worth noting that there will seldom be moment when at least one member of the group is not watching you. This scrutiny includes more time than the speaking time. From the moment you are in view be alert to the likelihood that you are being watched. At every moment you need to be on your guard by being at your best.

At a basic level, clothing will obviously be part of the projected image. Unlike your behavior, which can change from minute to minute, your attire is a fixed projection of you. As you realize, many employees in corporate environments are very attentive to appearance; this is particularly true in white-collar office settings. I defer to your judgment concerning your appearance because the situations will depend on your workplace. I can observe that being cautious on the side of conservatism is always safe, and dark basics in gray, brown, and black are common features of the "company look" if employees aspire to it.

It is important to look the part, which in a business setting means only that you wear normal office clothing. More important is the effort to avoid clothing that can be a distraction or that invites judgments. If you dress the role you will most likely meet the expectation that a group of people will have of you. In the closed environment of a company there are few surprises concerning dress practices. If you are an employee you will know the practices; indeed, at times there are even dress "codes," which make images unmistakable. Because a speaker wants to build confidence in an audience, it is wise to dress in a way that conforms to audience expectations. At times these expectations might suggest that you should dress as they dress. At other times you might, if you are supervisor, be expected to dress in an upscale image relative to the people you plan to address. These are variables that will be easy to judge on most occasions.

Team Presentations

There are times when two speakers are necessary to make a presentation. If you need an authority of some sort—perhaps a patent lawyer or an outside consultant or a specialist in research and development—the presentation can be made by several people. If you invite a specialist, do *not* expect the guest to do more than respond to your requests for information or opinions. You remain in control of the presentation and the authority is present as support or as evidence in much the same fashion as a visual aid. A guest of this sort is an aural aid. If, however, you are asked to speak alongside of a coworker, the circumstances are different. Then, you are *both* part of the presentation and you must decide who will discuss what and in what order. The advantage of this type of presentation is that you share responsibilities, and you might find this to be a comfortable way to handle certain public presentations.

Summary

- When you prepare a technical presentation, give due consideration to the audience profile. Design the presentation with the listeners' knowledge and interests in mind:

    What do they know?

    What do they need to know?

- You can also view audiences in terms of employment status, age, gender, and other considerations.

- Be aware of the possibility that the audience may have an attitude—positive or negative—concerning the subject matter of a presentation. This is particularly true if your presentation announces or encourages *change*—in the name of new and better technology, for example.

- Overcoming distance is a unique challenge for a public speaker even in the modest environment of a committee setting. Speak to everyone by turn and avoid focusing on a few committee members. Use the most distant person in the room as a measure of your efforts.

- Psychological distance can be reflected in physical distance. Fatigue is a general problem at any time. Be sensitive to both problems. Address the committee with enthusiasm and concern.

- Use a question-and-answer session if it is appropriate. Deliver the presentation first. Announce the time limits for the question-and-answer session and ask for raised hands.

- Visual aids are important in engineering technology presentations. The aids represent modeling and facilitate the discussion through a variety of common media:

    ✔ blackboards

    ✔ inkboards

- ✔ flipcharts
- ✔ posters
- ✔ handouts
- ✔ overheads
- ✔ slides
- ✔ digital projections
- ✔ models and samples

- Use visual aids to increase listener comprehension by presenting the physical dimensions of a technical discussion topic. Also use the aids to reinforce important points and to add variety to the delivery.

- Develop the aids with care and attention so that they look professional. The aids should reinforce your professional image.

- Give each visual aid a specific function. Do not make the aids overly simple or overly complex. If the aids contain graphic messages, decide how complex the graphic should be. If the aid contains text, avoid a single idea, but do not make the message too complicated either.

- Use videos and films with caution. These media can upstage a speaker and shift all the interest to the aids.

- Graphs are particularly popular in committee presentations. If they are useful, develop appropriate pie graphs, line graphs, or bar graphs.

- Your visual aids will help diffuse your anxieties about being in front of a committee. The aids help you, and they help the listener find interest in the delivery.

- Be alert to audience reactions so that you can build listener confidence in you. Rather than view your delivery as an effort to *tell,* look at the delivery as an effort to *share.*

- Avoid the dullness of the monotone and speak with enthusiasm and the correct articulation of your language.

- Dress the part.

Activities Chapter 10

Present a memo to your instructor that develops a discussion of a work experience that relates to the chapter. The memo should be 500 to 1000 words, typed. In it you can explain how you have related the text or the lecture discussions to some event you have experienced. This exercise will give you a better understanding of the material because your memo will explain incidents in terms of our perceptions of workplace structure and communication.

Select one of the following suggestions and develop an analysis that involves your current employment or a former position.

- *If you have any work-related speaking experience, explain the historical circumstances. Were there company expectations in terms of the subject matter, your image, intended audiences, and other conditions?*

- *If you have had supervisory experience—military or civilian—did you speak to your staff from time to time? Discuss the settings and the meetings.*

- *If you have had any civic speaking experiences, explain your interests. These can include service organizations, religious groups, action groups, community clubs, and other organizations. Discuss your speaking roles.*

- *There are other situations that are somewhat similar to committee presentations. Have you ever had to sell your services before a group? Have you ever defended your interests before an agency such as an engineering department? Have you ever been interviewed (perhaps a job interview) by a group? Discuss one of these events and your reactions.*

Work In Progress

11. Selling a Contract

Two days after my presentation with Tempo, I got a call from Simmons asking me to submit a formal bid for the company network upgrade. I called Randy and Cloe for a meeting at the local coffee shop and gave them the good news and told them to bring their calendars and prepare to cover a lot of ground.

First, we had to estimate the length of time the entire job would require and then add extra days to account for any unforeseen problems or delays. To do this, we did a "run-through" of the job with each of us throwing out comments and ideas about how things should be done and in what order. I wrote down ideas as they came up and organized them into general headings. We then went topic by topic, working out details on procedures and estimating time, materials, and the final billable cost to Tempo.

Each job we bid on is quite unlike any other, and there are almost always unique twists that we have not encountered before and that must be thought out. With Tempo, it was the lack of a ready-made wiring closet or area designated for one. From the employee poll I had taken earlier, I knew that the front-office workers were adamant about not wanting more computer equipment in their limited workspace. In fact, they could barely tolerate the machines already there. Randy's solution was to take a 10' × 10' chunk of space from the lunch room (the back wall of which bordered the front office) and turn it into a wiring closet large enough to house the new network equipment and some of the current equipment that was taking up valuable floor space. His idea would require us to build a 10' partition wall and run a dedicated electrical circuit back to the electric panel to meet the power needs the room would have. Cloe and I agreed with him that this would solve many problems and not add substantially to the overall cost to Tempo. We would do the construction ourselves (I am a former construction contractor) and hire an electrician to do the dedicated circuit run.

Since Tempo was on a heavy production schedule and only shut down after 7:00 p.m. and on weekends, finding a time to build the room and do the installation without interrupting production and disrupting office work seemed impossible. Cloe saved the day by reminding us that a holiday was coming up during which Tempo would probably be closed. Working straight through the holiday, we estimated we would have enough time.

Finally, we went back over our list of stated (and implied) needs that our contacts with Tempo had given us. We double-checked that our bid met all those needs and that we were slightly under budget. We like to use the difference to create a "screw up fund" to cover any overruns or underestimating we might have done. NetWorks NorthWest was now ready to create the actual bid document and send it on its way.

J. Q. C.

Sales Relations Basics

Sales Theory

Why would sales skills matter to engineering technicians? These specialized skills are important because most engineering techs, regardless of the career area, are marketing themselves and do not realize it. They may be selling technical products. They may be selling skills. They may be selling services. They may be at a city meeting trying to get permission for a tract home development. They may be new college graduates facing their first employment interviews. They may be computer system specialists explaining a LAN topology to prospective clients. In each case there is a product or service involved and the challenge is marketing. Consumer interest must be encouraged. The buyer must be made aware of products or services that will meet his or her needs.

If people are marketing and often do not realize it, it stands to reason that only a small percentage of people—the "naturals"—are going to do a very good job of selling themselves. The rest could certainly use a little training. Unfortunately this skill—sales—often goes unnoticed. It is even uncommon for a chapter such as this one to show up in a college textbook except in business programs. Perhaps Sales 101 should be required in a great many college programs since the "point of sale" is what business—any business—is all about. When you ease into a dentist's chair, it is the point of sale. When you pay the waiter, it is the point of sale. There is a buyer and there is a seller; people just do not notice. The patient and the restaurant customer are not likely to see the event—the point of contact—from a perspective they would call "selling."

Corporations spend hundreds of millions of dollars annually to *sell*. Selling is at the heart of success for any business large or small. This fundamental fact makes it all the more strange that most people look at *selling* with disdain—and the image of the "salesman" is not exactly princely. It would seem that all the golden plaques and sales service awards that top corporate representatives earn still do not convince the public that the $30 million-a-year real estate agent in California is not a salesman. Every book on sales strategy opens with the blunt truth: if you are in sales, you must always confront the suspicion of the adage "buyer beware."

In truth, the disreputable image was caused by a historical practice in marketing. The reputation is based on the fact that **sales pitches were usually conducted to sell the product no matter what—and many salespeople still honor that intent.** Although the jargon has changed, the sales mission remains a fairly timeworn activity for many sales people. Your company may pay $1000 to send you to a weekend seminar in "contact marketing," but the term does not change the world's oldest sales method: the face-to-face sales presentation. There are a host of conventional sales strategies, including the following well-known systems for contact marketing.

FAB

The features-advantages-benefits approach is popular. First, the *features* of a product are explained. Second, the general *advantages* of the product are presented. Third, the specific *benefits* that address the client's needs are discussed.

AIDCA

Attention-interest-desire-convince-action is another approach. The sales agent gets the *attention* of the client, builds *interest* and then encourages the *desire* to learn more about a product. The agent then *convinces* the client to *act*.

There are a number of variations on the theme, but in contact sales there are basically three ways to sell. You can sell the *product*, you can sell *yourself*, or you can sell the *client*. There are, after all, few forces at work at the moment of a face-to-face point of sale.

| PRODUCT | — | AGENT | — | BUYER |

The Focus Points of a Contact Sale

Until recent years, the time-honored philosophy that was encouraged by most company training seminars involved product orientation. These strategies involved a single concept—*sell the product*—that is evident in the focus of most of the traditional training procedures:

> **Basic product sales training involves—**
>
> - **product knowledge**
>
> - **market territory analysis**
>
> - **sales technique**
>
> - **power attitudes (can do/will do/did do)**

Sales representatives were expected to know a product line backward and forward and to sell the *qualities* of the goods.

> *Look at the wonderful palette of colors that Klein and Smith have developed for their carpeting.*
>
> *Hey, zero to sixty in 5.5 seconds!*
>
> *The price is a steal; it is the perfect location, and that acre on the other side of the road is included!*

The problems that can develop from such a sales strategy are mostly derived *from* the qualities, mainly because they are sold to the buyer as a bill of goods. The qualities become promises, and if the promises are not met—which can be frequently if you look at Detroit's recall practices—then the buyer is not happy. Hence, the result is a distrust of the seller.

The problem is that the motive for selling can have little to do with a customer's motive for buying. Sales agents often make assumptions. The hottest and fastest computers, for example, may not mean much to people if the speed of their five-year-old systems was just fine. Computer enthusiasts use speed as a yardstick as though it equalled horsepower. Many computer users are unconcerned—often unaware—of such issues (and perhaps drive economy cars also). The issue concerns buyer expectation. For example, Carol travels a lot for her college. She is a geographer. She brings back slides and videos of the various lands she visits.

I went to one of the big media stores to get a deal on a high-end camera, but the sales pitch was hype. When I'm in the field I want a tank of a camera. No pops. No whistles. I just want it to work in 120° deserts, and I don't want it to fog up in a rain forest. So these guys show me models with enough buttons to be an accordion. They can do thirty swipes and screen fades, and endless stuff I don't need. I couldn't even find the buttons I needed! There were too many useless features. I walked out. I had to mail-order a simple one.

The conventional product approach to sales followed a strategy suggested in the following diagram.

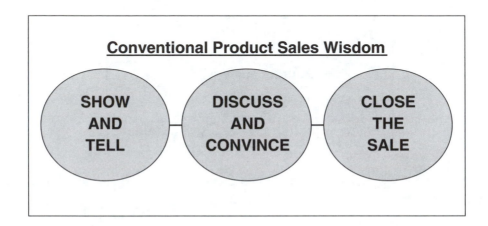

The Sales Agent

In recent decades there has been growing interest in looking at other options in sales management. The literature attests to the new trends, and behind the scenes are many companies from coast to coast that develop their own confidential sales manuals. Behind closed doors, sales reps attend training sessions, workshops, and seminars and they earn service awards or take home rewarding commissions if the job is well done. The newer strategies involve market savvy on a par with the sophisticated advertising and marketing packages that are designed with the help of consulting psychologists. These strategies have a strong focus on image building and people skills rather than on product selling.

As a result of the more recent thinking, many authorities came to see the sales representative as the selling force, not the product. Sell the rep, and the rep will sell the product. This approach meant that *hard sells* were out the door with the vacuum cleaner salesman. Instead, the seller was taught to build rapport and trust with the client. This idea shifted the sales training concept away from the old "believe in your product and you will succeed" to "believe in yourself and you will succeed" This is a very popular vision that is still persuasive today.

The attitude of the new century is certainly somewhat different from either of these two older sales ideas. As one young skeptic put it, "You can believe in the product *and* yourself and you can still go broke as fast as the next guy." There is truth to this observation. Every year I attend at least one county fair. I never fail to visit the tent with the housewares representatives: the reclining chairs that massage, the miracle cleaners, the vacuum to end all vacuums—and the especially gifted vegetable slicer demonstrator! Everybody stops to watch the vegetable demo. In a nonstop promotion that would leave an athlete breathless, the sales agent slices and dices everything short of coconuts. What sales skill! Those sales people certainly know the product. They demonstrate. They build trust. They are clean and well organized. There is no hard sell. And everyone walks away. What went wrong?

The Client

Current thinking suggests that the product and the agent are only two of the forces at work—and that the critical third force is the customer. If no one *needs* a vegetable slicer, then it does not matter how good the product is or how good a sales agent is at his or her job. What matters is what a customer wants. The sale depends on what a customer thinks he or she *needs*. The new look in sales is based more or less on finding that need, and the seller and the buyer are seen as partners in the sale.

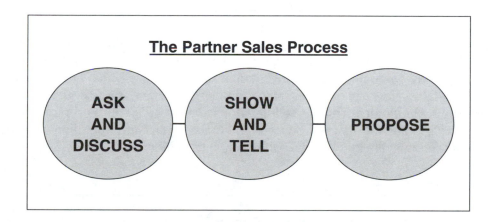

A sales encounter or a sales meeting is now seen as an opportunity for a sales agent to hear out a client's expectations, rather than the old-school approach of a customer hearing out a sales pitch from the agent. Typical of the "new selling" literature are the Miller and Heiman texts *Conceptual Selling* (Warner Books, 1987) and *Strategic Selling* (Warner Books, 1988), which are features of their consulting seminars for Fortune 500 companies and other corporations.

How did such a strategy emerge? You will not see a historical explanation in the literature, but it is very likely that the new strategy is tied to the shift in American manufacturing and technology. As many observers have explained, the second half of the twentieth century saw a dramatic shift away from manufacturing—particularly heavy manufacturing and smokestack industries. What emerged were high-tech industries of the Silicon Valley variety—and a service economy. Client demands shifted to demands for services.

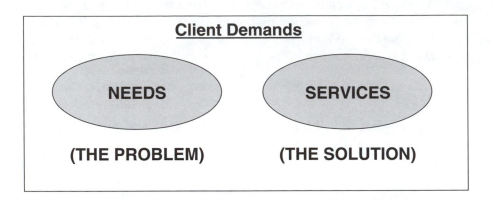

If you look at a service as a product—which is exactly the career potential of many engineering technology students—it is obvious that a service or skill is a more difficult product to market than a shiny product in a box. Marketing analysts have observed that "service marketing" is basically an "intangibility." A service is not easy to display. A service consists of skills, and if you want to sell those skills—let's say, as a software designer or Web page designer—there is more involved than product technology. *You* become the point of sale. This fact is perfectly compatible with conventional sales wisdom, which holds that contact selling (face-to-face sales) has one great asset: the personal touch. **Our service economy— at least on the technological end—is selling *knowledge*.** That is your end product. That is you. Your service potential depends on your ability to inquire after the client's needs and your ability to demonstrate that you have the capability to meet those needs.

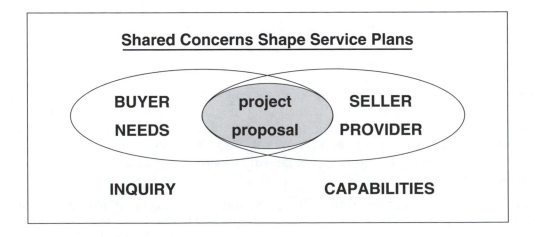

Selling Technology

Obviously there is a complicated issue involved here. I could say that *all* service contractors offer knowledge as a service, but the newer technologies are often complex disciplines related to electronics and computers. As an example of conventional technical services, consider Eric, a structural engineer. He meets regularly with clients. Architectural drawings are the center of his technology, but a contract proposal has to be negotiated with the client first. Eric has had a design and contracting firm in Boise for twenty-five years.

> *Fortunately I bought this storefront. It would cost me plenty now. It is very difficult to offer architectural services without an office setting. Clients call first, but I have to have the right environment where I can greet them. After all, I sell space. Designed space. That's all there is to it. I'm seldom here actually—except to work out proposals with clients. All my work is based on proposals. Joan is here—my secretary—four days a week. I'll come in to see a client or to work on drawings but I'm my own contractor also, so I have to be out with the crew.*

Tim, who designs Web sites, sells cyber space, which is certainly a less tangible market than, for example, an automobile showroom.

> *After I left Safeco Insurance I wanted to tap into this new market. Well, that was five years ago. I'm glad I got in early because I now have completed enough contracts to really demonstrate the product. I can show up at an office with a few disks, or pull up whatever designs my clients are currently using on the Web. To a lot of business people, this technology is a dead bore, and to them I sell the product—just the image on the screen. If they know their stuff I'll talk technology and sell myself instead. Of course the disks contain end products that go with the skills. It is selling for sure. I live my days with computers but I have to sell what I do. If you're not in my business it is hard to explain. Basically, I try to illustrate the services.*

The new economy is a service economy in which the products are technology and knowledge, and this has redefined sales strategy. The client has to be the center of such a complicated interaction.

Sales training priorities are unique for a technical service provider.

Technical Sales Training Involves—

- **client needs assessments**
- **professional service skills**
- **product knowledge**
- **sales techniques**
- **constructive attitudes (we can do/we will do)**

The face-to-face sale is now primarily a discussion of customer needs and expectations, and these interests are often technical and complicated.

Client Needs

To sell today's products and services—and at today's costs—a business person must precisely understand the customer's demands and meet them. In this respect the third factor in selling—the *buyer*—has become the focus of attention for people with technical expertise, and this includes both company employees and the self-employed. Rick, a photographer, explains a typical situation. He cannot ignore the buyer's needs.

> *Last week we received an 8″ × 10″ photo of a set of luggage for American Luggage. It was green. They phoned Hong Kong and changed the order to black. They needed fifty copies of the photo to get the sales reps going. But it was green. Fortunately I hired Jesse last year and bought a huge setup for Photoshop, the software application. He had it ready in no time, in black. I do what they expect me to do. I have to; they are my biggest contract.*

Rick met the client's demands with his technology. Often the demands involve extensive consultation. As I mentioned, Tim sells his Web designing skills, and his sites are the products. In his case the technology calls for extensive discussion.

> *They don't call if they aren't interested, but that isn't bringing me any business as such. I have to meet with them and discuss the whole issue of the .com world out there. Often they don't really have a clue and we have to define their plans together, because they have a vision, but I've got the technology. I have to bring the two together.*

Every successful professional is gratified by a job well done. **In technical services the path to a job well done involves the self-satisfaction of a contractor or service employee, but of equal value is the challenge to be certain you sold a client a product or service that rendered self-satisfaction for the buyer as well.** There should be no regrets for either party.

Essentially, the technical service provider is the center of his or her sale. There is no showroom floor, no sales counter, and perhaps no product to exhibit. In this case, the engineering technician is, as they say in the business, "the point of sale." *You* become the center of what you do regardless of your business setting. Service contracts then close the sales in a complicated manner that points to the uniqueness of technical service industries.

Types of Sales

Our focus in this chapter primarily concerns the self-employed technician—the entrepreneur—who has decided to hang out his or her shingle. There may or may not be an office setting or store frontage. Structural engineers and architectural design services usually have showrooms or formal offices for greeting clients. As commonly, the sales exposure for many technical contractors consists of business cards and Yellow Page ads. Technical contractors in either group depend on their telephones; the usual practice is to depend on customers who call for services or products.

There are a number of selling situations, each of which calls for a distinct approach to the client and the product.

- **There are client calls for proposals.** (customer originated)
- **There are sales calls to solicit business.** (seller originated)
- **There are walk-in calls for products.** (over-the-counter sales)

Each of these three situations is a selling environment in engineering that can involve company employees or self-employed technicians.

Client Calls

Professional contractors are sought after by their established clients and by referrals. Contractors—whether they design and install heating systems, audio systems, computer systems, or any other technical contracted activities—are expected to bid on contract opportunities. These technical contractors are distinct from specialists in other areas of engineering technologies such as communication electronics, digital electronics, or biomedical electronics where hourly costs are more common practices and the technicians are salaried employees.

Sales Calls

The more traditional craft of selling involved the *cold call*—dropping in on a *prospect* in the hopes that a new client could be developed from a previously unestablished customer-seller relationship. In traditional sales literature, the crack professional leader of a sales team often earned his or her stripes at cold-call selling. These professionals often showed their skills by taking over *territory* that other sales agents had covered with far less success. Rick, the photographer mentioned earlier, is one of Seattle's most respected professional photographers. I helped him design and build both his first and second studios.

> This current studio didn't even have that sign on the door until last year. I really don't need one. And, I took the ad out of the Yellow Pages. I have more work than I can handle. Originally, since I was new to the area, I hustled. I bought a $400 briefcase, built a portfolio of my commercial work, and went calling. Mostly I tried to work through friends of friends—which worked, and old high school alumni, since I went to high school here years ago. For example, I met one of my best clients, and also my wife, through an art gallery here that is owned by friends of friends. I don't know about other photographers, but asking around worked for me, but I still had to make cold calls to meet clients.

I assume that my readers are not planning on sales as a career. The self-employed technician will need a clear understanding of cold-call management but not because of a career interest in professional sales. The reason the new business owner needs to think about calling on potential clients is a bottom-line issue for a new business. Businesses seldom come out of the starting gate with more than promise—the promise of profit. Real profit is a different matter. The willpower, planning, foresight, and a bank loan will open the business, but it must be sustained, and that fact calls for strategy. Perhaps a few years later your business will be more than you can handle. Then the clients will come to you. The problem is that the first year or two are high-risk periods. Restaurant closures usually occur in that period because of high investment costs and low public awareness. Only 50% survive. The product did not sell, you might say.

Walk-in Calls

A third prospect involves over-the-counter sales, which I will include here, since you may be going through college while working at a related business such as a computer store. A case in point is Blair, who has been through my courses twice but never has had time to complete his degree.

> I work at a designer audio store. Our clients vary but they are overwhelmingly upper-income people. I guess I would have to say I sell over the counter, but with a salary plus commission I have hardly been able to put myself through college. Don't get me wrong; I make a very good living and have a home and support a family. That's the problem with college. I have no incentive to complete this electronics degree because of my income! I keep at it—sort of—out of interest in the field. Of course we're not talking Radio Shack. I could sell you an amplifier you could trade in for a new car.

The prospect of working in a salesroom is a practical option for engineering technicians. In Blair's case, the company installs five- and six-figure entertainment centers in homes throughout the area. Showroom exposure to learn how to market a technology is valuable. It may or may not lead to a sales position of long-range importance, but the people skills that can be learned while marketing a technology will help build a training background that will interest future employers. It will also be a valuable asset to the self-employed owner of a business.

Technical Sales Management

The Non manipulative Sale

Today's sales strategies are subtle. The strategies certainly make no effort to manipulate the customer. There are no mind games. The give and take of a meeting with a client is intended to help the customer and not to manipulate his or her interests. The challenge is to define. The challenge is to organize. The challenge is to listen and help articulate the needs a client perceives. I will discuss the basic strategies, but in among the suggestions you will find a hundred tips that help the subtle persuasion of the sale and that help the image-building process of projecting you as a professional.

Is there a conflict between a straightforward attempt to give a client a best-deal sale and an effort to sell yourself as a professional at the same time? The answer is no. **If you do not sell yourself as a professional while you help shape a potential customer contract, the effort will be lost.** In that sense you must take every opportunity to help create a positive image. This discussion will look at the obvious—never smell of tobacco (or conspicuous cologne)—and the not so obvious—never talk politics. To seize a tactful opportunity to image build—*and* to show your sincerity and your empathy for the client—is a healthy dimension of selling any product or service, and you do not have to stand in front of a mirror and practice (although sales people are urged to do so).

Image Building

Sales representatives are quite familiar with the concept of image projection. Endless sales meetings, incentive programs, training seminars, and retreats have focused on everything from mirror-image rehearsals, to videotape sessions, to meditation, to wardrobe consultants. Because you are involved in technical services, there are specific considerations that you should use to develop your professional image for clients. All the usual tactics will help: a little rehearsing, a little mind time, a neat appearance. You might even look at John Malloy's well-known book *Dress for Success* (New York: Warner Books, 1993). However, because you are marketing a technology, our client-directed sales approach signals a few specific elements of the professional image about which you need to be keenly aware.

Overall, you need to project expertise. You are in an unusual situation. The typical sales representatives offer no service whatever. They are not experts; they are salespeople. They offer products to sell; they do not engineer them. They may offer services, but other service providers show up to do the job. If you are offering a technical service, you are approaching this event from the other direction. **You *have* the expertise and you *will* sell it if you project it. Building a professional image is the challenge.**

First, recognize that the most common image-building strategy is the most prone to failure. Do *not* overpower your client with your knowledge. There is no stronger rejection you can invite than to parade a client's ignorance by celebrating your knowledge. A few visits to any audio store or computer store will usually turn up this type of sales tactic. The visit becomes a test to simply try to understand the sales pitch. Imagine the uninitiated client hearing this:

> *Yes, the Ethernet is usually a 10BaseT or a 10Base2. Cross-platform mass storage is what you need. Disk storage options could allow us to go to optical drives, DVDS or even SparQ, but let's decide on TCP/IP, Netbios, or RMON to get the protocols settled first. Just sign here.*

Regrettably, this approach is daily fare in both over-the-counter sales and sales meetings. Avoid it, but as I will often say, hope that the competition does a good job of it. For your part, *help* rather than *impress*. Focus on the *result* and not the product. You are offering a service, and it is a solution and not a box of pops and whistles.

Simple people skills and a basic understanding of sales tactics are all you need to be aware of in order to successfully have a meeting with a client that will create a favorable impression.

1) First, allow your reputation to precede you. Use established clients as first-choice service preferences. Use referrals as second-choice preferences. Deal with other prospects as time allows.

2) Second, during the first half of the meeting with your client, favor asking questions rather than providing answers. Do not evade answering questions, but favor asking questions yourself. If your questions are precise, they will reflect expertise. If the questions enlarge your client's self-understanding of needs, or precisely articulate those needs, the client will also recognize your professional skills at bringing the client and the technology together.

3) Admit it when you have no answer to a question. In particular, promise to get back with a response. This tactic shows humility, a willingness to be diligent, and sincere concern for a client. It even gives you a reason for follow-up contacts. By contrast, overconfident competition will, with luck, put a pin in their balloon and your client will see the bluff.

4) Have visual supports prepared. These aids are probably most appropriate in the second half of the meeting when you discuss the options you think will meet the needs of the client.

5) Be precise and technically up-to-date with your proposed services. Ask the client if your statements are clearly understood. Cover three areas in the second half of the meeting: *capabilities, specifications,* and *implementation.* Do not dwell on specifications unless the client understands them.

6) End the hour with a willingness to move on—either at a subsequent meeting or after submitting a proposal.

7) Offer to briefly present your portfolio samples, or perhaps letters from satisfied customers, or both. In the same sense that your history of good work resulted in this meeting, you can leave a "clincher" for the client to review. These documents will reinforce what the client heard about you and what he or she has just experienced. (Be sure the documents are copies and not originals. They could get lost.)

Notice that this is not a sales training routine. *Your image consists of communication management.* Sales tactics literature can be comically complicated. I would no sooner be able to apply the many tricks than I would be able to even remember them. The seven suggestions here make good sense for an encounter between a customer and a service provider. Preparation is easy and the meeting should naturally result in image building without manipulating the event. Since you are there to offer a solution to a problem, the correct procedure for the meeting is hard to avoid. (But hope the competitors gets it wrong.)

Contract Partners

It is important to see technical service sales as a special world in which technology is used to meet needs. This is not a simple "sale." The issue is mutual gain. **The service provider and the client usually have to work together to achieve their ends. They become *partners* to the contract by reaching agreement.** It is a kind of negotiated agreement similar to the negotiated settlements discussed on pp. 365–369, but here the bond that pulls the two partners together is technology. Simpler services—such as the offerings of a painting contractor—need nowhere near the amount of shared discussion that is needed by a technical contractor.

In the case of established clients, although the relationship is a professional one, there is a degree of friendship that will develop between buyer and seller. This relationship is a difficult one in the world of money. Friendship and financial dealings are not usually perceived as shared intentions, but because of the mutual understandings of contract agreements, the client base of a technical service provider is likely to be the exception to the rule at times.

The issues are many, the costs can be steep, and the liabilities are often quite defined. Though sales have not traditionally been seen as anything close to a *partnership,* the new sales concepts are unique. Conventional wisdom looked after the interest of the seller, often at the expense of the buyer, who for very real reasons, at times, could sense his or her role as a victim. Although this text strongly positions sales in a new light, be aware that thousands of companies continue to sell by "quotas," which is very old school. In a quota system the sales rep will not be concerned with clients per se; the rep will be concerned with meeting the numbers. In product marketing, perhaps this practice will never change, but in service marketing the practice can be devastating. You are not selling the smell of a new car. A service installation depends on a relationship. Without an understanding of client needs and provider capabilities, the customer and the seller are not partners to the contract.

Client Needs

If a client is to be viewed as a partner, you need to take a close look at sales meeting activities. The concern of the meeting is the client. Although it may seem apparent that a client will understand his or her needs, this cannot be taken for granted. The specific needs of a client should be the major subject of discussion during a sale. Many professional salespeople see a sales meeting as one-fourth talking and three-fourths listening. There are specific objectives that must be handled with a methodical intention each time a new client is approached.

1. Demands: The Perceived Needs

There will be an immediate need, perhaps even a pressing need, that the client will explain. Discuss the needs in detail; there may be many. Assuming that the desired product or service is technical, use a small notepad to quickly note the precise details of the needs. If you speak with several clients at once, be sure to use the notepad to avoid information overload. Technical specifications are your first priority. Conflicts in client perceptions must be noted if you have several clients present. Keep the notes brief. The notes can be revised later. Maintain the dialog and your eye contact; make your notes quickly so that neither you nor the client is distracted.

During the meeting, remember that the client, as the saying goes, is never wrong. As you listen to the expectations that are expressed, understand that a client is expressing hopes—and possibly fears—that cost money. **If a client is misguided, the challenge is to reconstruct the logic of his or her perceptions.** In real terms, clients can be very wrong, and when it comes to the desired cost on the bottom line of their imagined expenditures, many a sales rep is thinking, "They must be dreaming." The discussion that leads to a possible sales proposal is intended to unite customer and seller.

Win-win agreements—those in which both parties are happy with the result—are the secret to business success, and this is why the customer is never wrong—but might need help. In the process of hearing out the perceived needs, be supportive: "Good," "That sounds right," "Yes, a very common problem." But, when the ideas appear troublesome, steer the conversation with caution:

Well, there might be other cost-effective solutions.

That is one option.

Yes. I see. We could think that out carefully.

You certainly see some of the alternatives.

Return to the troublesome areas later once you have a clear idea of solutions.

2. Existing Conditions

Once you have a measure of the client's needs and expectations, discuss the status quo. How are things set up at the moment? For how long? With what effectiveness?

> *What strengths do you see in the current system?*
>
> *How does your staff like the existing setup?*
>
> *What made you think of the IBM thin-screen technology for this package?*

It might seem logical that the existing system would be the initial subject of discussion, but it is not likely. There is something to fix and so the remedy will be on the client's mind. He or she will want to discuss the *needs* issue promptly, just as a doctor's patient will want to discuss a bellyache first. But the doctor will soon get back to existing conditions and ask how much high-acid coffee the patient drinks. Similarly, you need to see what the practices are for the use of an *existing* situation or for a system you are asked to replace or upgrade. Is it popular? Why? What can it do? What can't it do? How *different* do they see new system needs? Do they want to feel all the comforts of the *old* system? You need to know the history and attitudes with which you will be dealing.

3. Discovery: Additional Needs

Part of your role is consultative. **During the discussion, you must be thinking beyond the client's stated needs and analyzing other needs that the client has not remembered, needs that the client was unaware of, new product support that could be of value, and similar matters.** After the initial phases of analysis are complete, move to the discussion of these additional needs by asking questions that will allow the client to "discover" any additional problem areas. You could bring up each issue and explain your perspective, but if you help the client see needs, the client will more easily be persuaded to think beyond the initial intentions. Since any additional features of a sale are likely to represent an increase in the cost of the sale, there is far less distrust of the suggestion if the client analyzes the problem you identify.

> *When did you last change the ceiling on your automobile insurance coverage?*
>
> *How would you handle a large claim against you?*

In both areas of need—*perceived* and *discovered*—your role is a professional one. You may or may not be trained "to sell," but for technical professionals it is your knowledge that will sell your interests. To the extent to which you can respond to needs with products and services, you will be seen as knowledgeable. Add to this your ability to construct the specific package that is seen as the *unique* need of a specific client; this is a demonstration of your technical skills.

If you convince the client—through discovery—that additional needs should and can be met, you add impressive insights to the business deal and show special interest in the client. There may also be a competitive edge here if you have competitors who did not see the additional needs. **The additional needs you can resolve become the *benefits* of accepting your particular proposal over those of your competitors.** As much as possible, you must personalize the bid or proposal so that it is unlike your competitors' offerings. Close this part of the discussion with a summary of what you understand to be the client's needs. Ask if your are correct.

The strategy of using the "extra benefits approach" to a sale allows you to build the likelihood of your sales success in several ways. By helping the client discover additional needs, desirable or undesirable options, or perhaps oversights, you demonstrate professionalism and build rapport. You also distinguish your bid from the competition's. The serious competitor will be meeting the same needs—if the competition is sharp—so you must demonstrate to the client why your products and services are superior. The client must see the additional benefits of *your* specific products and services. In every way that you can illustrate and define your *unique strengths,* you help create defined distance between yourself and competition in the eyes of the client.

Although it may not be the case in your area of specialization, it is even commonplace in certain sales areas to be prepared to indirectly help the client with his or her financing needs for a project by having brochures and appropriate business cards from loan service agents at banks and credit unions. It goes without saying that a client must be financially qualified to pay for your services, and in some service areas—real estate, for example—the sales agent is very involved in the actual financial arrangements.

4. Product-Service Solutions

During the discussion of client need, there is no reason to introduce and explain your products or services unless you want to briefly answer a question concerning them. If the interview is kept in compartments, there is a clearer understanding of needs, products, and services. **Keep the perception of the sale in the shape of a problem and a solution. The client identifies the problems; the service provider identifies solutions.**

Once the discussion of needs is complete, it is time to offer a solution to the problem. The second half of the sale is the presentation of your response to the client's definition of a problem.

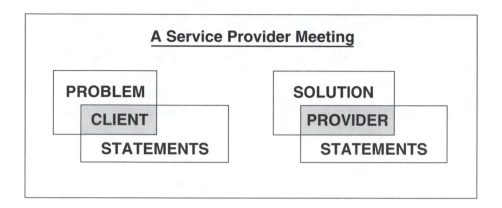

A Service Provider Meeting

PROBLEM
CLIENT
STATEMENTS

SOLUTION
PROVIDER
STATEMENTS

Assume that the competition is good, *almost* as good as you. All things being equal, you must make every effort to show that they are not! You must now match the clients needs: you must solve the problem, and you must be competitive. The discussion shifts to your side of the desk. You must present and discuss the elements of a tentative proposal. Typical concerns are identified here:

> **The Service Proposal**
>
> **Implementation phases to discuss:**
>
> - **technology**
>
> - **installation**
>
> - **training**
>
> - **trial periods**
>
> - **warranted services**
>
> - **costs**

Because the issues are technical and complex, considerably more discussion will continue, and very likely the *client* will now have many questions to ask. This half of the discussion is critical to the success of the proposal you will develop as a response to the meeting. Naturally, you must understand each other in order for the proposal to succeed and demonstrate your effort to resolve the client's needs.

The focus on services can be coupled with traditional sales concerns for products. Key products can be presented as features of your proposal. You can highlight relevant features:

> ## Product Considerations
>
> | Qualities: | Costs: |
> | --- | --- |
> | • capacity | • initial costs |
> | • speed | • maintenance costs |
> | • strength | • use costs/consumption |
> | • memory | • warranty |
> | • size | • product insurance |
> | • durability | • cost advantages |
> | • client dedicated | |
> | • unique features | |

Try to identify specific benefits that your products or services can offer the client. This is the *second* group of benefits you present to the client. (The first group consisted of your special insights into additional needs.) Specific benefits will depend on what your service offers, but there are predictable features or capabilities that are likely points of emphasis. The two big-ticket items are obvious:

- **Saving money with the product or service**
- **Saving time with the product or service**

The particulars will yield a list of other benefits you should adapt to your purposes:

- **Increases in sales or profits**
- **Increases in customer satisfaction**
- **Faster customer service**
- **Faster technology**
- **Decreases in material costs**
- **Decreases in employee time**
- **Better accounting**
- **Better retail/wholesale/ordering services**

Service Timelines

Your sales efforts have been successful if you are invited to submit a bid or proposal. That proposal is going to take time and energy, so *read* the client carefully at this point in the meeting. Are the client's comments energetic or tentative? Does he or she seem to have confidence in you? Is the body language showing energy that is positive? Do you see any sense of withdrawal? Time is money, and you can be generous or cautious with your follow-up efforts, but you must base the effort on your perceptions of the client's sincerity and interest. I have seldom been a victim of people who are just cross-checking bids before picking one they planned on in the first place, but this can happen. It is in the best interest of any customer to see what the market will produce, but for contractors who must submit proposals, it is time consuming to be misdirected.

Ask for timelines. Not only is this focus an easy measure of the client's sincerity, it also is the information you need to close the meeting. You need to match calendars.

> When would you like to see this facility in place?
>
> How do you see your staffing in terms of a calendar?
>
> Will there be a less intrusive month for the installation?
>
> If we could begin June 1, would that be helpful in terms of vacation plans?
>
> Is a three-week start-to-finish workable?
>
> Can we begin at seven?
>
> Can we have access to the building on Saturdays?

Note the answers to your dateline considerations on the notepad. Repeat your understanding of the timelines. Promise a timely response. A you leave thank the client for the opportunity to see him or tell her she will hear from you soon. Avoid any pitch lines about mutual satisfaction or similar sales talk. The job is not over.

The Close

Sales literature divides the parts of a sales presentation into specific areas or phases in much the same manner as our discussion. At this point, however, a professional in engineering technologies must look at the next phase of the sales routine from a unique position. In traditional sales, the craft of salesmanship is seen as a measure of the final act: the close. The close of the sale is a victory that demonstrates craft and skill. Even in the softest of selling strategies, where the customer is in total control of the buying motive and the sales agent simply massages the interest, the sale is the point of it all. However, for anyone in engineering tech fields, the point of sale is often not the sales encounter itself. The sales encounter, if successful, invites a bid or proposal. The sales encounter is only round one for us. Then the homework begins.

There is a unique element involved in technical service sales. Sitting squarely in this approach to sales is the contract agreement between the buyer and the seller: the *proposal*. There is very little sales management literature directed at this corner of the marketing game. There is precious little literature to help you because of the nontraditional "sales stall" in contract transactions that involve *proposals*. Sales agents are trained to keep a sales meeting in motion and to close the sale. Because the practice does not apply to a technical service provider, you must look beyond the traditional approaches to sales. The standard practices *are* valuable, of course. There are tried-and-true skills to selling. However, I will try to take this discussion beyond conventional wisdom so that you also look at the concept of selling a contracted agreement and focus on the client-centered strategies outlined here.

Notice the diagram on the following page. In the middle of this diagram of the sales process, you will find the proposal preparation. Obviously this is a unique situation that we must examine.

**The Service Provider
Sales Process**

THE FIRST STAGE

  • TELEPHONE PRELIMINARIES   **THE SALE**

    THE SECOND STAGE

      • MEETING PREPARATION

      THE THIRD STAGE

        • DISCUSSION MEETINGS

        THE FOURTH STAGE

          • PROPOSAL PREPARATION
           AND PRESENTATION

          THE FIFTH STAGE

            • CONTRACT DISCUSSION

            THE SIXTH STAGE

              • CONTRACT
                APPROVAL   **THE CLOSE**

Model 11.A (1)

NW
NW

NetWorks NorthWest
40031 Bluegate Blvd.
Everett WA 98000
June 6, 200X

Mr. Tim Simmons, President
Tempo Storage Design, Inc.
371B 8th Avenue
Seattle WA 98100

Dear Tim:

Enclosed please find my observations and recommendations for the Tempo Storage Design Inc. LAN upgrade. I have been in touch with Steve concerning his views on the general direction that the upgrade should take, and I would like to formally offer my services for all phases of the installation.

The attached proposal includes floor plans, installation drawings, and cost comparison tables. I would propose the following:

1) Transfer of the hub from the president's office to a new 10′ × 10′ "cabinet. The new room would be located on the east end of the staff lounge area.
2) Installation of a 24-port switch.
3) Installation of a Digital Subscriber Line.
4) Reconfigure the Intranet on IPX of NetBEUI for computers connected to the Internet on TCP/IP.

A cost analysis is provided as an estimate for goods and services.

We discussed tentative timelines that would be convenient. Could we hear from you by the tenth? That would give me enough lead time to meet your calendar.

I will call in a few days to see if you have any questions.

Sincerely,

Jane Q. Cauldwell

The Proposal and the Clientele

The Proposal

The proposal is the second step and represents your technical effort to secure a contract and close the sale. As I explain in *Basic Composition Skills* (the first volume in the *Wordworks* series), a bid or proposal is the expected protocol in many technical fields. A surgeon does not write a proposal for an operation, but the surgeon will expect a technical contractor to write a proposal if the contractor is negotiating to rewire the electrical system for the surgery theater. The proposal should be accompanied by a cover letter (see Sample 11.A) and attachments that will technically and vividly clarify the products that will accompany the service intended. The proposal should state the client's needs, clearly define all intended actions, list all product and service costs, and identify a time frame from start to completion.

Because of the thoroughness of a proposal, the initial discussion with the client is very important. If the needs of the client are misunderstood or are incorrectly assessed, the proposal will lose the second round of the sales effort by technical default. **The reason your interpersonal skills are so important to the sale is because the proposal can be only as accurate as your perception of the client's expectations.** As you will note in the following floorplan, the proposal is defined by many other communication activities. You should be able to outperform a competitor's lower bid *if* you have convinced the client that your proposal reflects a keener understanding of his or her needs.

A cover letter is standard practice in professional environments where estimates and proposals are required.

Model 11.A (2)

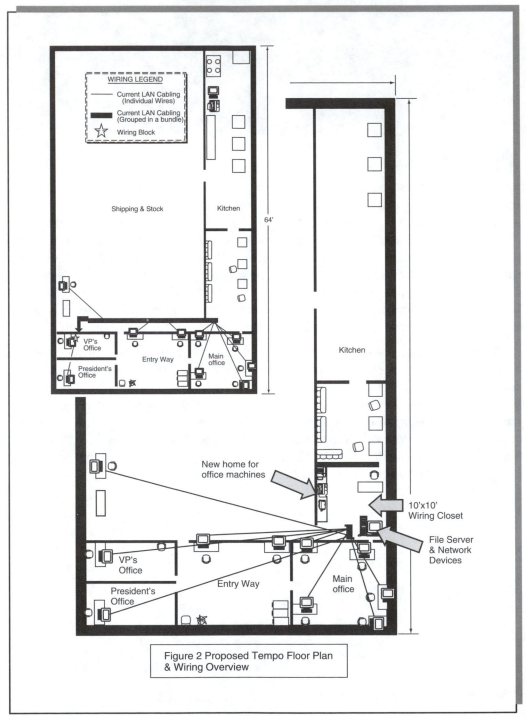

WIRING LEGEND

Current LAN Cabling (Individual Wires)

Current LAN Cabling (Grouped in a bundle)

☆ Wiring Block

Shipping & Stock

Kitchen

64'

VP's Office

President's Office

Entry Way

Main office

Kitchen

New home for office machines

10'x10' Wiring Closet

File Server & Network Devices

VP's Office

President's Office

Entry Way

Main office

Figure 2 Proposed Tempo Floor Plan & Wiring Overview

Be sure that the bid accurately reflects the time frames that were agreed on. Time is usually a very important consideration for clients, and they are likely to put it right up there with costs as a measure they will use to determine who will get the contract. **The timelines *must* be honored, so the time frame is as important to the success of your proposal as it is to the success of your future in business.** Timely technical and engineering contractors are much respected. The corollary is that tardy contracting is very bad business that will weaken your prospects for referrals. In fact, services that do not meet calendar datelines are the butt of gripes by business owners and employees alike, and word can get around. As I will later explain, some authorities warn that one bad egg can be a serious problem; certainly, a disgruntled customer can assert a negative force that is equal to the applause of your best clients. You have seen a car with lemons painted on it. Everyone looks to see what the model is!

To package the proposal, neatly design a touch that reflects professional care. (Observe the cover letter and proposal pages that accompany this section.) Many proposals are submitted in bindings. The bindings hold the proposal together and add a professional image. Spiral binding and similar bindings are available at copy shops, where you will also find plastic covers or card-stock covers and other materials that add a professional touch. You might also staple or glue your business card on the cover. Mail the proposal and the cover letter. It can be hand delivered but the fashionable way to show that your clients' time is valuable is to use FedEx or similar overnight services. In the cover letter, tell the client that you will call in a few days.

If you are encouraged to fax your proposal, great. Do not. Hope that your competitors are gullible enough to do so, because fax copy is awful—at least when you want to look your best. I would not, however, hand deliver the proposal. You need to give the client a chance to take a breath. Look timely but not overly eager. The sleek urban set here in my downtown area use a two-hour bicycle hot run to get just the right image. It has even become a local verb: "I'll Bucky it over." Off goes the Bucky bicycle service, uptown to the thirtieth floor. The bicycle express is not quite as fast as a fax, but the service shows concern, and the attractive proposal arrives intact.

During the proposal writing period, do not hesitate to call the client to request additional specifications. In fact, **at least one call a few days after the first meeting will serve to show your ongoing interest and your intention to promptly provide a bid or proposal.** This call may also slow down any efforts by the client to seek additional bids. You must judge what you consider to be a reasonable time for preparing a bid. For a small business that offers services to small businesses, perhaps a week is a practical timeline. For a civil engineering firm that will install water and drainage for a tract of 200 home sites, a critical path of activities will be needed just to build the proposal, much less perform the job! The task could take months.

Visual concepts can enhance a client's understanding of a proposal and play an important role in clarifying technological matters. Here we see a before-and-after drawing for a network installation.

Model 11.A (3)

LOCAL AREA NETWORK UPGRADE PROPOSAL FOR
TEMPO STORAGE DESIGN INC.

From

NetWorks NorthWest

INTRODUCTION

In this LAN proposal for Tempo Storage Design, there are specific upgrade recommendations based on observations and concerns outlined in our discussions. Tables and graphics show many of the proposed changes. Other tables show cost comparisons between items needed in the upgrade in cases where there are alternatives.

I will begin with the two floor diagrams and cabling schematics to illustrate differences between the current and proposed layouts. I divide my recommendations into "site improvements," and "equipment improvements," with subcategories under each.

For your convenience, network terminology written in *italics* appears in the glossary in the back.

SITE CHANGES

<u>Proposal 1</u> **Build a Wiring Closet**

A) The lack of a centralized area at Tempo for its networking concerns is arguably the single biggest problem in the current LAN setup. As shown in Figure 1, the present wiring setup brings all the cables directly into the president's office, where they are, in turn, connected to the hub. This arrangement is problematic for both the president and the network, since each has a somewhat compromised work environment as a result. The CEO can never really be sure that a network crash (and the resulting activity to fix it) won't come during an important meeting. The results would be frustrating and nonproductive. The network also does not benefit from this arrangement. The hardware and punchblock are located in exposed areas, where they could be bumped and jostled. There is the

By the standards of professional sales people, any business that must go through a proposal offering is operating at a handicap. Sales reps have many tried-and-true slogans and one that serves them well is the observation that timing is everything. Certainly this is true for a salesperson who can "close" some kind of sale or deal. Even long and drawn out procedures such as buying a house will not hold back real estate agents, who will always use the "earnest money" practice as a way to make clients see a two-month process as a decision that should be handled by tomorrow morning. Their traditional "close" is used to encourage the would-be home buyer into a financial commitment and a time commitment.

> *We could hold the house for thirty days if you want to put a thousand dollars down as earnest money.*

Unfortunately, for technical contractors there is no such strategy. You have to write a proposal and await a client's decision. You hope, if you talked out the bugs, that your proposal will be on the money and your competitors' proposals will contain discrepancies.

The Client Base

So far I have explained sales practices in terms of people skills and paper skills. There are other considerations. Since a service provider cannot easily close a sale at a meeting because a subsequent proposal is the final marketing effort, it is important to use every method possible to enhance the perception of the proposal. One important consideration is *predisposition*. There are different types of clients, and they should not be dealt with equally because the profits they can bring you will vary substantially. Three groups are of interest in technical services:

established clients   **(your existing markets)**

referrals   **(your new markets)**

new calls   **(your risk market)**

For many businesses, 80% of the financial activity is generated by 20% of the clientele. Established clients will usually generate the greater sum of profit. Not only will return business seek you out, but you will not have to invest the hours of time required for new clients when you deal with established clients. **The established client is your number one source of income.** It is often said that a customer is a long-range asset. The satisfied customer is your future. In addition, for both you and the client there is a comfort zone of familiarity: you depend on each other and know what to expect from each other. The clients know you will meet their expectations, and you know that efforts will be honored and paid for.

 A proposal should contain a brief introduction, which can outline the basic contract intentions or explain how the proposal is presented.

Model 11.A (4)

further consideration that the office is carpeted, which increases the likelihood that some element of the system will sustain damage due to a static electric shock.

B) Figure 2 illustrates my proposal that a wiring closet (or *MDF*) be created from a 10′ × 10′ section of the kitchen currently used as the lounge. Figure 3* shows the lounge space as it is now. I have heard concern that eliminating 10 feet of the lounge would be a hardship to the employees who eat lunch at the tables and relax on the sofas (since they do not have desks like the office staff). This is a valid concern, but happily, only 1 table of the 7 now there would need to be removed to make space for the wiring closet. If losing that one is not acceptable, other tables could be found that would properly fit the space.

C) The easternmost 10′ of the kitchen lounge area offers the most logical area on the Tempo site to locate an MDF for the following reasons:

- Cabling runs would be well within the 100-meter requirements of standard UTP Cat-5 cable.
- The existing floor is low-conducting linoleum, which is an ideal material on which to house electronic equipment.
- The main electrical panel is only 20′ away, which is convenient for putting in a dedicated breaker just for the MDF and relocated equipment.
- There is one existing doorway. Although there is currently no door, my measurements show that the door separating the kitchen and lounge and the western door into the kitchen (which is not in use) are the same size and can be mounted as the MDF's door. These doors are solid core and can also be keyed for locks.
- The ceiling is over 20′ high, allowing more than enough space for upper shelves to be built to hold networking and computer paraphernalia.
- Conversion to an MDF space requires few materials (*see Figure 7 for a table on costs*) and a modest investment of time and tools to frame in one wall, build shelves, and cover the plate glass window with plywood. The doorway mentioned above has mortised cutouts already in place for hinges.
- The area is large enough to house a standing distribution rack (*see Figure 8*).

** A number of drawings and tables have been omitted from the **Wordworks** sample.*

Established clients are also the basis of your growth. Advertising may or may not be critical for your service. You may be working for a company that spends many millions of dollars on product advertising. You may be self-employed and depend on business cards and a Yellow Page ad the size of a business card. In either case, **the underlying concept of person-to-person marketing depends on your direct relationship with clients who trust your efforts and judgments and who will recommend you to other possible customers.** There is a pyramid process here that is very good for business. If each of five major clients recommends you to one other prospect, you have the potential to double the base of established clients. Assume that you gain only two of the five referrals as new business. If, during the following year, the seven established clients send you referrals, you will again enlarge the base of established clients. And the number of potential referrals grows also.

The importance of these two groups—clients and referrals—is obvious to any self-employed business person. Employees depend on steady income. Self-employed people strive to maintain steady income in a world of wave-and-trough movements. The challenge is to balance a business and stabilize the profit flow. Feast or famine does not spell success. An established client base is very helpful. Returning customers greatly increase the stability of a business. Even with a dependable clientele, the business person must always be wary. For example, December is a soft month for many product and service sales sectors. All the potential clients are busy for the holidays. Winter can be a slump in many contracting services because April 15 distracts many clients who are concerned with their tax deadlines, and perhaps even the weather.

Now, if your proposal is sitting on the desk of any of the three customer groups, what can you expect? First, you have "sold" the established client who depends on you. In the case of the referrals you have a competitive edge because they have been told you are professionally and technically competent. The *new call* is your highest risk. If you received a call for a proposal and the call was based on, for example, your Yellow Page ad, then you do not have nearly the image-building leverage that you have with existing clients and referrals. In this case you must depend on the people skills and the sales strategies you put to work during meetings and phone calls, and you must depend on the proposal to be a winner.

Precise explanations of contracted activities are important. Detailed proposals protect the mutual interests of the client and the service provider.

Model 11.A (5)

Mr. Tim Simmons

June 6, 200X

Page 3

D) Creation of an MDF eliminates both the noise and space problems of the main office. Front-office workers have explained that they feel crowded by all the equipment, and drowned out by equipment noise when doing business by phone in the room. Removing these machines to a newly created MDF, as seen in Figures 4–6, solves both of these problems.

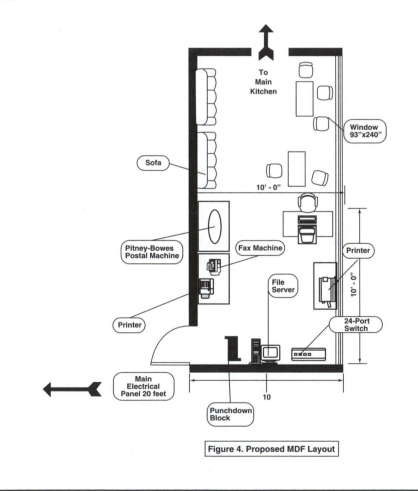

Figure 4. Proposed MDF Layout

Territory Management

When traditional sales agents refer to *territory* they might mean half of a city or four entire states. For a technical service contractor, the practical approach to the diameter of your service area is a personal issue. Usually, the desired distance will be little more than an hour's drive, since, even at one hour each direction, two hours of a contractor's time are lost in this kind of a commute. If you work an eight-hour day, two hours is equal to 25% of your contract time. Whether you work a longer day or include the hours in your eight-hour day, you still will lose ten hours a week. It is for this reason that sales management literature explores time management with care. Sales seminar leaders will analyze sales preparation time, commute time, lobby waiting, vendor contract time, paperwork, administration, and all the unnoticed ways that cost time—and money. Then there is the issue of real costs, from businesses cards to lube and oil. Of course, calls from established clients allow a contractor to avoid the wear-and-tear realities of conventional sales, but if any selling is to be done, time and money are considerations of concern.

My particular suggestion for new businesses in technical services concerns the use of the *market niche*. **To develop a market niche you must try to *specialize* in your technical service.** In other words, if you design small office telephone systems for small businesses, you may be able to develop a market niche by focusing on real estate offices, for example. If you can demonstrate that your business is knowing *their* business, real estate agents might select your bid instead of the bid of, let's say, a local phone utility that has no particular interest in special design considerations.

If you have established clientele and a referral base, you may not need to seek out new trade, but if you are just getting off the ground, consider looking at specific markets rather than broad markets. In the same sense that a sales rep marks off territory by geography, you can mark off a territory—but preferable one close to home. In my area there is a well-known art supply outlet. It is now one of the nation's premium suppliers of fine arts products. The owner started the business from scratch by selling art supplies to local schools. The schools were a niche market, and he cultivated it. Schools also happen to represent the single largest market demand for art supplies. Certainly, the company has been a huge success as a result of the original market target.

Graphics should be selected on the basis of clarity. This floor plan is nontechnical and was designed to create a visual image of a finished product in very practical terms.

Model 11.A (6)

Proposal *2* **Install a 24-port switch**

The choice of a switch over a hub is a sound one because switches can do more than strengthen a signal. They can also make low-level delivery choices, which make the network faster and more reliable. Figure 9 below is a table comparing the relative merits of different switches.

MARKET COMPARISON OF COMPARABLE 24-PORT SWITCHES

Vendor	Description	Ports	Model#	Approx. Price	Warranty
3Com	Superstack	24	3C16464A	$1,100.00	?
Cisco	Catalyst	24	2908XL	$1,720.00	1 year
Compaq	Netelligent	24	5708TX	$1,580	3 years
3Com	Superstack3300	24	3C16980	$1,800	?
Cisco	2924M XL	24	WS-C2924	$2,200*	?

Figure 9. A comparision of 24-point switches matching expressed TEMPO LAN needs.

Proposal *3* **Install a DSL (Digital Subscriber Line)**

Tempo has reached the maximum number of phone lines allowed into the building without incurring an additional $10,000 installation charge from the phone company. This presently makes the feasibility of Internet hookup for more than one phone line (as is the case now) impossible. Specific prices for a commercial *DSL* hookup will vary, but the ballpark figure I was quoted was somewhere around $250.00 per month. DSL service would resolve the current telephone limitations and Tempo could have the fast Internet access it needs without the tremendous investment in a major phone line expansion.

All prices effective as of this writing.

Ben finished a network technology degree several years ago.

> *Totally unrelated to the issue of computers, I've worked in three law firms over the years. When I decided to go into network technology I had to decide what to do with it; I mean you can't believe how big the market is. Junkyards are going to want them. Anyway, I only knew one business and that was law, so I went to law offices and the business is coming along nicely.*

Gloria is an interior designer. She does not see herself as any kind of engineer, but the story is the same.

> *The dog was sick. Kent was talking while he examined Chipper and I told him what I did. Well, in those days I guess I would say I told him what I would like to do. He was my first design contract—a vet clinic! Not your usual market for interior designers. Kennels. But after I did three clinics, things started to happen for a very specific reason: veterinarians talk to a lot of people! They were great references and I still do clinics as part of my business.*

Industry estimates rank sales-call costs at between $100 and $500 per call. These estimates are calculated as company costs where a sales force is at work. If you are self-employed it is heartening to know that you certainly are not costing yourself this kind of money. On the other hand, your time is valuable and you need a maximum return on cold calls, particularly since this type of selling is not for the timid. Because selling is only secondary to your technical interests, avoid trying to carve a wide path through a number of industries that could use your product or service. If at all possible, pick an area of familiarity and pursue it. Look up the specific business group in the Yellow Pages or in business indexes. It is old sales lore to *not* start with A. The A's take a beating; start with F or N, or try S.

Your credibility will be greatly enhanced if you know the client's business and you are then in a better position to help him or her get to know yours. You do not want to waste time or money. Joe Girard, author of *How to Sell Anything to Anybody,* fondly recalls going to football games with a brown bag full of business cards. During the hoopla of every touchdown, out would go a big handful of his cards to float down among the happy throng. Girard could afford to throw cards in the wind and could even afford to use dollar bills, except that his name and picture were not on them. He understood his market and was famous for it, and you must understand yours.

This proposal adopted a unique strategy by providing a host of cost options that were presented in tables. The strategy provided the client with various options, and the merits of the products were discussed in subsequent meetings.

Model 11.A (7)

<u>Proposal *4*</u> **Run intranet (connections within company) on IPX or NetBEUI and machines connected to the Internet on TCP/IP**

IPX, NetBEUI, and TCP/IP are Internet protocols. IPX and NetBEUI are protocols that run between machines in a limited geographic area, whereas TCP/IP is the dominant protocol of the Internet. It is a much more secure network that runs one protocol within its "walls" and another protocol outside the company environment. A hacker cannot, therefore, come into the network through the TCP/IP.

Also, if you are just starting out, consider joining business and professional associations. One avenue involves popular organizations such as the Kiwanis or Rotary (which are not only for men, by the way). Another consideration involves your professional specialty. There are hundreds of *societies* that serve the interests of hundreds of engineering and technical specialties, and many of them meet regularly in large metropolitan areas. I am regularly asked how a writer can break into technical writing. The first thing I do is hand the interested person a membership form for the Society of Technical Communication, a national association of technical writers. The local branch is quite active and has frequent meetings, dinners, banquets, and other activities. Such specialized gatherings are the perfect place to learn business and do business. This fact is one of the reasons everyone attends. For contacts in your specialty, consider joining a professional society. There is no pressure to get involved. For more information see the *Writer's Handbook,* Chapters 9 and 10.

Competition

This entire chapter can be interpreted as being a *soft* selling strategy. I now want to temper that perception by taking a forceful stand on the issue of competition. You are making every effort to respect the client's needs and you must make every effort to match his or her expectations. Of course, there is a profit motive that is part of the goodwill, and that motive will help you be wary. Your product or your service must outsell the competition. The competition is equally determined to outsell you. In this dimension of market competition the game is clearly adversarial.

Fortunately we are not gladiators. You will never meet your competitors head on. To the contrary, you must anticipate the strategies—costs, for example—that competitors will be using. Obviously you will try to be cost competitive. Then again, maybe not. If, for example, your discussions of client needs revealed $10,000 more in desirable features of a service installation that had been left unspoken in discussions with other bidders, you are in an excellent position to outbid the competition and come in substantially over the competition in gross estimates—and you could win the contract. How? Because of superior image building. The client will respect your efforts to help design the desired system in as thorough and perfect a manner possible. Yes, it is $10,000 over the top. The client still might take your bid over the others—even if he or she backs away from the desirable extras. Why? Because you showed more professionalism at defining need and more technological skill in perceiving the need and providing the service.

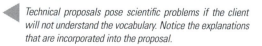

Technical proposals pose scientific problems if the client will not understand the vocabulary. Notice the explanations that are incorporated into the proposal.

Model 11.A (8)

Cost List for Tempo, Inc. Network Upgrade

Materials

Item	Vendor	Price	Total
Main switch (1)	First Source	$XXXX	$XXXX
Catalyst 5509			
Secondary Switch (1)	All Net Direct	$XXXX	$XXXX
Catalyst 500 Fast			
Ethernet 100BFX 12 Port			
Backbone module			
Hubs for office areas (3)	PC Connection	$XXXX	$XXXX
Router (1)			
Cisco 2600 Enet/TR router	Cisco Systems	$XXXX	$XXXX
Cabling - Copper-UTP			
CAT-5 (1000')	PC Connection	$XXX	$XXX
Cabling-Fiber			
(200')	PC Connection	$XXXX	$XXXX
(Includes termination)			
Subtotal for materials			$XXXXX
Installation costs			
XX Hours		$XXX	$XXXX
		(Hourly Rate)	
Subtotal for installation			$XXXX
GRAND TOTAL FOR UPGRADE			$XXXXX

As a service provider you must approach the competition in the spirit of competition. Outsmart them, not your client. All the old-school, sell-the-product types will be immediately left behind because the *service* is the real product. The commodities (computers, for example) are secondary to the event. You sell the service by selling yourself—as a network systems designer, for example. You sell yourself by serving the client. Untrained competition will not understand these subtle workings of the selling process. Too few technical service providers learn to sell, even though the small-business owner of a technical service should realize that the contact point for profit is the sales effort. Who is then left among the competitors? The serious contenders remain, but you will be among them.

Contenders are your serious competitors. Repeatedly, you may see their cars pulling out of the lot or arriving after you. In several trades I have watched them stop to talk to each other because the selling markets are so highly defined that they are bound to see each other. If you sell and install X-ray facilities, the market is highly specific. You and the other local rep will both be standing on the street corners opposite a new medical clinic that is under construction. Expert contenders have to be very smart. They will know their business. They will try to underprice your proposal. They will know much of what you are reading in this chapter and the next chapter. What they will not know is *all* of what you read here. Nor will they know your own discoveries for sales success; nor will they know how you sell your image or what exactly you do with a client. These factors are keys to winning over competition. There is no win-win relationship between competitors.

One of the reasons I place considerable emphasis on a base of established clients is not simply a steady-state issue of economics. Of course, it is ideal to have clients beckoning on a regular basis, but there is more to their utility. **Established clients are past victories in you efforts to serve customers for a mutual gain. They are also much less vulnerable to the onslaughts of competitors.** I have had many clients who admit they do not even try to get other bids. In fact, I have clients who do not even ask me for a figure anymore. "Just do the job" is the attitude. Established clients represent a tactical victory over competition (although sales strategy books always warn readers not to rest on those laurels).

Referrals, though not as solid a foundation as the established clients, are your second tactical strategy. They emerge from the established client base. Referrals are a preferred sales target because they come to you with a *recommendation*. The sale began without your meeting the referrals! Your reputation precedes you. Your reputation is a combination of your technical skills and your people skills for handling clients. As you can see, it all comes together. Selling is a very large and dynamic process to understand, but without these practices and perceptions, the service provider, particularly the small-business service provider, is going to risk the business investment. The service must be *sold* to yield a profit.

 The cost analysis is usually a separate part of a proposal. It is itemized with care and should thoroughly reflect all costs. It can, of course, be negotiated in subsequent meetings with a client.

As a final note, avoid comments that criticize your competition. If a client offers criticism of other providers, you can show interest—"Oh really?"—but do not tag along and involve yourself in scuttlebutt. In my contracting experience, I seldom knew the competition in any case because there were too many. If you provide a service that is very limited in the service area, you may get to know the other technical service business people. You do not serve your purpose or theirs by commenting on their skills in a negative manner. You might, however, compliment them: "They are very good. I just hope I can beat their price. We are smaller and usually have a faster response time."

Summary

- The traditional "hard sell" was focused on the value of a product and the closing of the sale.

- The current thinking for technical services is that selling should focus on client needs and the sale should close only if the technical service is appropriate to those needs.

- There are three distinct types of sales situations:

  ✓ client calls for proposals, which are customer originated

  ✓ sales calls to solicit business, which are seller originated

  ✓ walk-in calls, or over-the-counter sales.

- Image is important and will develop as an outcome of a sales encounter. Be prepared, and conduct the sales meeting in two halves: the discussion of needs and the discussion of the options you can offer.

- Handling questions with skill is one key element of success.

- During the first half of the meeting, help the client articulate the needs that prompted the invitation for your visit.

- Discuss the existing conditions in terms of strengths, utility, staff perceptions, and other matters that will give you an overview. Do not assume that a need means that some process or structure or system is unpopular or ineffective.

- Identify additional needs if you hear the client speak of them or if you perceive such needs.

- In the second half of the meeting, offer the appropriate product-service solutions.

- Be very supportive and enthusiastic rather than technically overpowering. Keep the technical vocabulary within the range of comprehension for the client.

- As you begin to define solutions, establish service timelines as well. What can be done when?

- For your own purposes, ask for contract-decision time frames also. When will the client decide on the bid?

- The "close" on a contract proposal is not a sale; it is an invitation to submit a proposal. The final decision rests with the client after receipt of the proposal.

- Prepare the proposal in a thorough and knowledgeable fashion and present it in a professional looking manner.

- Established clients are the most important element of the clientele.

- Referrals are the best time investment for new business.

- Competition must be considered at all times. Established clients are the least vulnerable to enticements from your competition.

- Develop a market niche. It will allow you to focus your business, which can be a strong tool in maintaining a solid image by defining unique expertise. The market niche concept also helps eliminate nonspecific competitors.

Activities Chapter 11

Present a memo to your instructor that develops a discussion of a personal experience that relates to the chapter. The memo should be 500 to 1000 words, typed. In it you can explain how you have related the text or the lecture-discussion to some event you recall. This exercise will give you a better understanding of the material because you will explain incidents in terms of your perceptions of workplace structure and communication.

Select one of the following suggestions and develop an analysis.

- <u>Commissions</u> *If you have had any experience with commission structures, explain the system and your experience in the sales environment. Were you successful? Were other agents successful? Examine the sales strategies and explain the situation.*

- <u>Sales Training</u> *Have you had sales training? Explain the learning procedures. What were the company expectations? How did you perform? What was the attitude of other employees regarding the sales training and the actual day-to-day work involved?*

- <u>Customer Satisfaction</u> *Compare at least two encounters you have had with sales agents. These can be positive experiences, negative experiences, or preferably, one pleasant encounter and one unpleasant encounter. Define and explore the concept of customer satisfaction and/or customer dissatisfaction. Explain your encounter with superior or inferior service and recount the details. Explain your reaction, as the customer, to the considerations you identify.*

- *If you have appropriate technical work experience, can you start to construct a portfolio for a sales presentation? Include thank-you letters, sample bids or estimates, a resume, licenses, and certifications. Develop a package.*

- *Develop an analysis of your anticipated sales territory and explain the logic behind it. Develop a strategy for opening your business and discuss your advertising and marketing plans.*

Sales Strategy

Work in Progress

12. Sales Methods

Writing a bid proposal for a six-figure contract like the one for Tempo is a lot like writing those papers I did in college. The topics Randy, Cloe, and I had discussed somewhat casually at our coffee shop meeting were now fully written out and annotated. I included printed forms of the Visio diagrams I had used in the presentation and had overlays made to show how things would change with our proposal. I almost didn't include the overlays thinking they might be too slick, but I put them in at the last minute with the hope that they might win us a positive vote from Sharon. I had fully annotated all expenses and equipment in the back appendices and also included a possible work schedule (using the holiday period Cloe had remembered). I had Kinko's bind the bid and add a heavy cover made from my original title page. The finished product was impressive, and to add the final touch, I sent it to Simmons via FedEx.

The following week, I got a call from Tim. He was friendly but a bit distant. He had gotten the proposal and found everything to be in order, yet there was an awkward pause and a hesitation to commit one way or the other. In my experience, this usually signals some concern over money or financing, so to nip a premature rejection of our bid, I suggested a meeting the next afternoon. Tim agreed and we set a time.

The following day I met with him and he started the meeting with sincere compliments about the professionalism in style and content of our bid. He continued by saying that the one sticking point was the creation of the wiring closet area. Apparently, his nephew, Steve (remember him, the computer buff, from Chapter 6?) had studied our proposal and flatly vetoed the wiring closet. Tim shrugged apologetically and let me know that with that exception, the bid seemed just fine, but he didn't see any way around the veto. I sensed that my original assumption that the objection was strictly monetary was wrong. This was about company politics!

I have to admit, it was hard not to want to throttle the nephew after all the work NetWorks NorthWest had done on the bid, but instead I asked if Steve had any alternative suggestions. Tim said he did. Steve thought the equipment should go in Tim's office, where the current hub and punchdown block were, and Tim agreed. Privately, I imagined what a mess it would be to have a network center in the president's office, where business meetings and network needs couldn't help but get in each other's way.

I gave Tim an abbreviated view of my concerns, but I suggested that I work up another bid that would not include the wiring closet and that would use his office as the network center. I let him know clearly that I did not think it would be a good idea, but I was willing to amend my design to fit these new parameters. He looked visibly relieved at my suggestion, and I promised to have the second bid to him by the end of the week.

J.Q.C

Meeting Manage-ment

First Impressions

Opinions vary, but authorities suggest that you have little more than several minutes to create your initial impression during a business-related encounter, and at least one observer cut that time down to twenty seconds. Because you cannot accomplish more than a handshake in twenty seconds or a minute, a first impression is a visual impression. The impression you make is one that is totally in your control: *seeing is believing.* What you *appear* to be is what you *will* be to the client. It is important to construct an appropriate image that represents your specialization in a professional manner. The image initiates the client relationship and is an important first step in the development of trust.

Begin with larger environments. If you show up at a client's place of business, you might assume that you are in good shape if your shoes are shined. Suppose, however that your car is in the shop and you drove an old beater that you keep for such emergencies. It is not washed; your hubcaps are long gone; the back seat is filled with empty paint cans you forgot to take to the disposal site. As luck would have it, you pull in just as your client pulls in. He sees you and greets you and sees the car. Not cool. You just lost the challenge of the first impression.

If there is any chance that a client will see various environments such as your office or your car, or your van, be sure that these spaces look the role of someone who is a caring professional. You do not need a new car; you need a clean and orderly car. You do not need an oiled walnut desk, but offices can make powerful impressions if a client visits you. If you are self-employed, perhaps as the owner of a structural engineering and design firm, you have absolute control over the office image, and you can construct a clean, orderly, and professional space. Even if you are an employee with an 8′ × 8′ cubicle in an open office situation that holds fifty such cubicles, you still have control over your 64 square feet. It should be clean and orderly, and a small space can still signal professionalism with certificates and awards appropriately positioned on the wall.

It goes without saying that your physical appearance will matter. This is not always a suit-and-tie image. I mentioned a design firm before. If an architect is going to welcome a client, he may have to hunt to find a tie in the office—and wear it with jeans. When I visited clients to bid a contract, I wore flannel shirts to create a specific image that fits the mold for one particular group of general contractors. An insurance sales rep, male or female, will wear a conservative suit. Real estate agents will cover the full spectrum of possible images and drive the cars that match the image—from a mouse-gray Ford to a full-featured Mercedes sedan, perhaps with a wet bar. Why? Because it is an old idea in sales that your image should match the image of your clients. You want to appear to be their kind; the kinship builds trust.

Match the image that you think is appropriate. The image must be professional, but you can adapt it to the client also, particularly in the case of existing clients who you sense will notice subtle image-cutting details. Your briefcase or any items you carry will be part of that image. At the conservative end of the scale you will often see sales professionals dressed in dark clothing, usually grays. Women will often wear black, although the blouse is usually white. The briefcase tends to match the image.

If you study your presentations you may find that briefcases are a necessary feature when you call on a client. If you watch another sales representative in action, or see a video clip of a sales effort, notice the gyrations that are part of the game of handling briefcases. As you bend over to fetch one thing or another, your voice trails off in a muffle. By the third time your shirt or blouse is starting to pull out of the back of your waistband. Each time, the client sees you disappear in front of the desk and then bob back up. Many sales people use a briefcase that sits upright on the floor like an accordion folder. The compartments are organized front to back by the phase of the sale. You reach down to knee level and never need to bend over or root around. The upright briefcase is probably the better product for your work.

Office Rapport

Another impression that you make is the result of the manner in which you deal with staff. Sales management books explain the point succinctly: a salesperson is *always* selling. The point is not that a pitchman never shuts up. In fact, it is important to avoid such a cliché of an image, since one secret to sales success is never to be seen as a seller. However, in one sense a salesperson must be selling goodwill at all times.

It is important, for example, to get to know every secretary of every client you have. If you deal with small businesses—medical offices for example —there will be at least one secretary. If the setting is larger, there may be a receptionist and perhaps another secretary or a bookkeeper. Corporate environments may be even more demanding. Sales reps do not always have positive feelings about secretarial staff because the client's personal secretary is a gatekeeper and, as such, is an obstacle to doing business. Phone calls can be blocked, drop-in calls can be blocked, mail can be censored. For these reasons alone, selling never stops. **The client's secretary becomes an important aspect of a sales call. *Everyone* is part of the sale; everyone is part of the support you seek.** The perceptions that staff members have of vendors, sales agents, or service providers are very public stuff; the minute you are gone there may be a comment: "Nice woman. Always friendly." Or perhaps: "Pushy sort of guy." These images obviously help or hinder.

The well-known issue of remembering clients by name is made more difficult in this case. You would not show up for a call if you did not know the client's name, but do you remember the name of her secretary? How about the receptionist? Professional sales reps will keep files of this sort of information. They reason that friendliness is important to building trust and maintaining clients.

> Hello Sandy. I haven't seen you in awhile. How is that boy of yours doing at Howard School?

As an opener, this rep is referring to her notes from her visit a year ago. At that time the secretary, Sandy, was deciding if she could afford a private school for her son.

Managing the Sales Inquiry

Regarding sales management, we need to finely tune a few important features of the discussion in the last chapter. Additional observations will be helpful.

Probing

The easiest way to help a client discuss needs is to ask questions. Because you are in the process of developing a potential sale, the questions must be managed with skill. Your client is aware that his or her needs also represent a type of vulnerability to the buyer. Always in the environment of a sale there is the potential for distrust. The customer will be thinking,

Is it overpriced?

Is the product a good one?

Will the service be properly performed?

Since your questions can answer these unspoken fears, they will serve to develop trust, but *only* if they are managed with skill, because questions *ask* for something. Questions are inherently aggressive because they demand response. Since our *theory* of service sales is based on the idea that you respond to the demand of a client, the actual *practice* of compelling the client to respond to questions can reverse the sale. If a client is in any way intimidated, the sale has been reversed. The client must be in control of his or her needs, and your goal is to understand those needs. Nonetheless, since you are in the professional capacity to help clients articulate expectations, you must ask questions—perhaps plenty of them.

Open Questions

Focus on questions that use the keywords *what, where, when,* and *how.* These words will ask for information without cornering the client. One particular open question—*Why?*—should be avoided. This question can easily put a client on the defensive. The basic open questions—*what? where? when? how?*—will probably lead to an explanation of why a situation is as it is without your involvement in an overly direct inquiry.

When did your staff first experience a problem with the system?

What will be the final number of employees you intend to train for a new system?

How did you attempt to adapt the system to your growth?

What does your staff need?

How do you feel about the new state requirements?

What do you think?

Closed Questions

You can ask closed questions but they are far less exploratory than open-ended questions, and they can blunt the conversation. Closed questions begin with such words as *can, would, do, are, will,* and *is.* Because the questions can be answered with a *yes* or *no,* they do not invite discussion without other questions. In addition, if someone is asked a yes-or-no question, the person is being asked for a decision or a judgment, no matter how minor it may be. Again, there is the risk of defensiveness. **It is better to ask questions by which the client *explains* rather than *decides.*** The client's confidence in you is all-important. As long as the client is comfortable with the conversation, you are winning the sale, at least in terms of image building.

Support

At all times maintain an upbeat and enthusiastic attitude. Your client's problems are your problems; you need to understand them. Your problems, however, are not your client's problems and must never enter the picture even with long-established clients. During the initial minutes of a sales effort, you may want to use a few personal touches, but the photographs in your wallet will not be necessary. If you portray some aspect of yourself—perhaps you own a sailboat—you will color a picture that has the ability to show you in the proper light. Say nothing negative; do not talk politics. Be agreeable and energetic but make no effort to be witty—and *never* tell a joke. Jokes have the ability to offend, and people who tell jokes will bore clients who do not tell jokes.

As you realize from earlier discussions, a sincere smile is very supportive. In the intensity of a technical discussion it is easy to frown while you are in intense communication. Remember to use the smile to acknowledge good ideas and to keep up the goodwill. Also use the traditional supportive gestures of nods, comments such as "That's right," and rewarding remarks when the client significantly contributes to the development of the discussion: "Fine idea!"

After the first few minutes, focus your support on the client's responses. The supportive devices we discussed at several points in this text will serve you well.

- *Body language* Listen with care and do not interrupt. Maintain ongoing communication while you are listening. Use your eyes, your face, gestures, or similar signals of support. Show interest. Show attention. Show poise.

- *Paraphrase* Restate key comments that you hear.

- *Repeat* Precisely reiterate technical demands that a client expects of a contract with you.

A sales meeting has a tendency to put a service provider on edge. In traditional sales relations, the *client* usually feels a little uneasy, and the challenge for the salesperson is to put the customer at ease in order to sell the product. The strategy is undoubtedly helpful and you, too, want to keep your client at ease. However, because you are in the client's territory, you are the party that is likely to be a little nervous. You will also feel that you have to produce, both a sales image and a superior product or service. *There is no hurry.* The rush is simply an anxiety that all but the most professional sales professionals will experience. Pace the discussion. Be self-conscious of rushing or talking too much. *If you use your questions to examine client needs, the meeting will slow down as soon as dialog begins.*

In addition, allow both yourself and the client time to think. The only way to achieve thinking time is to stop the discussion. Moments of silence are valuable. After a key question, silence is a good sign. Perhaps the client is thinking out his or her needs; perhaps you are thinking out your response to a request for an alternative to an idea you had.

These should be useful moments and not awkward moments. If there is any sense of awkwardness, busy yourself with something such as taking notes while you await a response. You do not want the moment to be a pressure point.

Dining the Client

The restaurant routine is not an important consideration as a general rule. You might invite established clients to lunch to keep the contacts fresh, but the expense is otherwise not usually an expectation. In addition, for technical presentations you need space. You may have visuals, brochures, or larger layout drawings. You need room for a briefcase, and you will need your notepad and possibly a calculator. An office is a better setting—or perhaps a meeting room is even better at times. Besides, the polite task of assuming the cost of the lunch is yours. You will save money by avoiding the issue.

If you do lunch with a client, avoid alcohol. Some shrewd sales representatives offer their clients alcohol. Alcohol can dull the client's perceptions, and by today's selling standards that is not part of the game plan. Coffee? Surely you can drink coffee. Yes, in the restaurant but avoid it in the client's office. In the office, thank the client, but explain that you just had a cup. Why? Murphy's Law. Three weeks ago a woman came to see me about a course. She sat in front of my desk, placed her coffee on the coffee table and began to put down her books and pack. Over went the coffee, all over the books on the coffee table, all over the floor, all over Mary. There went that discussion.

Basically, technical services are probably dull business in the scheme of things, at least in terms of sales strategy. You will not have giveaways: balloons for the kids, key rings, pens, and office paraphernalia with your name on them. You will not have local newspaper ad copy, much less radio and television spots. Contact marketing is a people skills issue and depends on only one person: you. Even if you are a company representative with an expense account and a company car and reservations at the Hilton dining room, the basic reality is not any different. Although your backdrops may be rather elegant at the corporate level of one-to-one selling, the basic contact marketing reality depends on the contact between you and your client.

Rehearsals

Although professional sales people will attend seminars where simulations are part of the action, these people are probably the least likely to need a dress rehearsal. It would be the inexperienced person who would need to practice a presentation. In my experience, simulating a sales encounter is particularly difficult. It reminds me of fishing in a stocked pond where you can rent folding chairs and buy odorless bait. Working in front of a mirror is probably not much better. However, if rehearsals seem ineffective, preparation is a different matter. Preparation is a must. Reading the chapters you are reading will help prepare you for a sales encounter, but you need to sketch out a few pages of specific intentions before a meeting.

The initial phone call is the most important dimension of preparation. Without it you cannot anticipate the event, but with it you can gather enough information to shape your strategy. Frankly, preparation *is* a rehearsal; it is not staged but at least you have the script. You cannot memorize the script in any case because you do not know precisely what directions the meeting will take. **However, with preparation you can help guide the meeting. What matters is your initial effort to approximate the probable situation that is about to occur.**

You can, for example, design a list of questions that you can use. I would suggest that you develop two lists. One list will consist of whatever standard questions you have found useful in your previous meetings with clients. The other list will consist of exploratory questions to examine the current situation of this particular client. In addition, if you have a sense of what the needs will be that you know you can meet, you can begin to write out product options and service alternatives to discuss. You might even have an idea of *exactly* what the solution is, but do not present such a solution abruptly. You do not know for certain that your ideas are ideal, and you do not want to be overconfident. You will not know any details of the client's thinking until there is considerable discussion. Nonetheless, the point of the exercise is to *prepare* ideas and products and services that might get the job done.

There is no reason not to take your questions with you as a resource. The same is true with a list of possible options, although you should not title the options with the word "proposal" because you are not ready to draft a proposal. Use the preparation to help you during the meeting. Have the material positioned in your notebook. Do not follow the list rigorously unless there is a technical reason to do so. Follow the natural course of the dialog with the natural pattern I identified earlier: problem to solution. Discuss subject areas in order, but allow the questions to develop by circumstance.

First half of meeting: Client inquiry

- **Needs**

- **Present condition**

- **Additional concerns**

Second half of meeting: Service offering

- **Capabilities**

- **Products**

- **Services**

Consult your list of questions as needed and briefly survey the entire list before you move on to an offering in response to the problem. As a part of your response, consult the alternatives you anticipated and discuss them. The meeting will doubtlessly have generated additional needs you must respond to at this time. You might need research time, which is a realistic consideration at all times. You can see that preparation will help you manage both halves of the meeting. In addition, anticipation has additional utility. **You need to determine what sales literature is of value and take brochures, product descriptions, and specifications.** Perhaps you have completed similar contracts and have floor plans, schematics, and other materials on hand. If you anticipate what materials will help the meeting, you can be well prepared. Just as you had prepared the script (your questions and suggestions), you can also have the props ready, even if there is no rehearsal.

Problem Areas in Sales Relations

Sales Resistance

Selling involves buying. If a buyer decides not to buy, the sale is off. You would think that this situation is clear enough. The sales rep should pack up and go home. Interestingly enough, traditional sales practices are extensively dedicated to the "control" of the conversation and the skillful trouncing of every possible objection. Because a sales rep doubtlessly hears all the usual rejections within one year of practice, it probably is a matter of skill to overcome every objection in some way. As Henry Ford is reputed to have said in the early days of automobile manufacturing, "We have every color you want, as long as its black." In this sales environment the goal usually seems to be to sell at all costs. Undoubtedly there are industries that believe in this approach to this day. If you are offered the wrong color on a car, you will resist. If the dealer throws in free options you might change your mind. It is Henry Ford's legacy, and the car is being sold, by using whatever strategy it takes to get it off the lot. There is, however, a way to involve a client in the ongoing discussion even when there are objections. Open discussion can help to keep client objections from becoming hard-line conditions.

In Japan there is a tradition that is well known among American business people who work in Japanese markets. An objection or a no is always politely avoided in favor of a yes. This famous courtesy confuses business meetings at every turn, but the message of politeness is an excellent idea, if only the objection were also as clear. In English we have a perfectly polite way to say "Yes, you have a point, but hear me out," or "Yes, you are right, but what about . . . ," or "Yes, your suggestion is a good one but so is this alternative." The phrase that will involve your client in positive discussion and keep your ideas in a favorable light is simple:

Yes, but. . . .

Any variation on this structure is respectful but moves on with the business at hand in a very polite way. Sales tactics try to overcome resistance in order to sell the product. The resolution of conflict in contracting a service is critical for technical reasons having to do with the complexities of service needs. The issue for you is the fact that the client and the contractor must agree to the terms of a contract, or else there is no sale.

You can avoid certain cost objections by carefully respecting a client's price range—if possible. In conventional sales, the effort to get a customer to "buy up" is common. A more expensive model, options, attachments, and other elements can involve considerably more money than a customer intended to spend. In an expensive service contract there is something of an opposite concern: can the client afford the cost? Whether the fee is $5000 or $50,000, the circumstances do not involve inducements to spend. Even the benefits that you promote are enhancements and *not* hidden incremental costs. There can be *no* hidden costs in a contract proposal. **Rather than create an unaffordable situation, one challenge is to carefully discuss costs, both those *cost expectations* of the client as well as the *cost realities* presented by the seller.** Depending on your technical service, suggestions concerning financing might or might not be part of your presentation. The discussion of financing might involve considerable research time in any case. As I noted earlier, in some cases you might carry the business cards of financial officers of banks and lending institutions who know your specialty and are familiar with the business finance involved.

You do not want to squelch any objections that could have been foreseen before a contract starts. That much is good business for both parties. Second, the objections that occur should be discovered during contract discussions because they will work on the client. If they remain unresolved when the proposal arrives, additional roadblocks then separate the proposal from acceptance.

Two areas of concern for the seller are easy to zero in on. *Listen* for hesitancy and *watch* for any body language suggesting discontent. If the client purses her lips over a drawing and starts tapping a pencil, acknowledge concern.

> *Does that floor plan reflect what you had in mind?*
>
> *This is tentative and I want you to indicate anything that doesn't look right.*
>
> *I placed that additional terminal on the second floor. What do you think?*

Enthusiastic clients who jump in with changes are strong prospects because they are open in their expectations and are showing a willingness to work with the seller.

Current thinking suggests that it is not smart to block or try to overcome objections. This may be particularly true for contract proposals. Unlike most sales, contract proposals are delayed. Point of commitment is *not* point of sale.

> *I love it. Write it up and lets see what it costs.*

This time lag provides a window on the world that conventional salespeople do not have to handle, since they get to hear the wonderful music of such lines as "Sold" or "I'll take it." Client hesitancy must be negotiated out of contract writing to close the sale.

In a sales proposal situation, *strong* client objections are a much more serious issue than hesitation. Minor sales *objections* can be overcome; sales *barriers* are a more serious matter. **If a client is setting up strong barriers, the contract provider should avoid the contract.**

> We can't afford it.

> I really won't need the installation until the new insurance regulations are in place.

It stands to reason that a client will be hesitant both because a service is intangible and because services can represent a very considerable cost that is not in the normal budget of day-to-day profit and loss for any business. Typical challenges to costs with suggestions for overcoming them are listed in the following table.

COMMON OBJECTIONS AND STRATEGIES FOR OVERCOMING THEM*

Objection	Strategies
"Your price is too high."	Don't even try to compete on price alone. Talk to the prospect only about value, quality, satisfaction, profit, prestige, or service.
"I can't afford to buy it at that price."	If you qualified the prospect properly as to ability to buy, his objection simply means that you have not made the prospect want or desire your offering. Show why the prospect needs or should want it and how it can be obtained most easily.
"Your price is out of line. I can buy something like it for less money."	Very few products are exactly the same. Avoid argument and concentrate on distinguished features.
"The cost is far over our budget allowance."	Walk the buyer through facts and figures, demonstrating how your product will solve a problem, increase profits, or cut costs, and point out how much it could cost not to have it.
"We're really doing fine with our current version."	Point out how much better the buyer can do with the new version.
"I don't like the color (or style or quality or features) of your product."	Acknowledge concerns and express willingness to convey these thoughts to development staff, but steer conversation back to features the buyer does seem to like.

To suggest how thorough the literature can be in detailing strategies for sales trainees, I have included this sample from the text, Marketing, by Keegan, Moriarity and Duncan (Englewood Cliffs, New Jersey: Prentice-Hall, 1995).

Traditional sales strategies jump at the opportunity to challenge objections and barriers, and these strategies encourage the sales rep to take a deep breath and keep up the chase. *Don't*. Contracted services call for a different approach. Client *objections* are part of contract adjustment strategy. *Barriers* are a different kind of doubt that probably should *not* be encouraged to the point where a wary prospect is signing on the dotted line.

Prospective clients who are argumentative should be appreciated. At least they show their colors. You can act accordingly. If you do not, there can be problems. Nick is a custom cabinet maker who had an encounter of this kind.

> *We should have known better. He argued with us from the start. At least we had the wits to go hourly or we would have lost the shop. He wanted this classic Airstream trailer restored and fitted like a walnut-paneled law office. Nothing was square; all the corners were rounded. A shipwright could do it but we had to fit and fit and fit to get the cabinets in. He won't even speak to my father anymore. They were friends. He still complains about the job to people. And guess what? He is a millionaire.*

Sales literature generally takes a strong stand on standing firm. Sales is not for the meek as we all find out sooner or later in the contracting fields; however, your client base and your referral base are of the utmost importance, and prospects *can* and *should* be sacrificed from time to time. On the other hand, your best clients will disagree with you at times, and it is important to be able to argue your technology and defend your interests or haggle over your prices. In the discussion here I could not begin to suggest the details that involve cost disagreements. You will learn what your margins are by experience, and every contractor learns where the bottom line is after taking a loss on one contract or another.

As I explained in the last chapter, the best advertising short of free news coverage, is a satisfied client. A happy customer is the ideal reference who can quickly broadcast your merits—which is why you will see successful business people leave their clients with five business cards. An established client does not even need one; it is already in the card file. A disgruntled customer does not help your client pyramid. *Do not try to close a sale if the customer is, in your mind, a risk*. The client is not likely to be happy. You are not likely to be happy. A lose-lose situation could be the result, but worse is the potential criticism you do not need. A seasoned, hard-nosed salesperson might say "You let yourself get stiffed," but by avoiding a risky contract that could make both buyer and seller unhappy, you created a win-win situation *because* there was no effort to sell a proposal.

In a study from the White House Office of Consumer Affairs, the guesswork about customer backlash was turned into blunt statistics:

- **96% of all disgruntled customers never address the problem with the seller.**

- **91% will not offer return business to the seller.**

- **The average annoyed customer tells nine people about the event.**

- **One in ten of these annoyed customers goes out of his or her way to complain to as many as twenty people.**

(Robert Miller and Stephen Heiman, Conceptual Selling, New York: Warner Books, 1987)

A promise is a promise. The contract you agree to can create problems that can be costly, because, as a professional, you must honor your commitments. Objections and barriers must be handled before a promise is made. In Lloyd's case, the misunderstanding on a contract cost him a week's labor, which was not too severe.

> I'm an electrical contractor. I bid a kitchen job for a hood fan for a stove. This wasn't the recirculation filter type; I ran ductwork into the attic and out the side of the house and wired power from the basement because the kitchen was already pulling too much power from the kitchen circuit. Well, the bid was the only bid I never put in writing. The guy was a teacher and knew my wife. When he got the bill he faded like a flower. I gave him a figure for time and materials when we originally agreed on the deal. All he remembered was the materials! He thought that was the total. So there went my time because I didn't want any disagreements.

Client Profiles

Sales people are sometimes seen as hypocritical, and you may think that keeping file cards on clients is overdoing it. Professional salespeople look at the issue differently. Their client base puts them in contact with too many people. They cannot remember much more than anybody else remembers. If your daily world involves twenty people, you do not need file cards. But what would you do if you had two hundred clients? Your family doctor is a case in point. Observe medical office practices carefully. If you have an appointment, your file will be sitting by the receptionist when you arrive. The nurse drops the file in the in-basket for Exam Room Three. The basket is mounted on the door. You are escorted to the room. You hear the doctor approach. There is a minute of silence but you hear her pick up the file. In she comes: "How is that thumb?" Thumb? What thumb? Even *you* forgot you stubbed your thumb a year ago. Perhaps a thousand patients have come and gone— and more than one thumb. This is not hypocritical; it is professional tracking— and good business.

A very professional way to deal with the scope of your activities is to develop a system so that you can quickly reacquaint yourself with each of your accounts. This can be a simple matter of having an account history file, and this should, indeed, be a practice you should consider. A surprising number of business people bury their accounts with their annual tax accountings. These are "dead files" that are kept for legal reasons, and they are organized by the year. If a client calls and you need to recall the contract details of your work for her, not only will you not be able to promptly do so while you are on the phone, you will have to go hunt the material—if you remember the year.

If you copy account records for a *client account file* you can quickly produce the history of your transactions. It might be useful to keep the account files within reach of the phone. If your memory of the client *appears* to be sharp, you impress the client at the outset, even though you are in fact opening his or her file folder.

Sales representatives have additional strategies to offer you that might be equally impressive. Since sales reps are professional salespeople, they deal with far more customers than you or I would ever want to handle. If only because of the sheer numbers, they look for additional ways to help their business. One practice that could be of use to you is the *client profile*. It is one thing to keep a client account history; it is another matter altogether to remember the client's business setting. A client profile can help. What matters is that you return from a meeting and record what you might need to know in the future. Here is a sample.

Client Name: Business Name: Address:	phone: FAX: E-mail:
Business type: Number of Employees: Previous Service Provider: Technical Problems:	Owners: Date:

The client profile can be enlarged if desired. It would be appropriate to review the stages of the contracted sale and jot down highlights that might be of value during a future sales situation with the client. If nothing else, it helps to remember the successes and failures of a contract history.

> **Meetings:**
> Phone problem. In the future ask Bill to ask Cheryl to take the calls if possible.
> Didn't have extra table.
> Clear headed and quick.
> Will never go with it on day one.
> Seems to talk it all over at home, and then with younger brother.
> Cautious but pays what it takes.
> I like him. Good all-around client.
> **Job History:**
> What he liked—
> 　　　Quiet work crews, courtesies, no smoking.
> 　　　Loved the printers and 19" screens.
> 　　　Attention at the discussions.
> 　　　The proposal and color blurbs.
> What he disliked—
> 　　　Fussy about power tool noise.
> 　　　Wanted dust vacuumed constantly.
> 　　　Very worried about incoming clients.
> 　　　No parking in his lot!
> 　　　I stuck that new tax on the bill and it wasn't on the proposal as a cost. Didn't sit too well.

With this sort of information you are well armed. In five minutes you can reconstruct people and events that might date back five years or more. Consider this your medical history for the client. You should, of course, design a form that meets the concerns of your industry. These profiles could be 5" × 7" cards or a template design kept in your computer.

Consider taking the patient history a step further. Because sales representatives have a large public, they will not remember all the "people details" that help support their sales efforts. I, for example, am terrible with names. I score no points whatsoever by calling a client by the wrong name. Sales reps sometimes keep an additional file—*a client data file*—that helps them reconstruct a client's setting before making a call. It might look something like this.

> **Client Name:**　*Dave Mortenson*
> Knows what he wants/wife is half of decision/incorporated/SeaFirst Bank/will dicker/avoids bids/likes golf/sails—uses Southend Marina/doesn't eat red meat/follows college football/2 kids/Catholic schools (one child—married in Phoenix)
> **Employees:**
> 　　Karen (the short one) does accts/divorced/one child/has a time share in
> 　　　　Cancun/Jeep/cooks and swims/athletic
> 　　Mitch/ok/parents are local/father is Navy retired/Mitch did Navy 4 yrs/has dogs/runs and
> 　　　　bikes/(a little cynical)
> 　　Cheryl/secretary/always in yellows/grows roses/all biz/is important/opinionated/make
> 　　　　the effort

If the idea of "showing you care" seems a little sentimental to you, you can use an equally impressive strategy, which is to show them you *remember.* It is good business and saves little embarrassments such as asking Cheryl about the dogs she never owned.

These files should *never* leave your office and you should avoid serious value judgments in writing. Also, if clients call at your office, the files should not be consulted or used while they are present. There is nothing secretive here; everything you write down consists primarily of what clients and employees have told you. However, "personnel files" are a fragile matter in any company and they are often confidential. Remember that a client profile is *not* invasive. It identifies people by their traits and interests and records public life-events. You would not be writing most of the information down if you had not been told the information. The intent is not personal or secretive, but if your memory is anything like mine, you can easily lose points by remembering nothing about clients, *including* their names!

Datelines for Contract Decisions

Trying to pin down a decision on your proposal is a difficult matter. Although very little of the sales technique literature that I have on my shelves discusses the delayed-purchase response involved in contract proposals, you should think about the peak-risk period that awaits you while the client decides to buy or not buy. In traditional sales, every effort is made to close a sale while buyer energy is up. This is the psychology we see underway when dozens of salespeople work the feeding frenzy on the showroom floor during a one-day sale or the December 26 blitz. Keep it hot. But nothing is slower than the proposal as a sales device. What to do?

There are few strategies available to you. If the proposal goes to a *committee,* you cannot even use your client rapport, because the committee may set the meeting dates, determine the selection criteria, and decide on contract approval dates. Committees establish datelines, but they are self-serving and meet their own needs. Be sure to ask who has final authority for accepting the proposal. If you have only one party to deal with (or perhaps two are working with you) the options provide a little flexibility. If you run a small business that deals with small businesses, you are in luck. Then, your client is the decision maker. **If you have a client or two involved in a potential agreement, ask them if they would establish decision datelines.** As I explained earlier, you want to be sure to get a sense of the client's time frame (and sense of commitment) before you leave a meeting and agree to write a proposal, but you also want a commitment to a final decision-making action if possible.

Will you be moving on this soon?

When will we know if you have approval?

Can you present the proposal to the bank by the tenth?

Will you be able to respond to the proposal by the fifteenth?

Can we set a decision date?

Can we trade datelines here? If I meet your deadline by the tenth could I hope for a decision by the twentieth? It is complicated to calendar everything and I need lead time. I know you want this done by February, so I can't offer any surprises to my vendors.

I can lock this in only if I know by the fifth.

These samples move from questions to a mixture of questions and observations to a specific declaration. You may or may not want to use one or any of the tactics. In the same sense, your cover letter to your proposal can be specific about timelines if you want to be or have to be.

I can guarantee to meet your timelines if I hear from you by the third. I hope this isn't an inconvenience. This is proving to be a busy season, but the calendar is still looking good. Call me if this won't work for you.

It is only fair that clients should respect your efforts. Your time has been provided to meet with the client, to analyze the situation, and to develop a proposal. There are costs involved. In return, a measure of the goodwill of the client will be to let you know where you stand, to sincerely follow through on the intentions that brought you to the meeting, and to provide lead time for you to schedule the job while adjusting your other commitments as well. Rather commonly, the contractor defaults on the deadlines far more often than the buyer, so also honor your half of the agreement at all times.

The Postsale Period

Assume that the contract was awarded and you completed the job. After the sale, the issue of trust takes on a new meaning. You must stand behind the service you provided. For example, if the client encounters problems and finds that product liabilities are not being handled properly by vendors, you will have to jump in and expedite matters. You will have to get the product repaired or replaced or install a temporary model. You want to help the client, who will certainly respect, if not praise, your assistance. The headaches can interfere with your work schedule and can involve temporary costs at your expense, but the reward will be a faithful client and perhaps an excellent referral. Mark secured a contract from a state university. It was a network installation. The university is in a small town in the farming region of the state.

The server was down before I got back to Seattle! I called HP and there wasn't another one in the state. The problem was the three-hour drive over the pass. Snow or no snow I went back over the pass, got the server at the campus, and drove back to Seattle to take it to HP. They were going to run it back to the campus after they tracked down the problem but I wanted to be the guy to put it in and make sure we we're "a go." That was embarrassing. Since then there have been no problems. I call to check once in a while. They appreciate that because the campus is a hundred miles from a major computer service—well, make that exactly 131 miles door to door. Each way! I know.

Motivation and Long-Range Planning

I am omitting certain features that loom large in sales management literature. At least a third of each of the books is usually devoted to goals assessments, motivation analysis, performance analyses, and long-range planning. The focus is usually self-improvement, and the catch word is "excellence." To the last, these discussions take on a can-do attitude where the sky is the limit. More than one of them shows the businessman or -woman, briefcase in hand, striding to the top of a mountain. There are charts, tables, questionnaires, and plans. There are one-year plans, two-year plans, and five-year plans. This material overlaps with business practices, time-management studies, accounting, and many other issues and disciplines that are beyond our scope. However, I will end the chapter on an image that might be worth thinking about.

When you drive by local barber shops in your neighborhood, what are you likely to see between 10:00 and 3:00 in the afternoon? In all likelihood you will see a fellow sitting in a chair reading the daily paper—and of course that fellow is a barber. As a second image, think of that dream vacation in Belize for two weeks of scuba diving. Think of putting two children through college. As a final image, think of your success and ask yourself if you can achieve it without selling.

I am sure my readers always begin this chapter with the traditional misgivings about people who sell. Now you probably realize that it is one of the unspoken keys to success. Unfortunately, few people have been trained for sales skills, though some variations of those skills are part of the path to success—particularly for small-business and professional practices such as one you might soon own. If you see that selling—in whatever fashion or manner that suits you—is the path to profit, you can put good business practices to work. The million books that Joe Girard sold to explain how he sold his way to millions of dollars in profits contain no charts or tables or questionnaires. Joe sold cars, like no one ever sold cars. His motivation analysis consisted of one word: *want*. The want motive is very simple. If you want to be in Belize, you cannot sit in a barber's chair. You have to sell what it is that you do—so that you will be what you *want* to be—and where you *want* to be. The idea is simple, but doubtlessly valuable. What you want is motivation—near term and long term—for what you do.

The Drop-in Customer

A few of the sales strategy details in this chapter can be applied in over-the-counter sales and showroom situations. The overall concept of a client-first strategy can also be applied, and it is the key to the sales successes of some very large businesses—notably, for example, the Nordstrom department store chain. Just as commonly, a manager may hope to move quotas of products, and a client-first approach will be discouraged. If you are working in a salesroom environment, consider those elements of discussion in Chapter 11 and Chapter 12 that can have value in your sales efforts. A few additional observations might be helpful.

The browser is often browsing with an interest. If someone walks into your sales environment and is "just looking," consider the person a prospect. The browser often knows exactly what he or she is actually looking for and wants to see what is available and at what prices. Your role is to explain the options and investigate the customer's needs. A little positive thinking is involved here, but the logic is also obvious: someone just drove five miles in the rain to "just look." The person is being cautious. You can respond to that cautious attitude with a helpful offer and a polite withdrawal. Offer your business card if the opportunity is right.

If you are working in a facility where you are engaged in showroom sales or over-the-counter sales, try to use your time to greet customers and build your sales percentages. This goal may seem obvious, but occasionally there is a workplace where cronies tend to hang out rather than work. Avoid the sales-agent gatherings. The cluster of employees is not going to impress a supervisor, and incoming prospective customers should not have to wait for attention or feel that the alpha wolf has left the pack and is headed in his or her direction. There is plenty to do: stocking products, hands-on learning of new products, reading new marketing literature and instruction manuals, and dealing with customers. Remember that this sales environment may involve a salary, but the effort is the factor of success—which is why commission structures exist. Even without commissions, supervisors will reward successful employees with other perks or promotions that indirectly increase salary. Many goal-orientation and motivation seminars will encourage sales agents to seek new heights: "Go for the 5% you need to top your best quarter." Without moving the goal post, you can usually improve your production with improved time management, which includes the suggestions you see here.

Sales Supports

Checklists and Visuals

Avoid the comedy of errors at all costs. The most likely annoyance at a business meeting will be what you forget to bring. Because you will be carrying a briefcase of important items, carry a checklist so that you can be certain that you have the correct items. Understand that each client will provide you with a set of conditions that you must perceive as unique. A number of items that you regularly will carry along to a meeting will be in your briefcase at all times. A number of other items can be collected in response to the contact phone call during which you gained some preliminary understanding of the particulars for a specific client. Certain items can be removed from your briefcase to avoid clutter and confusion; certain other items can be added to enhance your prospects at the meeting. A basic list might include the following:

notepad	questions and proposal ideas
graph paper	product promotions and visuals
pens/pencils	product specifications
calculator	business brochures
tables of product costs	business cards
sales tax tables	references (copies)

Other items can be added, and although I do not discuss demonstrations and simulations, you might from time to time have reason to roll out a product and demonstrate it. Anything that is important should be on the list—from tools to your reading glasses. The meeting should be seamless, but everything will go wrong without the checklist—and the glasses.

Many of these materials play a role in the second half of a meeting when they help support your service presentation. You must continue image building with aids that visualize the product or service you offer. The use of aids involves a little rehearsal time and an understanding of the psychology of the sale.

Visuals, simulations, or actual product demonstrations are traditional sales features that are used to prove that you are getting the whitest laundry or the brightest car wax finish, but they will also show up in the best executive suites when a CEO uses a *PowerPoint* presentation before the shareholders of a major corporation. The utility of these visual aids or demonstration techniques is based on research findings that an observer will retain 20%

of what is heard and 30% of what is seen. These numbers are modest, but the research suggests that the percentage can come together: an observer can retain up to 50% of what is seen and heard.

Remember that the use of any visual aid can also work against you. **Visuals complement an explanation but they must not be allowed to distract the client.** The number of visuals should be limited. Too many visuals or too many samples are awkward to handle and can confuse a client. Dwayne keeps 3000 samples in his van!

> More than that probably, but a painting contractor learns pretty fast that color wheels confuse people. A color wheel can have hundreds and hundreds of color samples and I keep a lot of them in the van for color matching. All I usually show a client is the current recommended selection for the year—maybe three display cards with thirty colors on each. Then the decision is easy to make. Even white is confusing. Technically there are hundreds of whites. Thousands. I use a display card with about a dozen popular shades. That's all.

Sales management literature usually suggests that you should not hand descriptive brochures to the client during the presentation. The literature can distract the client. Carry a complete assortment of literature in a briefcase. If you have product supports that will be a predictable part of a sales package, make visuals out of the brochures by enlarging important photographs. Suppose, for example, that you have twenty or thirty brochures or pamphlets concerning computers, modems, servers, and similar products for a network installation service that you are developing. If five or six of the items are the usual features that a client desires, reproduce a full-color glossy image of the items on 81/2″ × 11″ paper at a copy shop and have them laminated. Mount each item on card stock. Use these pictures as demonstration cards.* They will be suggestive and supportive, but without the written blurb that was in the brochure they will not be distracting. Do not, by the way, place any on the client's desk unless you are invited to do so. Hold the cards; point to the item and speak highly of it.

If the client asks to see one or more of the cards, he or she might then place it on the desk. In this case, your discussion unfolds and shapes itself for the client. In the sale of technical services, you can anticipate the needs of the client and have floor plans or schematics or drawings of sample installations prepared. These can be the layouts of former contracts or sample models that are intended to illustrate the concepts. Again, these items could be presented as glossy laminations for a professional appearance. If you use large drawings in your presentation, you can carry a tube of drawings, but unless you are in a field such

---

* Stationary stores have useful materials for your presentations. Your visual aids can be kept in an elegant ringbound folder for example. The folder is a little better organized than loose cards but not as quickly adapted to the needs of the moment since there may be aids that fulfill the needs of one setting and not another. The ring binders can be held in front of you. Visit a large, professional camera supply store and you will see a wide variety of frameboards, photo mounts, cases and other uncommon supplies that can be used for presentations. Large art supply stores also carry unique items that are useful, such as large portfolio cases and matboard.

as CAD drafting or structural engineering where oversized drawings are the norm, these drawings may prove to be a clumsy distraction. If, for example, you visit an architect's office with a tube of drawings, the architect is familiar with the medium and probably has a table on which they can be placed. If you visit a law firm with the same drawings, the lawyers may find the material clunky, and you may all end up standing around the drawings, which will end up on the floor.

If a layout or floor plan or similar feature is an important aid in your sales, think in terms of a small display board. Can it be kept at a reasonable size? Perhaps 11″ × 14″ or 11″ × 17″ are practical sizes. Larger drawings can be carried in a special portfolio case. They can be dry-mounted on foam-core board at a picture frame shop or laminated in glossy stiff plastic by a local plastics laminator (look in the Yellow Pages). Depending on your career specialty, you might find that there are helpful software applications. *Visio,* for example, can be used to design vivid layouts and schematics of computer network systems for office facilities. These visuals can greatly enhance a presentation by creating an *image* of concepts under discussion. Similar drawings, ones that respond to the specifications of a contract, are then drawn at a later date to meet the *precise* intent of the client.

If you are invited to a meeting table where there is room to spread out, ask the client if you may use the tabletop. Put the briefcase and other carryalls on the floor and remove items as needed or else set up the presentation you will make. If you do not have the good fortune of a table, beware. The client's office desk is sacred territory, and if your first display knocks his daughter's picture off the desk, you are not doing too well. Avoid the desk if possible. As I mentioned, hand items directly to the client and invite him or her to look at them. The client will probably place the items on the desk, which is fine. In addition, do not lean on the desk and do not go around the desk for any reason. The client expects you to be in the seat that is there for your convenience. This is a territory issue that is subtle, but violations are noticed. The issue is easy to handle by remembering that your reaction, in your office surroundings, would be similar.

Advertising

There are million-dollar advertisement packages available for Monday Night Football games. It may cost a lot less to rent the streamer flown around a bowl game by a little propeller plane. It is cheaper yet to take a client to a local game. Or, you could offer your business card to clients and ask if they are going to watch the big game on television. Advertising is what you make it, and certainly, for every ad on television, thousands of business people are handing out their cards on the same day.

For the small business, advertising considerations are modest. As I explained, a high opinion of you among your clients is your best advertising. In practical terms, you will need to make a modest expenditure for business cards, and you must establish the habit of carrying them *and* distributing them. Freshen the ones you carry in your purse or wallet from time to time. Wrinkled and smudged old cards are very unimpressive and almost defeat your purpose. And if you change a phone number or address, print new cards. Cards that you have to revise immediately suggest that you are not on top of things.

If you are involved in showroom sales or over-the-counter sales, ask your supervisor about the policy concerning business cards. If the management will pay for them, order 500 for a start. Make a habit of using them in these settings, particularly if you receive commissions. Even if you do not receive any financial incentives, make the effort to help customers and hand out those cards. With the right approach you can increase sales and be branch manager soon enough.

For self-employed service providers you will need a Yellow Page listing. If you can afford the charges, do something more than list your name in fine print. Underlying this chapter on sales is the basic idea that you *must* do something more if you want to get out in front and stay there. The telephone company will explain the fees for various advertisement blocks and the color options. Be aware that there are strict datelines for these ads, since phone books are printed only once a year. If you miss the dateline you miss the entire year of listing and the ad copy exposure.

One other avenue to explore is the brochure. The typical full-color, three-fold brochure is an 81/2″ × 11″ sheet of glossy paper designed with two fold lines. It is a popular promotional format for business and deserves your attention. A small company can make a big impression by having a brochure that explains services. In addition, you should have letterhead stationery (many people design their own) that you can use to present additional information. If you present a client with a brochure and several pages of letterhead information, you will make a strong image-building projection of yourself. If desired, these items, along with a card or two, can be mailed the very day you meet the client on the telephone. The literature can serve as part of the ongoing sales effort. The client will receive the material somewhere between the first call and the first meeting.

Apart from the image that stationery projects, everything you commit to paper should be carefully prepared. I always made it a practice to submit all proposals in a typed format, knowing that my competitors would often use a form of one kind or another that was basically a checklist of services. The well-prepared document is critical to the image of a contractor because contract agreements are very involved, and they are always in writing. Neatness counts in preparing such a document.

You will easily match the challenge of documentation if you review the advice concerning proposals in *Basic Composition Skills* another volume in the *Wordworks* series. In Chapter 14, I point out that your file copies of former cover letters and proposals constitute the basis of upcoming letters and proposals. Modifications of existing documents are a ready source for new material. Modify existing contracts or other documents to meet the needs of new contracts. The work will look professional and attractive, but the labor involved will, with luck, be modest.

What will not sell? It is possible that slide presentations and *PowerPoint* presentations may be inappropriate expenditures. These sales tools are extremely popular for boardroom presentations but may not return the investment for a small-business service provider.

There are additional marketing considerations you could consider, including free handouts such as the ever-popular desk calendars and wall calendars that are used by businesses ranging from dry cleaners to banks. More appropriately, greeting cards are often a part of the annual marketing routines that companies and professionals will use to spread goodwill. Among the card possibilities, the thank-you card is the most appropriate item that will be good business in the technical service area. Once a job is complete and the client has become a satisfied customer, it is perfectly appropriate to write a thank-you note—with or without any reference to your hopes of doing business with the client in the future. This is considered a "postsale" strategy to maintain business relations. You might design a card specifically for this use, perhaps in the shape of a greeting card with your company logo on it. Avoid a printed message and write a brief note inside the card. Enclose a business card. I received one recently after buying a stereo component from a store that prides itself on customer attention.

The Telephone

Contact marketing with telephones has been and remains big business. Phone canvassing is called "telemarketing," and from the local funeral director in a town of five hundred to the cellular phone companies in Chicago and New Orleans, companies of every size try to sell by phone. For a business in technical services, telemarketing may or may not be critical, but for a small company—perhaps one you might own—the decision is yours. This text does not discuss this highly specialized industry (there are telemarketing consultants available to serve you). Our concern for the telephone concerns contract negotiation for a technical service proposal.

You receive a call from a prospective client. She would like you to drop by her office and look at the business setting. It is a three-story facility with the manufacturing facility on the first floor. It is an old and venerable local bakery. You know it well; everybody does, because of the slogan on the back of the trucks: "Drive safely, buns on board." They want a LAN system—maybe. This moment is important. It is the initial contact. You want to politely and enthusiastically handle the conversation. Use the opportunity to make a few upbeat comments about her industry, or her company in particular if that is possible. Ask a few primary questions.

> If you have a moment, it would help me if I ask a few basic questions before we plan to get together. Is that okay?
>
> What is the current situation?

What computers are you using?

How are they linked?

How long ago were they installed?

Is there a dedicated software system? Is it a popular business program or is it specific to the bakery industry?

With a little background you can arrange a meeting. You have shown interest and gathered enough information to begin to tailor your preparation. As you realize, the five-or ten-minute phone conversation is a valuable tool for initiating what you hope will be a relationship with a new client. The question is, what do you know about bakeries? Call the local community colleges and see if they have a culinary arts program. If they do, visit the campus library for a few hours to see what you find. Or if you think this is overdoing it, be sure to look around the client's office while you are on your first call at the bakery. If she subscribes to trade journals, you might show an interest in them. The interest is a good image-builder. The competition will not notice the magazines and you may get to see the commercial lay of the land if you are invited to borrow a few copies.

Subsequent to your business meeting with the client, you will draw up a proposal if it is requested. You might have to return to study the building and the floor plan. Perhaps you should call during the following week to ask a question or two and to show that you are working on the proposal.

Once the proposal is forwarded, allow for a grace period before you call. You might call the client's secretary to be sure the package arrived. This shows good form, and the client will hear that you inquired after the package. A cardinal rule in sales is to not let the client's interest cool. Equally obvious, the books do not explain how or when to act. I, too, will waffle on the issue. Is there a committee or executive board involved? As I noted earlier, that alone makes it impossible for you to hurry along a client. I would probably call after five days and hold out hopes of a decision or a dateline commitment. As I explained, you might suggest that you want to calendar the contract as soon as possible so that your own schedules can go forward—but be subtle. This is, by the way, a very honest concern for most contracting services. In my local area, the current boom economy has many contracting services backed up six months to a year or more and clients often go begging for quality work.

Notice that the phone plays a critical role. In every phase of sales when a contract is involved, the telephone is the beginning, the middle, and the end of the sales contact in its own humble way. Granted, the meeting or meetings get most of the work done in developing an agreement and drawing up a successful proposal. Nonetheless, the telephone is a valuable element in sales at the one-to-one level. Telemarketing it isn't. Marketing it isn't. Communicating and understanding it is, and this is at the heart of successful client relations.

In your use of phones, be alert to the fact that electronic conveniences can function as obstructions. The hang-ups on message-machine services make the point quite clear. There is no way around the need for message machines unless you have staff to take messages. One device, however, can put off clients and should be avoided: call waiting. If at all possible, avoid call-waiting signals if you have an important client on the phone. That same client might have the courtesy to tell her secretary to cancel all incoming calls while you are in a meeting with her. This courtesy is impressive, indeed, and you can return the favor with appropriate use of your telephone.

If for some reason you have to answer a call, make it short. Ask for permission to interrupt the discussion.

> *Would you excuse me a second? I have a call. I will be only a moment.*

When you return to the conversation, begin with a courteous apology.

> *Thank you. I'm sorry about the interruption.*

If at all possible avoid a repetition.

Not surprisingly, employees of every stripe are being encouraged to use transit time as office time: the cell phone office concept is in full swing. If you are an employee of a company, it takes no math whatever to realize that the concept involves *your* time and not company time. The reasoning here seems to be that if you have a cell phone you might as well be working as you pull out of the driveway of your home. The employer will realize most of the profit of this concept instead of the freeway telecommuter. Unless you are self-employed, the cell phone is intrusive and diminishes your per-hour value relative to your salary. There will be less stress without the device, or without the company's knowing you have one. Perhaps you should put the account in your spouse's name and have it unlisted.

The self-employed worker, the technical service provider, and the small business owner are another matter. As I noted elsewhere, apart from meetings, your phone is the link to your clients that you will use most of the time. It is the link to your vendors and other support services *all* of the time. You may find it immensely valuable as a time *saver* to own a cell phone in this case. Ralph was a building contractor who specialized in custom summer homes that were usually in fairly remote locations:

> *I bought one of those phones the first year they came out. They weren't called cell phones at first. All the contractors wanted them because of the milk runs; I'm always going or coming because we need materials. It made a big difference and made things a lot more efficient at the job site. If I lose it or drop it and damage it, all my headaches are back. I mean real headaches.*

Portable telephones are time-saving tools that will be worth the costs involved.

Summary

- First impressions count. Look the role that meets the expectations of the client.
- Put your social skills to work as soon as you encounter the first staff employee.
- When you interview the client, use open questions with a focus on the key words *what, where, when,* and *how.*
- During the meeting, follow the basic procedure of dividing the meeting into two distinct areas of interest.
    - **Client inquiry**
      Client needs
      Present conditions
      Additional concerns
    - **Service offering**
      Provider capability
      Provider products
      Proposed services
- Objections are part of the reality of sales and help define buyer-seller understanding.
- Serious barriers indicate that a contract should not be drawn up.
- Established clients are your best advertising.
- Establish a file for client profiles.
- A simple file will focus on former meetings, job histories, and account files.
- A more thorough client file can include client and employee sketches to help you remember details about the various people you have encountered. The file is strictly a reminder so that you appear sociable and thoughtful instead of forgetful and indifferent.
- In over-the-counter sales, the customer deserves all the attention and concern that is used with clients in service-contract selling.
- Prepare the visual aids, literature, and portfolio that will help construct the sales presentation.
- Keep a checklist of everything you need and check the inventory before you leave for a meeting.
- Study the most effective manner in which to present all the materials.
- Avoid the client's desktop and avoid any materials that will cause clumsy moments.
- Study your advertising needs, which, at a minimum, involve business cards and a Yellow Page listing. Your stationery and your message machine are adjuncts to the advertising strategy.

Activities Chapter 12

Present a memo to your instructor that develops a discussion of a personal experience that relates to the chapter. The memo should be 500 to 1000 words, typed. In it you can explain how you have related the text or the lecture-discussion to some event you recall. This exercise will give you a better understanding of the material because you will explain incidents in terms of your perceptions of workplace structure and communication.

Select one of the following suggestions and develop an analysis.

- *If you have had both sales experience in stores and contracting experience, discuss the differences in your strategies in the two environments.*

- *Examine the contracting services that are familiar to you. What is the sales process used in these technical service areas? Discuss the typical time frame from contact to close.*

- *Consider your reactions as a sales representative if you have sales experience. Compare several encounters that were rewarding or stressful. Discuss how you have handled annoyed clients or customers. Explain company policy concerning complaints if you were working as a sales employee. Explain your own concept of how to manage such situations if you were self-employed.*

- *Discuss the salesman stereotype. What images do you have of salespeople in different sales areas? Explain how you came about the stereotyped images.*

Work in Progress

13. Looking for Contract Work

Shortly after I resubmitted the bid to Tempo, I received a call from Tim saying that NetWorks NorthWest had won the bid—the one with the equipment room! My employees and I were excited. We had worked hard to get the contract and it was gratifying to see that the effort had paid off. Partly, I wanted the contract because it wasn't too big for us to handle. It was just the type of contract I knew we would enjoy.

As we got ready to do the Tempo installation, we spent a lot of time in preparation. I realized just how many different pieces of the puzzle must fall into place in order to create a winning bid. The first was being able to project a true and accurate image of my company (and myself as its representative) as capable and confident. During my initial interview, I needed to convince the people at Tempo that NWNW had the technical expertise to do a fine job and also that we had the sensitivity to ensure that our work would cause as few disruptions as possible.

Another factor was being able to understand the customers and what they are really saying they want and need as opposed to what you think they want or need. I think it has been obvious throughout these chapters that the key has been to communicate clearly with the customers and make sure at each stage along the way that you understand what they mean and vice versa. Nothing sours a business relationship faster than miscommunication, and when you are in business, it can mean losing contracts!

The sheer number of phone calls and meetings I had with Tempo all throughout the process might appear on the outside to be excessive or perhaps even nit-picking, but it was not. Preparation is the name of the game in winning bids (and in a lot of other work-related projects). I firmly believe that the contract is won or lost at this level. Poorly handled bid documents or a poorly prepared presentation sends a definite message to your potential customer that you are either not taking his or her bid very seriously or your company itself shouldn't be taken very seriously. Either position is a loser. It really does pay to be prepared.

Going through the rigors of contract bidding is very much like interviewing for a job and requires a similar mind-set to be successful. All the same warnings and observations apply. Again, preparation is the key. Knowing a little about the particular business or company gives you fodder for intelligent questions and a way to show your interest. Being clear and confident about yourself and the abilities you have, and dressing and acting the part, will take you a long way in either winning that bid or landing that job.

J. Q. C.

Selling Your Skills— Plus

The interview is a decisive tool used by all manner of employers in every size of company in every industry. You will probably be interviewed before you are hired, and you are likely to be interviewed for promotions. In terms of day-in, day-out communication, the interview is obviously only a fraction of your work-related activity—a very small fraction. The interview looms large in this discussion not because you will be involved in interviews often (none of us usually are), but because of another consideration: the power the interview has over your future, and your income.

The employment interview is a unique communication experience and is a commonplace tool in companies large and small. Most hiring practices involve interviews. Many promotion practices also involve interviews. A business will hire a new employee to fill a certain occupational vacancy that is defined in terms of skills and knowledge. However, the actual way in which interviews are conducted clearly suggests that there is more to selecting an employee than professional background.

The function of the interview involves an interpretation of much more than your professional capabilities. For the same reason that *you* do not want to be judged solely on the basis of your application, the employer usually does not want to depend on the application either. Employment applications simply save time by weeding out the unqualified and the less qualified candidates. The *real* issue involves the qualified candidates and which one to hire.

The hiring process is a selection process by which a company determines the person who is *best qualified* to be part of the *team*. Notice two criteria here. You might think that the application for a position says it all, so if you are a qualified candidate, why bother, for example, with a resume? For the same reason the company bothers to interview you. You are *more* than your skill or your knowledge. You are *more* than your application. Send a resume with the application if for no other reason than to show that you go the extra mile. The candidate *with* a resume can make points at the expense of the less savvy candidates. Be qualified, but be more; show them you should be on the team.

You send along your resume to try to stand out among the qualified applicants. This is your way of trying to reveal your *extra* potential. And this is the exact reason why a company conducts interviews: to see where there is *extra* potential among the qualified. In the case of promotions that involve interviews, the evaluation of the additional assets of candidates can be particularly critical because the selection of finalists eliminates all the also-rans before the selection process even starts. The additional assets *are* the measuring stick.

Interviewers look for skill potential, *and* extra potential or skills—plus. **Since companies look for skills plus, a smart interview strategy is a plan *based* on skills—*plus,* a plan *based* on your capabilities, *plus* personal energy and ambition, *plus* any supporting talents that can be identified.** For example, showing a "can-do" attitude may make all the difference. Energy matters. Energy means profit. Enthusiasm means quality. An extra skill such as software knowledge, foreign language proficiency, or previous work experience will also be a plus.

For example, if you finish a drafting program and head into the job market, you can be certain that your degree will help you find a position. If you have the kind of energy and ambition that will interest an employer, you have an additional asset. Suppose that you also are gifted at freehand drawing and can handle watercolor drawings for an architectural firm. Your degree skills *plus* your additional assets might prove to be the significant key to being able to put you well above the competition and into exactly the position of your choice. Every architectural firm needs one draftsperson who has the artistic talent to draw or paint the all-important conceptual graphics for the big day when an architectural firm submits a proposal. You could be that special person.

Motivation Analysis

Can you "prepare" for an interview? Simply knowing the "skills plus" concept is a good start because it will help you market your skills. There are, of course, entire books dedicated to the art of the interview—all of which argue that preparation is possible, and necessary. There are a number of practical and well-known approaches to preparation. The first task should be a self-analysis concerning your goals. If you are finishing a college degree you are well on your way to fulfilling the goals you have mapped for yourself. **Clearly understanding what your goals *really* are is a strong headstart in preparing for a confident interview.**

You are likely to be asked about your "goals" in an employment interview. The question is quite difficult for most people to answer probably for the same reason most people never really squirrel away much money for retirement. It is difficult to take control of the future. Even the *idea* of the future is confusing or unsettled at times. For the sake of the interview you will need to organize a response to a question about your future plans and goals. Since you may have difficulty deciding whether to take a particular job, take the plans and goals issue seriously. It will give you the criteria to judge the job prospect itself.

Divide your goals into short-term goals and long-term goals. Decide what you mean by "short" and "long" before you go further. Use paper and pencil and build the two lists as a record of your ambitions. If you are married, it will be easy to create the list if your spouse helps out, and a spouse is very useful in pointing out where your own ambitions may not be the only plans in the family that have to be considered. In fact there can be outright conflicts in ambitions at times.

Our ambitions or plans or goals are rarely in conflict with our personalities. Expectations are usually somewhere between realistic and possible rather than between unrealistic and impossible. Your lists might reflect personal plans, family plans, and monetary plans. Your lists could include realistic short-term and long-term interests of a financial sort. If you have the disposable income, where will it go? House? Children? Car? Vacations? Investments? Education? Boat?

The word *fulfillment* is one way to create the lists. Know what you *want;* know what you want to *do.* Or think out what you find fulfilling or what your long-range plans are for a successful future. Do not, however, simply write down "a good job" under both lists. Nor would you tell an interviewer that your "goal" is a "good job." The motive is not false or improper, but the word *good* does not tell anybody anything. What is a good job to you? Break down the real fulfillment or pleasure you have taken in a job. Those fulfilling elements probably are your overall criteria for a "good job."

The two lists—short-term and long-term goals—will serve two practical purposes. First, during a job interview, if asked about your goals or future plans, discuss the plans that will affect the company if you work for them. Carefully select goals that coincide with the mission of the company. Explain that you want to be a field service representative or explain that you want to be an engineering aid in research and development. Perhaps you want to be a sales representative and you look at commissions as the key to your success. In addition, if you want a certain style of working environment, explain your needs. You may seek a small company environment, you may depend on teamwork, you may need a desk job. Perhaps you are quite the opposite and you may love the bustle of the big time, thrive on independence, or enjoy plenty of outside activity representing a big corporation.

Be careful not to describe yourself in terms that conflict with the employment picture of a company. Some firms will not look at future schooling as desirable, for example. Other companies will virtually pay the bill to send you. Be honest but be realistic. Judge *which* goals you might discuss, and omit your financial needs. It goes without saying that we all have financial goals. The question is, what ambitions or plans do you have that *fit* the interest of a company?

The second purpose of the short-term and long-term goals is to examine, at *home,* what plans you have that fit the interest of a prospective company *and* what ambitions you have that do *not* fit. This is the kitchen table summit meeting that will help you decide how to deal with the right dollars/wrong job problem and the wrong dollars/right job problem. Your lists of goals will allow you to

- tell a company why you want to work for them

- and tell *yourself* why you want (or do not want) to work for a company, which is the *real* acid test!

In this text, we are generally looking at your role in working for a *company.* At your kitchen summit meetings you should try to look at the company as though it were working for *you.* It might be to your advantage to do a little homework and learn as much as you can about a business before you interview.

When an executive makes the decision to move from one company to another, he or she will study the prospects from a great many angles—including a close look at whatever information is available concerning the financial health and financial practices of a corporation. How were gross sales last year? What was the net profit? How did the company look over the last five years in its financial performance? Is it in debt? And so on.

You may not plan to be in the executive suite any time soon, but you also should find out what you can. Is there flextime? Do they pay college tuition? Is it a sweatshop or a popular employer? The pay? The benefits? Find out what you can before you interview. Are there big layoffs? Frequent layoffs? Huge production swings? Unions? Your own interests are obviously of prime importance. The company will judge you as a candidate; you must judge the company as a candidate.

Basic Interview Issues

Many people are afraid they will not provide the "right answers" during interviews but such a worry is probably a misunderstanding of the interview as a type of communication. An interview is *not* a test. There may not be many "right" answers. With this said, there may be confusion about what to say and how to say it for other reasons. If an interview is not "testing" you, what are you supposed to be doing? First, let's look at two problems about interviews that you may or may not have thought about.

Many people overlook a simple fact about job hunting: you are trying to shape your destiny. Your future depends on your efforts. An interview for a job takes on an importance that affects your future plans in a total way. We dream of owning a home. We dream of the perfect job, high pay, and fulfilling work. No other moment is likely to shape your future in less time. It takes a half-hour to get married. It takes a half-hour for a job interview. Enormous decisions lead to marriage; enormous fates evolve from the decision. In the same way, you prepare for a type of employment by going to school for years, and the interview determines the outcome—in less than one hour. The interview is a critical moment. You hang on the edge of your future.

At the same time, interviewing can be difficult to do because it is artificial and conflicts with lots of natural concerns we have about meeting and talking with people we do not know. Even the best conversationalists and the healthiest of egos may be uneasy in a job interview. Such concerns may seem odd at first; after all, isn't an interview just another form of conversation? **If you are fortunate, an interview *will* be a conversation—meaning a natural and possibly a pleasant discussion of mutual interest between you and an interviewer.** On the other hand, interviews are likely to be anything but natural, and for good reasons. In the first place, the conversation is very serious and often in depth, and possibly somewhat private, and yet the interviewer is someone you do not know.

The result is a very rare combination of a first-time meeting and some amount of overexposure on your part. You are not likely to be so open with life's details—even just your professional concerns and career interests—in any other first-encounter conversation, particularly a formal encounter arranged by appointment. In reality you are talking to a stranger in a manner you would normally avoid, and you are having to explain yourself and *sell* yourself. The setting is unique and not very natural.

An interviewer asks a great many questions that demand an equal number of answers. However, do not think of the interview as "answers" just because all you hear are questions. If the interview is going to be successful it will depend on your presenting yourself in an honest way. **An interviewer wants to see who you are, not just what you know.** We can call the task of presenting yourself in an honest way your "projection." We project ourselves with most clarity among our friends and close coworkers. The more removed someone is from your personal world, the less of "you" you are likely to care to project. In a strange situation you just "won't be yourself" as we say. The interviewer has no social importance to you, except that you may want a job that has put you in contact with the hoops of the interview process. Your reaction will be to provide information because you want the job, not because you want to talk. In other words, you will not be too inclined to come forward with the kind of relaxed confidence that could help sell you.

Projection

There is a trap in this situation that you must face because the interviewer does want to have an idea of who *you* are. If you were going to hire someone, you would want to have a sense of the personality of the person you hired. Information, however, can be presented with or without much personality. **If you project yourself honestly, the sincerity and enthusiasm will say a lot more about the information you are providing.** There are specific ways to open up and maintain a natural discussion. Trying to "relax" is one strategy, although relaxation does not indicate how you should project yourself.

Your ability to project yourself is partly in the hands of the interviewer, but it is difficult to construct a general description of an interviewer in a company setting. Larger corporations may have staff in personnel offices who do nothing but interview. The professional interviewer may have a solid educational background in psychology. On the other hand, a small company may have no office staff and you find yourself being interviewed by the "boss," who has never seen a day of college, but he knows what he needs for a successful business. One interviewer may have a long list of questions and follow the list rigidly. Another interviewer may have a few key questions on paper and little more. If you are really lucky you will run into someone who hasn't a single question written down but who knows how to hold a conversation. In this last case the interview will be spontaneous for the interviewer, and this means that genuine conversation will "drive" the interview instead of questions.

Interviews often tend to result in a fairly ho-hum presentation of ourselves for obvious reasons, as you now realize. We go home and think back at what we *could* have said or how nervous we were, or how shy, or how slow. Is there a solution?

Think of the information you provide as facts. The facts are all very "black and white" as the expression goes, rather like early television. Real conversation turns on the color. You must color your information. You want to *paint* yourself into the interview.

The key is projection. Because we shrink away from such an unfamiliar or uncommon activity, you must make your first task the effort to uphold your self-image. People *really* shrink: they sit low in the chair, put their hands between their legs, pull in their shoulders and neck, look at the floor, talk quietly—and run the risk of not getting the job because of their appearance. The employer may see this behavior as the key sign of a poorly motivated applicant. You do not want to run that risk.

By contrast, some people act out an interview as though they were on a stage. This is likely to be difficult for most of us and might convey a false sincerity, which creates another set of problems. You must not overact. What you need to do is be yourself, and there are ways to do it.

Natural Behavior

First, be *natural*—but not informal. Be a neighbor for example. Or accept the interviewer as your coworker. Or tell yourself how hard it is for the interviewer if people like you do not help. In other words, try to overcome the strangeness by accepting the stranger. Give the person a chance. You also want to be *formal*. Whether you wear a suit to the interview or not, pretend that the situation deserves formal dress. This will give you the right control over little errors you could make, such as slouching, chewing gum, getting palsy, or even swearing. There is a balance. You can be natural *and* formal at the same time.

Next, have *confidence,* but do not exaggerate. If the truth were known, we would all probably love to be highly appreciated, but we are usually too "cool" to tell the world what it is they should admire and respect about us. Many of us would love to brag, but would not dream of it. To the contrary, we do not promote ourselves. Confidence is less *what* you say than *how* you say it. Confidence is self-assurance. Confidence has humility. The promise "to do my best," for example, does not brag and it does not promise. It simply suggests faith in ourselves and the courage to try. Confidence is the courage to try.

Respond with *color,* but do not act. Be conscious of your tendency to shrink. Stay poised (not stiff). Look at the interviewer. Speak in your normal voice volume. Remember that you are potentially accepting this stranger as your future coworker. Talk to your new neighbor in your normal voice. Avoid monotone answers, but do not falsely tone for effects. You can energize a sentence with a lot of color and excitement without acting. On a scale of one to ten, a monotone is a one. The way you normally talk is with plenty of tones and pitches and words that color your voice at levels between two and eight, let's say, on our scale of ten. Go into that interview so that when you come out you can say, "I gave it a six."

Finally, try to relax. The jitters will be a fairly natural problem for many people during interviews, although a great many people have no problems at all. Nervousness is most likely the result of a lack of confidence about being interviewed and lack of familiarity with the process. No one can tell you to relax and expect you to do it. Many people overcome their anxieties by charging into the interview process to learn what it is all about. When the mystery is gone, the fear is gone. You might get right at the chore and start interviewing with companies—perhaps those that come to campus—to build familiarity with interviewing and to gain confidence. Those who do seek out practice interviews are very likely to be the people who have few fears in the first place, and if you are a little timid, you need the practice more than they do.

Sell Yourself without a Word

Let's begin our discussion of specific strategies with your image. Obvious points can be earned without a word being spoken.

Be prompt. You may not know how many points you will make by being on time for your interview, but you know exactly how many points you score if you are late. Even if you have to get there two hours early, be on time. If the rush-hour traffic is going to be impossible, you must leave early. No excuse will *ever* work. You can always sit in a coffee shop nearby (but probably not the company cafeteria or any part of the company grounds). Also, call the company to ask how easy it is to locate. Some office parks are impossible to find, and "suites" are even more impossible to locate inside the grounds complex. If you suspect such a possibility, go for a Sunday drive and locate the company days before the interview.

Be alert. We discussed poise, and we talked about eye contact early on in this text. If you sit up straight you will look alert. You probably *will* be more alert, and good poise may help you be a little more animated because your body is not slouched.

Be neatly and cleanly dressed. We are living in a time when weird is cool—at least until you look for work. This shift from counterculture to mainstream concerns only a few people. Handle these matters in your own way. There is, however, a more common problem that we need to look at. If you see yourself as an average or regular sort, you may be prone to making dress mistakes nonetheless. Let's note that we suggested that you should talk as though you were wearing a suit. In reality, then, interviews are often fairly formal.

How formal? Jeans are out. T-shirts are out. Sneakers are out. What not to wear is obvious. What to wear is a more difficult question. You might "dress the job," or a little better. In other words, you can often dress as you would for the employment, or a little fancier, since the interview is a special occasion. Example: You apply for a service tech job that calls for a company uniform of dark trousers and matching shirt. You notice also that the clothes are always well pressed and never soiled. There is the image. Suggest it: wear a plain shirt and dark pants well pressed, and maybe a sports coat to add a touch.

I say "dress the role" or "dress the job" because a tie or heels may not be appropriate and may suggest the wrong image. You probably do not want to leave the impression that "She doesn't want to get her hands dirty." On the other hand, we have all seen the copy machine service technician in a suit and carrying a leather briefcase—full of tools, of course. For this reason, dressing for interviews in the technologies always puzzles men and women because the positions are professional, the education is college level, and yet our hands and even the knees of our pants can end up dirty. We often wear a blue-collar shirt for a white-collar job or a white-collar shirt for a blue-collar job. Look at the job and dress for it, but dress for an interview also. In other words, upgrade the image if you think it is appropriate.

Know the Company

You can avoid a few more surprises by showing interest in the company that has chosen to interview you. See what you can learn about that company *before* you show up.

Some awareness of the company will help distinguish your interview. Show a sincere interest in the company. An interviewer will gain confidence in you if you ask questions about the company. You will gain further confidence if you explain why you want to work for the company. **The interviewer might be impressed if it is clear that you even know a good deal about their products and services or related matters.** If you have investigated a company to your personal satisfaction, do not conceal your knowledge, since it will show a serious commitment to a potential employer.

Put Your Mind in Control

And so you arrive and the interview begins. Begin by synchronizing your thoughts with your comments. In a spontaneous situation, people will make comments they do not mean or would not say. The problem is pacing; you must take the time to think out a question. People *rush* their interviews, either because they are nervous or because they think the fast pace suggests an energetic image. Usually the results are not particularly effective; sometimes the answers do not even correspond to the question.

Do not rush. Focus on understanding a question first. Pause it you need to. Ask for clarifications if you need them. Do not pretend you understand when you do not.

Frankly, if you ask questions of the interviewer, you have still another tool to help create conversation out of interrogation. This strategy will slow down the pace of your answers. It will also build your interest in conversation so that you are less focused on the "right answers." With questions you invite interviewers to explain and talk *with* you. And explain they will. In some situations, the interviewer takes over and does *all* the talking, which is no more desirable than your doing it all. Your questions will usually help create a balance in conversation.

Meanwhile, you gain clarification, and you gain time to think. Do not rush. Think, but do not leave them waiting either.

Guard Your Image

Who, you might ask, is going to put themselves down in front of an interviewer? It is probably rare. In fact, overselling is far more common. Too much confidence is the more likely problem. I take the viewpoint that the *major* barrier is the challenge to act natural in an uncommon situation. Given that awareness, most candidates can make the effort to sell themselves. They will project skills plus. However, there is one exception.

People will and do criticize themselves. This must be avoided in an interview. There is a common habit among recent graduates, for example, to accidentally criticize themselves, probably because they see the college experience as only a fraction of the experience they would like to have. With little or no work experience, job candidates may express a lot of insecurity, such as in a comment like this:

We only worked on the IBM-PC.

This graduate of a computer repair service program has just created a tremendous tactical error in the way he presented himself. The error is exactly one simple but very powerful word: *only.* Forget the word exists.

You have enormous control over the way people see you. Use it wisely. The interviewer might review your file and recall "only IBM-PC experience." This particular student could have said this:

The program at the college was very strong. We focused on the IBM-PC because it dominates the corporate market in purchase demand and follow-up maintenance. Besides, with the cloning that goes on in the industry, if you know the IBM-PC inside and out, you pretty much know them all inside and out.

Forget words like *only* and *just.* Your program was not "just" six quarters or "just" four semesters. You have spent two entire academic years completing a highly specialized degree in the field of your choice.

Be the Applicant You Would Hire

It can be helpful if you reverse your role and decide what kind of employee you need. This tactic can help you define a specific image in an interview. Essentially you impress the interviewer by showing how you are just the stuff for the job.

Suppose *you* are the employer looking for talent to help the productivity of your company. Would you hire yourself? Are you a smart investment? Are you worth the money? There are simple ways to evaluate yourself in monetary terms if you look at profit both as a quantity (of a service) and a quality (of a service). These values may not seem to relate to an employee in simple terms but the employee is hired for his or her skills. Profit is seen in our proficiency and our efficiency. You then realize that the *level of proficiency*—how good you are at what you do—is one obvious concern you would have in hiring anyone. Skill is the key word here, but "*measure* of skill" means even more. Lots of people share a given skill; the question is who is better or who is *best.* **The first-choice candidate will be the one who has the skills that are needed but also the one who is better at those skills than other applicants.**

Employers will also look beyond proficiency. Products, services, and profits depend on a well-oiled work setting. The teamwork of the employees builds the other trait you want to look for: *efficiency*. Translate this as the search for a personality that fits the company. Team spirit and cooperation are critical needs of many offices if they are to be efficient. For other companies, an efficient employee must be a self-starter and must have the ability to work alone.

You can avoid surprises if you look at the likely demands for skills and traits before you go into an interview. Emphasize the talents you have that seem appropriate. Also, listen to what the interviewers say about a company. The talent they need is no secret. It is likely to be explained to you.

Avoiding Surprises

The following list of questions includes some of the obvious ones that might be presented during your interview. Be aware of them and think about your answers. Better yet, also try to anticipate the professional or technical questions you are likely to be asked in *your* area of specialization if you want more confidence in your interview. Preparation *is* possible, since certain questions are predictable. Even the questions regarding your technical skills are predictable.

1) Tell me about yourself.

2) What work experience in your background will be useful to you if we hire you?

3) Did you enjoy your last job? Why did you leave?

4) What training have you had in school?

5) Do you have any particular reason for wanting to work for our company?

6) What are your career plans?

7) Are you willing to relocate?

8) How do you feel about overtime and working on weekends?

9) What specific skills do you have?

10) Do you think your background has prepared you for this position?

11) What salary were you thinking about?

12) How is your health?

13) What do you do in your spare time?

14) Do you enjoy working alone?

15) Do you enjoy taking charge?

16) Have you ever had problems handling your workload?

17) Do you prefer to work with others?

The Question Shapes the Answer

During an interview, it is obvious that questions drive the entire discussion. Study the questions. Construct the answers. A rule of thumb would be to say that your answers to questions should be specific and thorough and that you should not drag out responses. What an answer should be—accurate and complete—often conflicts with deciding how specific and how thorough you should be. The key is in the question.

The question shapes the answer. When you pass a friend in the cafeteria and say, "How are you doing?" you expect a response on the order of "Fine." Even a disconnected response such as "Hi!" is appropriate, because you really were not asking a question as such. But you *were* calculating a response, and if that same friend plops into the chair next to you in a state of despair and tells you how she *really* feels for five minutes, you certainly get more than you bargained for. Hear the question. It controls your response. You will look your best if the response is appropriate.

Let's look at questions in very simple terms. Questions frequently shape very simple yes and no responses. Sometimes called "closed" questions, as I noted in an earlier chapter, questions that call for simple answers need not be discussed. What more do you need to add to your age when you are asked? What else is there to your address?

The opposite of a *closed* question is an *open* question: "What do you know about studio amplifiers?" More open: "What do you know about digital electronics applications?" And there are comments that are amazingly open: "Tell me about yourself." Notice that this one is not even a question. It is a command. The point here is that you want to close questions that are closed. You want to open and explain your answers to questions that are open. Where do you stop? Stop for a breath of air. People always shift discussion on a pause; it is our natural habit and a courteous one. **If the interviewer wants to hear more he or she will smile or nod or ask you to continue. Interviewers will change the subjects when they are ready.** You will quickly understand at what length *they* expect you to answer the question. Be sure to allow them to cut you off.

There is no "right answer," but there is certainly a right length to an answer—one that will leave you and the interviewer feeling very good about the conversation. If you "yep" and "nope" the interview, it will be unsuccessful. If you say too much about too little, you will be equally at risk. The middle is a wide path and it should not worry you, but be aware. Interviewers will hint at the right pace by moving the questions, watching the timing, and closing parts of the interview. The length of the answer should be in the interviewer's hands; that is the correct length. Your job is to be *ready*, especially for long comments if they are requested or suggested. But show a willingness to cut if off.

There is another way to look at questions. It makes sense that there are *key questions* and then lots of questions that we might be tempted to call *minor*. Questions could even be ranked from most important to least important. The logic of the open questions would suggest that there are important answers to important questions—meaning, be specific and thorough. There is a problem here, however, in that "importance" is more of a value judgment than the idea of open questions.

I suggest that you learn to answer important questions before the interview. There are two practical tactics you can take. First is the scenario method. Sit down and anticipate what questions you will be asked. Divide the areas of discussion and rank them, perhaps in this fashion:

- **professional skills**
- **professional education**
- **professional interests**
- **previous work experience**
- **other education**
- **other work experience**
- **personal information**
- **health**
- **personal interests**
- **service record**

Write out key questions you might be asked in each category. Talk out your answers—preferably out loud. Focus on strong selling points, but consider weak spots, also, so that you do not stumble on answers.

Anticipate the important questions of concern to the company. Build your answers with precise *nouns*. Resume books place great emphasis on what are called "action words" or "action verbs." In truth these verbs often say very little: *plan . . ., direct . . ., deliver . . ., establish . . ., control . . . .* Be specific by using numbers, nouns, and proper nouns. If you can troubleshoot computers, name the brands and the model numbers and where you did such work and when you did it (the years and months). Use dates, names of companies, names of schools.

Avoid generalities because the impression you will leave will be vague. Build your image, and carry a resume. If a resume is well done it will remind you of dates, companies, institutions, skills, equipment, and so on—*if* it is in your lap, which is one of the best uses for a resume. Having your resume with you is like having your script ready—unless you just gave your only copy to the interviewer. *Always* take a resume along for yourself. It will stay calm and collected when you are searching for a date, and you can refer to it when in doubt. Two sample resumes follow.*

There is a second strategy for determining important questions of concern in your field. Go to job interviews. This is obvious enough. If you do not find your responses to key questions to be adequate, you will then know what to work on.

*For a complete discussion of resume writing, see Chapter 11 of the Writer's Handbook (another volume in the Wordworks series).

Sample Resume

Dennis Bell
630 Sunset Ave.   (206) 323-0000 Home
Seattle WA 98112   (206) 243-0000 Message

CAREER OBJECTIVE

Seeking responsible position in electrical power related technologies.

EDUCATION:
ASSOCIATE OF APPLIED SCIENCE DEGREES

Electrical Power and Control Technology Degree: North Seattle Community College, Seattle WA. Anticipated graduation: December 2000. Program Content: Electrical Code (Residential and Commercial). 3-Phase Power Distribution Systems, Dynamos (all types). Programmable Controllers and Interfaces, Electrical Controllers, Servomechanisms (Hydraulic and Pneumatic), and Electrical Design. All subjects included hands-on experience will the equipment. Familiar with applications of digital gates, analog circuitry, and servovalves.

Digital Electronics Degree: North Seattle Community College, Seattle WA. Hardware and Software of Mini-Computers, Pulse and Linear Circuit Design, Trouble-shooting, Boolean Logic, Assembly Language (Motorola and Intel microprocessors) and Analog Circuitry. Have built an 8085-based computer and designed an operating system.

Educational Honors

Remained on dean's list for high scholarship for eight quarters. Elected to Phi Theta Kappa national honors society.

Additional Education

Highline Community College, Midway WA.
Blueprint reading and drafting, math, general studies.

EMPLOYMENT

Apprentice Electrician, Westco Electric, Seattle WA. 1996 to present.
Residential wiring.

Machinist Specialist (Hone Operator), Kent Machine Products, Seattle WA 1992–1996

Machinist Trainee, Air Tech, Kent WA. 1990–1992
Manufactured parts for electronic and aircraft industries. Received letter of commendation from Hewlett-Packard for work done on computer printer rail guides.

WORK-RELATED ACTIVITIES

North Seattle Community College Computer Group: Member of Computer Maintenance Committee.

REFERENCES
See Attached

Sample Resume

Eva Smith
33948 93rd NE
Bothell WA 98011
(206) 485-0000

CAREER OBJECTIVE: Seek responsible position as an Engineering Aid in Research and Development or Calibration and Testing.

EDUCATION: ELECTRONICS TECHNOLOGY, Associate of Applied Science Degree, North Seattle Community College, 9600 College Way N., Seattle WA
- Graduated: June 1999

An intensive program that addresses the fundamentals of electronic theory. Additional course work included:

- Computer Programming in Basic, and 68 HCII Microcontrollers
- Soldering and Handtool Practices
- Precalculus Mathematics
- Certified First-Aid Credentials
- Technical Report Writing
- Electronics Drafting
- Digital Logic and Computer Fundamentals

(Transcripts available upon request)

ADDITIONAL EDUCATION: RIVETER/MECHANICS VOCATIONAL TRAINING, Boeing Aerospace Company, Everett WA. Graduated with certificate (160 clock hours.) 1994.

ADVANCED BLUEPRINT INTERPRETATION, Boeing Aerospace Company, Everett WA (240 clock hours.) 1994.

WORK EXPERIENCE: ELECTRONICS TROUBLESHOOTING LAB ASSISTANT, North Seattle Community College, Seattle WA 1997 to present. (part time)

HEAVY STRUCTURE MECHANIC, Boeing Aerospace Company. Everett WA. 1994–1997.
747 and 767—Work involved joining of body to wing sections. (Received certificate acknowledging performance on 767 wing assembly of airplane #1)

MAINTENANCE ELECTRICIAN, Proco, Inc., Everett WA 1990–1994
- Developed preventive maintenance program.
- Developed troubleshooting procedures.
- Retrofitted programmable logic controllers to electric and natural gas furnaces.
- Upgraded primary cooling water system.
- Installed nitrogen gas distribution system used in vacuum furnace operation.
- Installed and maintained hydraulic and pneumatic lifts, elevators, and conveyors.

REFERENCES: Available upon request

Loaded Questions

There is another type of question to add to your list. We noted that questions indicate whether the answers should be detailed or brief and whether the answers are major or minor. There is also what we might call the distinction between *loaded* or *neutral* questions. Most questions are informational and probably develop very neutral responses. **Be careful about questions that move your responses into opinions.** There is no way to be certain about how someone will take your opinions, so be very cautious. An example is criticism of your former employers. The stronger you emphasize the matter the more risk you run, even if you are correct in your judgments. Why? Because the interviewer may instantly suspect the reverse and think that you are a potential troublemaker.

Seemingly harmless questions may be "loaded" because of the concerns a company has, but interviewees are not likely to sense the criteria. Do you see yourself as a self-starter? Are you a team player? Questions of this sort are difficult because there is likely to be a standard a company is after and you cannot be certain about what they want. Be honest. What will matter is a contented employee. Concern yourself with what *you* want.

One loaded question that you cannot miss hearing is "Do you smoke?" Employers know very well that smokers are expensive people. They are more costly to insure. They take more time off for illness. They waste more time at work. The *Wall Street Journal* trumpets these statistics regularly. So a yes answer to this question is a big problem. What to do? Be honest. You will not be happy in their office anyway if you are a smoker.

There is not much point in fearing loaded questions, since they may be hard to recognize. In the end, even recognizing them does not help if you see a conflict between the question and who you are. Overall, you never have reason to worry because you have a keen instinct for what is right. Here is a loaded classic, but I have all the faith in your answer, and no one should have to tell you how to deal with it:

Suppose you see another guy in your office steal something; what would you do?

The Prepared Response

There are also questions you, alone, fear. A very common dilemma for an interview is any insecurity a person has about employment prospects because of some element in the background that could be troublesome. Perhaps he was fired once or twice. Perhaps she had too many jobs and thinks the track record looks bad. Or perhaps the record is spotty because she was raising children, or had health problems, or worse, he had legal problems. These difficulties vary from large numbers of people with lumbar injuries, to a smaller group in alcohol and drug rehabilitation, to the rare case of a prison record.

If you see a smudge on your background, decide how you plan to deal with it before the interview. You can take two approaches: omit or explain. Obviously, interviews are intended to be truthful. You must never lie about anything. That is not to say that your life is not private, and obviously you can omit concerns that you do not think are relevant. A person with diabetes is not going to discuss the condition during an interview. There is probably no reason to. A drinking problem that was cured ten years ago is obviously long gone. If, on the other hand, your concerns are recent or very obvious from the look of your application or resume, think out the issue and determine your answers. Do not try to conceal the obvious.

Disabilities are a worry to many college graduates. Vocational programs bring in many rehabilitation students who are retraining for desk jobs because of injuries—often from the building trades. They are very worried about their future; they feel insecure about college, and they are very depressed about the job market for, let's say, a guy with a disability from a lower-back injury. Attitude is the key. First, being injured on the job is like being under fire in the line of duty. Be proud. Second, disabled people qualify under affirmative action guidelines and are sought after by industry nowadays, so be shrewd. Third, figure out how to make it clear that you are a healthy worker, regardless of the disability. For example on your resume you could include a "personal" section that might say

- **Health:**   **excellent**
- **Disability:**   **partial**
- **Rated**   **S.3**
- **No work restrictions except weights over 25 lbs**

There is no conflict between fine health and a disability. There are millions of us. So the third point is simple: **have a confident, upbeat explanation ready for the questions you worry about the most.**

What about questions you do not like? What right does a company have to ask personal questions? What are your rights and how do you stand up for them? You must follow your better judgment. If you want to work on some black-box project that is in the Stealth bomber program, you will need a clearance, perhaps a top secret clearance. You may be asked a *lot* of questions. You may be asked if you will take a urinalysis. An exciting job or excellent pay may conflict with your values from time to time. Frankly, the event will be unlikely at the technician level of most industrial sectors. Electronics technology will be the likely exception.

Larger companies will be familiar with affirmative action guidelines and personal rights issues. The smaller the organization, the less likely it is to be familiar with its legal rights and your legal rights about interview questions. The issue is more a matter of your integrity, and you will stand by it.

If you are lucky you will not be forced into a lot of yes and no questions that limit your answers to a word or two. I noted before that if you are fortunate, someone who does not have a single question written down will interview you. In this case the interview will not be "canned"; it will be as spontaneous for the interviewer as it is for you. The result will be fairly close to a genuine conversation even though the interviewer is doing the driving.

It is true that interviewers are in control because they ask the questions. However, the real hire-or-don't-hire factors have to do with the employee that a company is looking for. The questions reflect the silhouette of the desired employee. The questions see if you fit the outlines that define company needs. You must hope you do fit the need and that the quality of the conversation helped you appear to be the most qualified candidate. If the interview is a fairly natural conversation, you cannot ask for more.

Depend on Preparation

Plan on a little preparation, in the hopes that the competition will not. There are a great many do's and don'ts we could apply to interviewing. In a nutshell, the major tactics depend on a little homework, and maybe a little practice:

- Try to learn something about the company. Look at the company in terms of services and products. For your own information, look at its performance, ownership, and financial practices. Try to determine your job responsibilities, perhaps by requesting a "job description" from the company. (A company may or may not use job descriptions.)

- Know why you think you want to work for a company, and remember that needing a job and needing a good income are not reasons of interest to a company. Do not bring up the obvious.

- Be prepared to discuss you short-term and long-term ambitions.

- Have two resumes ready no matter how many applications and electronic resumes you filed. Offer one to the interviewer and keep the other one in hand for your use. (See the sample resumes on pp. 343 and 344.)

- During the interview, speak with numbers and nouns, dates, and brand names. Be precise but judge the length of your answers by the interviewer's cues.

- Ask about your job responsibilities when you are asked if you have any questions. If you are asked if you have any questions, "nope" is not an answer.

- Do not discuss money and benefits too much but do not hesitate to ask for printed information about medical plans and the like. If certain concerns are very important to you—overtime, college compensation—perhaps you should ask about them.

- Try to turn the question of salary around. If asked what pay you expect, reverse the question and explain that you hoped to hear what salary they offer.

- Take along a portfolio or samples of your work if you are in a career area such as drafting. If you are fortunate enough to have a product to sell, by all means sell it. Show the interviewer your work.

- Look for the opportunity to talk casually about something of interest to the interviewer. Perhaps some topic unrelated to work is best if the interviewer brings up something to "chat" about. This is your chance to be more relaxed and show yourself by your first name rather than by your last, but the opportunity rests with the interviewer.

- Be cordial at all times, but be rather formal.

- Remember to think in terms of the skills plus strategy. There will be a lot of talent in the waiting room.

Don't Shoot Yourself in the Foot

There is no end to what we could talk about concerning what you should do and what you should not do. How-to books about interviewing and resume writing list hundreds of tactics and taboos. We can sum up our observations with a very few highlights you must remember concerning your manner during the interview.

Do:

- Listen to the interviewer.

- Take time to think.

- Ask questions, but not too many.

- Let the interviewer explain.

- Show an interest in the company. Show an interest in the job.

- Be positive about yourself.

Do Not:

- use slang (You are not "into" electronics and computers are not "cool.");

- talk more than the interviewer wants;

- oversell yourself;

- get emotional or opinionated;

- be witty;

- fidget or be stiff;

- lean on or otherwise use the desk in front of you;

- state a salary unless you are familiar with the pay scale;

- criticize yourself!

Right Salary, Wrong Job

A first job offer after college graduation is hard to pass up. You will not be discouraged from taking a job simply because the commute is a long one, but if you and your family have misgivings, you should study the problem. An offer of employment is a great temptation and may sweep away serious problems that will surface later. Perhaps you will be away from home too much. Perhaps overtime is too common. Or maybe you concealed the fact that you smoke and will find the going difficult. Talk out whatever the problems are—at home. List them on paper. Weigh the pros and cons of not taking the job.

We all have seen the results of a company-employee mismatch. It is no different from what you see in college. Many of the bright hopeful faces you met during your first semester are long gone before you graduate. It is very difficult to work against our own wills. Students drop out of college because the programs and the students did not match. The problem with a job offer is the same, except a little more confusing because there is the offer of employment and a paycheck—and the fear that you might not have another offer soon.

If you are unsure about an offer, you can also take the job for a time to see how it works out. This is an option. If you take the job you may solve your problems, or you may create problems. Most likely you will simply struggle along between these extremes. The *options* are clear. The *choice* is the difficulty. Given the opportunity, always take the path you find most fulfilling or rewarding.

Wrong Salary, Right Job

In the real world, this is the more likely problem: most employees would probably like to see much fatter paychecks. You may know a few people who make a lot of money doing something they dislike, but you probably know a lot more friends and acquaintances who enjoy their work but feel that the career field is underpaid.

Once you become involved in the job search, you should be able to determine fairly accurately what your "worth" is, not as you would like it to be, but as a practical expectation.

Any job offer within that range of expectations is fair game, and you can then look at the job from other points of view: work expectations, coworker lifestyles, personal benefits, commutes, and so on.

If you are anxious to get that first job experience, perhaps the decision is an easy one. If you are in a hurry to get back into the job market or if you are anxious to pay off college loans or debts, you must decide whether you should hold out for higher returns for your investment in college. Remember, however, that it is *while* you are in school that you should get to know the job market and understand what you can expect in return for your education and talents. And expect entry-level positions to be modest in pay. The simple reality is that the bottom of the ladder does not have much of a view.

If You Were Not Selected

The interview is over. It proved to be a worst-case scenario. Not to worry—you were not meant to have the job.

However, if you are confident that you were close to getting the position, but that perhaps there was too much talent, you might be persistent. Begin by sending a thank-you note to whoever interviewed you at the company. Ask to be informed of any new openings. Show your continued interest. This strategy may cut down on the competition, particularly if you did not get a job in May or June because the graduating talent pool was too big. After all, it is at that time that corporations visit campuses all over the country, precisely to select the talent they want. Count on the *next* interview as an opportunity to approach the job with more confidence and, possibly, less competition.

You might call the interviewer to see what the company is looking for in the near future. Provide the opportunity for tips or even criticism, but be subtle. Interviewers are often frank and helpful. In other instances, candidates never get honest answers about their failure to find jobs.

Because of your follow-up, the interviewer has been given a different perception of you. You have shown continued interest in the company. Notice that it is honest. It is not a sales pitch. A sales pitch will not work. You might get a call.

Summary

- Employment interviewers look for "skills plus," meaning that they look for desired skills and extra potential that makes a candidate stand out.

- You should identify your short-term goals and your long-term goals so that you understand your motives and so that you can target companies that appear to match your interests and goals.

- There are no "right answers" that a candidate should rehearse for an interview, and there is no way to anticipate the thinking of the interviewer. However, you can anticipate the questions and tailor your answers to the likely needs of a potential employer.

- You should *appear* to be the right person. Be natural, confident, and properly dressed.

- Respond to questions cautiously rather than rapidly, and provide precise information. *Explain* your technical programs as you *describe* them. *Discuss* your former employment as you *describe* the jobs. Be sure to sell these two areas of interest.

- Do not criticize yourself or describe your limitations.

- Offer supplemental skills or knowledge whenever possible.

- Expect to be asked questions in any of the following areas of interest:

professional skills	**service records**
work experience	**other education**
professional education	**professional interests**

- Closed questions call for precise and brief answers.

    When did you begin your college program?

- Open questions invite discussion.

    What did you think of the program?

- Prepare a few questions you can ask if you are given the opportunity (but the salary is not usually one of the questions).

- Prepare those answers that respond to troublesome matters. In other words do not volunteer the fact that you may have been fired from a job, but have your answer ready in case the problem comes up.

- Be upbeat, enthusiastic, and poised.

Activities Chapter 13

During the course of several weeks you will complete the following exercises:

---

1. A writing project *in the form of an interview script consisting of twenty questions you will use to interview* another *student.*

2. A performance project *consisting of you and your teammate interviewing each other.*

3. A critical evaluation that will be a self-analysis of your presentation.

4. An observation form *that evaluates another interview from among those you will see. (You may be asked to complete several of these forms.)*

These four tasks are explained below.

As you read Chapter 13 you will be asked to prepare an interview that you will conduct with a teammate. That teammate will later interview you as the second half of the activity. These interviews will be videotaped so that you can review the activity.

---

1. *To prepare to conduct interviews, pick a class member who shares your technical specialty, but pick someone other than the close friends you have made in the program. You will have less trouble simulating the interview if you are not too familiar with each other. The instructor will then ask the teams to get acquainted and get a feel for each other's job plans and technical interests. Then you will be asked to build the first project, an interview script of twenty questions to ask your partner in a brief interview (a ten-minute segment).*

---

2. *The interviews are the second exercise and the performance will be videotaped. The primary focus is the person being interviewed, but the role of the interviewer will also be a constructive experience for you.*

3. *You will have the opportunity to watch the videotaped interviews with a small discussion group consisting of the people who conducted interviews in the same session. Your third major task is to then complete a self-analysis (see the appendix, pp. 415–416) of your handling of your responses when you were interviewed.*

4. *Finally, you will complete an observation form that evaluates one of the other interviews you will see. (See the appendix, pp. 413–414.)*

Work in Progress

14. People Dilemmas

The installation at Tempo began without a hitch of any kind, and I knew it was only a matter of time before something went wrong. I have never been on any installation where there weren't some glitches, so I wasn't surprised when, after all the cable had been pulled, and we were beginning the install of the workstation PCs, we discovered that the vendor had not given us the specific network interface cards we had ordered. Since we were doing the install over a holiday weekend, chances appeared slim that we could get the right cards in time. Randy panicked and stormed around muttering while Cloe quietly went about her work looking pale and worried.

I called the vendor 24-hour order line and spoke first to a very unhelpful operator. She checked our invoice and insisted that the mistake had been ours and therefore she couldn't help us except to take our reorder for new NICs. Now I looked pale and worried! I asked to speak to her supervisor and found a much more sympathetic audience with him. He apologized for his operator's lack of helpfulness and explained that, due to the holiday, they were using newly trained staff who didn't know how to deal with a situation like ours. He assured me that since we were a commercial account in good standing he would send out the replacements within the hour and credit our account for the cost of shipping them back.

The NICs arrived as promised and all was well until Randy knocked a prized plant off the graphic designer's desk and it broke the roots. If he had tried, Randy could not have picked something worse to damage! The graphic designer was going through a nasty divorce and was touchy to begin with. When she came to work Monday morning and saw the plant with the note explaining what had happened, she exploded. We had anticipated this, and while she ranted and raved at me, Randy slipped out to our van and brought in two replacement plants that were actually much finer than the one he had damaged. She stopped in midstream and almost cried. Randy felt like a hero, and I was glad that disaster had once more been averted.

Later that day we were finishing up small touches and double-checking configurations when the office manager stopped Cloe to complain that the new PCs were 4" too high, and everyone was unhappy about it. Cloe, Randy, and I had a quick huddle to decide what do to. Cloe suggested mounting the drives under the desks to lower the monitors. I thought that was a great idea, but we would lose a half a day in labor charges. We decided it was well worth it to have happy customers.

The Tempo upgrade was finished, and all parties were happy with the job and pleased with all the new features the new LAN gave them. Tim Simmons and Sharon Hartell promised us great referrals and promised to spread the word!

J. Q. C.

Resolving Conflicts between Employees

In any workplace environment there will be workplace politics. The human community, from the family unit to the nation-state, illustrates this unique trait of humanity. The inner workings of all our relationships with one another are held in a balance. The balance is fairly delicate and if the assumptions of that balance change for any reason—if the balance of the scales tilts—there will be conflict.

Conflict is commonplace. There is little reason for alarm if conflicts of interest appear almost daily in the workplace environment. One employee enjoys the radio in the office; another employee is distracted by it. If conflict is balanced with resolution, there is seldom a difficulty. Unresolved conflict is the more problematic situation.

Because the business world is seldom structured for stable-state economics, the climate and relationships of employees in the world where we work is not as stable as it looks. There is constant change. The capitalistic concept is caught in the old idea that growth is good. Changing the term to "sustained growth" only provides a politically correct word for more growth. Growth, for the employee, means change. We are caught in a biology of struggle. We adapt to survive. Conflicts at work frequently reflect change. Resolution of the conflicts reflects adaptation. The process is continuous. Conflict is, in a word, natural in the scheme of things in today's work world.

Consider the high school you attended as a case in point. Schools are major employers. The schools appear steady and stable to the adolescents, but there is always a need for adaptation. The schools have to keep up with the times. When American schools and colleges struggle with economic and social turmoil, they too adapt. Sputnik, the first Soviet satellite, revolutionized the American education system. Television created a different kind of dilemma for the education system. Computer technology later emerged and gave a different sort of television screen a useful status and constructed the "new literacy" taught in the schools. Racial conflicts altered the school system structure yet again. In the near term the social picture may appear to be stable, but there is a constant adaptation to circumstances, often technological, that is under way in the workplace.

One source of the conflict of change is technology, and this is an ongoing dilemma that every engineering technician must face. Rapid change is very demanding of our willingness to adapt and our ability to learn.

> Andy was a typesetter. Computers put him out of business. Now he is studying *Auto-CAD* at an area college and he has also mastered *PageMaker* to put himself back in business. His conflict is resolved. Once he realized that his technology was obsolete, he decided to retrain at a community college and reenter the workforce. A common workplace conflict for a technician is the confrontation between old and new technical skills in a world of changing technologies.
>
> > *I was almost forty when I found myself out of a job. Worse than that, I was outdated and there were no other printing companies that would hire me. That was a grim year until I started college. Now I think I will be in good shape. I can do better publishing design work electronically than I did manually, although the job prospects in CAD interest me a lot now that I also have had plenty of experience with it.*

Regrettably, we are not usually trained to resolve conflict even though the culture is in such constant transition. The results of change often create relational conflicts that disturb the lives of many employees. Coworkers often quit a job to resolve such conflicts. Julia was responsible for accounts payable. She was forced to take over payroll—*and* accounts receivable! This created a serious shift in her job expectations. Then her boss complained about her overtime. She quit. The changes in her employment created one conflict. The resulting conflict

CHAPTER 14 CONFLICT RESOLUTION

with her supervisor was the other problem. Leaving unpleasant employment is one strategy that will stop conflict, although it is not exactly a resolution to the problem.

Apart from change, the other conflict that can affect you as an employee is the issue of employee relations. Employees do have conflicts with one another at times. **Conflict is usually interactive; that is, it occurs between people who have a relationship to one another—such as employees.** In the workplace community we will often see an existing balance shift and create conflict. There are many theories about the actual issues of the conflict. We should examine at least three: *territory, power,* and *self-esteem.*

Territory

No one wants to lose *territory.* No one wants to lose power. No one wants to be shamed. Julia quit her job because her boss had redrawn her territory. She had been perfectly happy with the existing definition of her job and she was quite satisfied with the quality of her work. In her mind she was being absolutely abused by having to cover all the new ground. What is important to understand is the point we made in several other chapters. Real territory, for example, may be the issue but, what really matters is *perceived* territory.

Power

The struggle to maintain control of your work role is a *power* issue. Real power may be the issue, but what really matters is *perceived* power. In other words, whether Irene, another employee, really experienced a salary loss because of a complex change in a union agreement somewhat does not matter if she *thinks* she took a beating on the contract. She felt that her status was weakened and that her earning power was threatened. Knowing the truth—the reality—will help, but *understanding the perception* is the real key to conflict resolution.

Self-Esteem

Ken is a supervisor who is proud of his territory and his power but his *self-esteem* took a setback in October. Ken's reaction to being denied a promotion was a surprise to everyone. He resigned. This seemed to be an overreaction according to his staff, and they did not want to lose him. They had trouble talking him into staying because he felt insulted. Not only did he not get the promotion, but the two supervisors who denied him the promotion were his best friends. He absolutely could not believe that someone else got the job, because he perceived himself as the best candidate. Conflict.

Let's assume a conflict is initiated at work. It will usually be due to change—or perceived change. Someone you work with is upset. What can you do? First, remember the issue of perception. Accept the conflict as real, even if you know better. This will help you see the "other" point of view of the employee. It will help you understand. Second, let's draw on some of our resources elsewhere in this text. Be aware that there are phases to the conflict, or at least *parts.* Authorities on this issue seem to imply that there is an order to the way the conflict takes shape.

What matters is that you see the conflict in terms of our two concepts of social maintenance function and task function. In other words the conflict exists emotionally *and* rationally. You will have to deal with both. How? Again use the resources from other chapters. *First, be supportive.* Work toward resolution. It may not be best to encourage people "to get it out of their system" in a big emotional way (although you could under certain circumstances suggest therapy and other techniques that call for professional help). Instead, **the easiest path to emotional balance is to deal with the rational issue in a rational way.** *See the task function as the resolution.* Help analyze the problem, and study the alternatives. Help someone in conflict see the options and look for avenues that offer best solutions. The task is the primary focus of attention—but do not overlook support needs. Remain very supportive.

Third, work toward a *win-win* resolution. In other words, there can be no loser. Only a philosopher will be able to shrug and say "Win a few, lose a few, some get rained out." Most of us lack that maturity. Nobody wants to be a loser. There is seldom a good loser, and you should help people resolve win-lose situations if at all possible. The key word is "balance," not victory. Irene wanted to quit the union. Her friends helped her resolve her anger by getting her to run for office in the union! She won and was empowered. The scales must be restored somehow. Trade-offs are the usual solution at work.

When Judith was denied a new computer for her office she had an argument with her supervisor. The truth was that the order was strictly a special deal on IBMs and a LAN system that was to be integrated for electronic mail. Judith wanted a new MAC and didn't care about e-mail. The order was a volume order based on a discount from IBM. Never mind the truth. The perception was the issue. She felt that her interests were being ignored. Conflict.

> *What bugs me is that some contractor sold a bill of goods to our executives and they didn't consult with us first. If I'm supposed to receive a new computer, shouldn't I be asked what my needs are? No, these characters saw a deal and that was that. It didn't help that Jamison, the new guy in administration services, is big on IBM.*

Resolution came a year later when her MAC order went in. A healthy resolution *could* have restored the *balance* in the first year, had her supervisor been supportive and had he promised her the MAC as soon as he could finish up the electronic mail commitment to his supervisor. The gesture would have helped resolve the conflict that she thought she was second fiddle to everyone else. Unfortunately, he could hardly deal with her esteem needs because he was angry with her and added to the conflict! He let her stew over the issue for a year rather than seek resolution.

I am assuming that you can counsel someone through a conflict. Actually, there are also other roles you might play in a conflict situation. In any conflict, an employee must examine where he or she is standing in relation to the problem. *You* could be the injured party—the Judith. We all *are* at times. Or you could be the perceived *cause*—the manager or employee who is not accommodating. Or you could be the mediator between two angry employees. Regardless of your role, you need to use your skills for being supportive and understanding. Much of the literature concerning conflict resolution seems to address the third-party role of the mediator. There may be none! In fact, *you* might be holding down either side of the scales of balance; *you* might be seen as a threat, or *you* might be perceiving the threat yourself.

You have the skills to mediate conflict. You often do mediate conflicts. Suppose the familiar case of alcohol abuse becomes a problem at work. Here the conflict originates with the employee. The scales of the balance tip. Projects are late. Routines are being violated. The drinker becomes defensive and perceives a threat. His supervisor is also defensive and perceives a threat to the company. The track record of this particular conflict is a poor one in most situations. Companies often lose employees for lack of resolution. In this setting, what can you do? Be supportive. Work on the task, in this case the conflict between substance abuse and work ethics. **Being supportive can mean a lot of different positions or attitudes.** There are many options. You can help the person resolve the conflict.

There are very practical efforts you can make to try to help:

- Try to understand the person.
- Talk to the person.
- Be honest with the person.
- Think about what to do.
- Try to persuade the person into a resolution.

Unions, civil rights acts, staff psychologists, medical aid plans, and any number of forces have somewhat helped construct mechanisms for conflict resolutions in the workplace. If an employee wants to help resolve conflict from the bottom up, the *best* mechanism is the awareness of conflict and the need for resolution—and the desire for community. *Organized resources* can be helpful and represent one plan of action. Get the community involved.

There are, of course, a great many ways to help, including leaving the person alone! The preceding list might suggest that the strategies are all lacking in authority or discipline. **If you are an employee, but not a manager, your efforts to resolve conflict should focus on the word *help*.** Once a manager is involved in a conflict he or she must address, there might be a call for disciplined responses, or even disciplinary action.

If you are a manager the approach to conflict is going to be more structured. Authoritarian responses may not be ideal, but they certainly are real. In industry there is only so much time for resolution of conflicts. Industry also believes in rewards, and by implication, the denial of rewards—or punishments. Conflict can cost an employee his or her job. The employee is removed from the community. Time is money and the system, as you well know, is usually topdown.

In your efforts to be personally helpful, adopt a *task-oriented approach* as a first strategy and try to resolve conflicts by reasoning out the issue. You can use emotional appeals if they are needed. In the workplace, blunt realities often demand that conflicts be resolved. The typical options will be familiar to you but it will not necessarily occur to an employee or manager to use these devices systematically.

1) The promise of reward

If you get compliance you provide a reward.

2) Threats

If Joe doesn't shape up, he ships out.

3) Systems knowledge

***You* know what will happen, come what may, and you inform her.**

4) Friendliness

Kindness is always one path to resolution.

5) Debt

You hired him. Todd owes you the solution to the problem.

6) Moral appeals

Everyone else gets to work on time. It is only fair to share. Doris should be on time like everybody else.

7) Appeals to dignity

Use pride and shame as a motivator.

8) Community

The team will pull for Brian if he adapts.*

Although we could examine each of these strategies in context, a few illustrations should serve to illustrate the practical nature of these approaches to conflict resolution. The "psychologizing" involved in these tactics has more to do with common sense than with any deep-seated psychological matters that have generated the conflict, so these helpful approaches are well within your grasp in your efforts to deal with conflict.

Based on the work of Marwell and Schmitt. Cited in Joseph Folger and Marshall Poole's text Working through Conflict, Glenview, Illinois: Scott, Foresman and Company, 1984.

Managers have access to a unique resolution to conflict: financial incentive. If the funds are available, certain conflicts can be resolved without compromising the values of any involved parties. **The ability to "buy" a resolution to conflict will depend on a very specific solution that is not always available: financial incentives or promotions or the like.** Of course, there has to be an understanding that a financial adjustment is appropriate; not every request for a pay raise is greeted with a signed check. If an employee is overdue for compensation, the employee is very likely to begin to show the conflict in work behavior. The conflict, which is often anger, is one of the most serious reactions you will see in the workplace. If a financial solution is available, this conflict can often be readily resolved.

Paula was being paid to work as a keyboard specialist for a small publishing company. She was paid for hourly services for four years until the publishing company was awarded a contract for a large project that involved several books that were to be distributed by a major New York publisher. The artwork estimates from Hong Kong were huge and Paula was asked to accumulate her work on the two art books as a separate freelance contract—payable upon publication.

> *Actually the first year wasn't too bad. I was excited, and Lesley guessed that it would take only another six months. All those extra hours looked like a nest egg. In the second year the whole project bogged down, the spirit at the office wasn't the best, and I became doubtful about the project and I was doubtful about recovering my time. I just seemed to find less and less time for the project, much to Lesley's frustration. We weren't getting along too well at that point.*

> *Then Lesley finally gave up on me and went to see the owner, T.J. T.J. stepped in and made the rest of my work on the project part of my regular work week—and gave me a raise. The raise wasn't a big deal but I was happy to forget about overtime, and I enjoyed working on the books again. And sure enough, it took three years before we saw the first printing, but then I also got my investment back from the first year of freelance time. That was great.*

"Fairness" is a useful appeal to put to work in situations where conflicts may be the result of self-interest. A very common headache for managers is the tardy employee. Absenteeism is a similar difficulty. The motives behind these problems can be many, and a moral appeal may not always prove to be effective, but troublesome employees must be made to understand that they must carry their weight.

Andrea was a senior technician, partly because she had been hired away from Challenger Technology, where she had years of experience. Whether her track record was strong at Challenger, Hugh didn't know, but at Trident she became a problem because of frequent week-long absences. The problem was that other employees had to cover for her, and this created an annoying amount of discontent that Hugh had to deal with. If he had one problem employee, the issue would not have been critical, but because Andrea's important workload had to be picked up by a number of technicians, the grumbling numbers grew.

> *I don't know if she is a hypochondriac or what. I didn't want to see the medical records even though she offered them to me. This problem went on for two years and involved one- to three-week absences every couple of months. Well, I happened to get the right opportunity to deal with it. We lost a guy to cancer. He took all his sick leave and never came back. Then Rosemary had a kidney problem and she was out for two months. These were real serious. Andrea was getting the cold shoulder anyway, but with serious illness and work backups in the office, she shaped up. Everybody made her realize she wasn't being fair. And her excuses looked lame next to the real illnesses. I think she figured that out.*

The issue of "community" is highly important in certain career fields where teams work in tightly knit groups. The fire service is a particularly unique team-effort business because lives depend on a concerted effort to act as a highly skilled single entity. If one member of the team is out of coordination with the group, lives are at risk.

Brad was a very popular firefighter and paramedic. He had ten years of service with a Portland area fire department and he was as comfortable with the fire-fighting teams as he was with the emergency medical aid units. He found the work exciting and humane. He was promoted to a position as lieutenant and was prepared to apply for an upcoming position as Captain. Brad also had a substance abuse history. He had been in clinics twice—once before his promotion and once after he became a lieutenant.

> *I have had problems getting off the stuff. The first divorce got me back on drugs and then there was another crisis involving my father. The department always finds out because we are a close group. The threat of being demoted didn't do much to threaten me, and people with drug problems already have dignity problems, so that threat won't work. What brings me around is always my fear that I could be the weak link and get somebody in trouble on a response call. These guys have stood by me and I'd never forgive myself if anyone got hurt. I'm off the stuff again because of that, but the problem cost me a shot at being a captain also.*

If we look at conflict as a response to perceived change in our power or our territory or our self-image, we realize *adaptation* is the key to conflict resolution. Whether we are threatened, or appear threatening to someone else, we need to look for adaptation. Similarly, if we stand outside the conflict and try to mediate it, we can seek compromise among others and try to bring about adaptation.

These are practical tactics for one-on-one conflict resolution. Being aware of the reality of perception is, as we have noted, the most important point. People fear what they fear. People are threatened by threats they think they see. Your role is to help resolve the threat of conflict as best you can, as thoroughly as you can, as quickly as you can. Your efforts to resolve conflict are clearly task-oriented. Seek a well-reasoned solution if possible. On the other hand, conflicts create unfavorable behavior that can be a negative influence in the workplace. **Supportive maintenance skills will probably be a necessary tactic as well, since you will need to encourage behavior modification.** At the least, you need to encourage a positive attitude. Social maintenance, after all, is the not-so-easy task of constantly building community in a world of change.

Resolving Conflicts between Groups

Our previous discussion of conflict established basic strategies for dealing with conflicts between employees on a one-to-one basis. Let's now look at the usual approaches used to resolve conflicts between groups. At the outset I would suggest a paradox. You might think that the difficulties of conflicts multiply with the number of people involved. This is not necessarily the case. Group conflicts do not grow either arithmetically or geometrically in group negotiations such as in labor-management agreements, federal mediation, or civic organization conflicts. In fact, group conflicts may be *more* resolvable than one-to-one conflicts.

A conflict between two groups of people can be efficiently brought to resolution if we recognize that there are two features unique to their numbers. First there is the risk of war. The entire community can be destroyed. When President Reagan refused to resolve the strike of the radar guidance union some years ago, the "community" was destroyed. The union and all its jobs were eliminated. There is great risk in conflicts that involve large numbers of people. War, which risks the destruction of community, can result in defeat. There are no winners. There is no victory. It is better to resolve the conflict.

The other force at work in groups is flag carrying. Whatever issue is in conflict between groups is a flag. There may not be the visceral involvement, the gut reactions, of a one-on-one conflict. The issues are often less emotional if there is leadership, a platform, and rational issues to discuss. In other words, **the focus shifts strongly from social maintenance to task orientation in group conflicts.** You do not have to *like* the union in your company; you simply have to *agree* with it. The union does not have to like the CEO; it simply has to *agree* with her. Disagreement is the source of the conflict. Agreement is the resolution.

The first issue in working toward a solution between groups is the need to perceive stages in the path from conflict to resolution. In the same sense that the dialectic process we discussed in an earlier chapter is quite universal, so too is the path to problem solving. The method that you use for structuring the agenda of a task-group committee is identical with the path of any logical decision-making process, *including* conflict. Be aware of these stages and use them, as *necessary* guidelines, for a path to resolutions.

Identify the problem.

Analyze the problem.

Offer solutions.

Choose the best solution.

Remember that the best solutions in conflicts between people are win-win outcomes. Resolution must not be viewed as "unfair" by conflicting members of a group or by a group that is in conflict with another group.

If a group is experiencing internal conflict (perhaps a committee on which you serve as a member), there are any number of paths—positive and negative—that can be taken. Suppose Roberts and Stevenson are the two managers on a committee. They gathered the group to discuss a problem. A realignment has been proposed for your department that will promote a few people and demote a few others even though most of the employees are left as they were. These are *changes*. There will be conflict.

The *first* strategy could have been to not have any such meeting. Roberts and Stevenson could have sent a memo with the new guidelines and simply said "this is how it is going to be." *At one extreme, conflict can be ignored.*

Since the managers called the meeting, and showed up, we realize that they are willing to discuss, and to listen—to some degree. At this level compromise might be at work, but there are different kinds—*extents* is perhaps a better word—of compromise. *A second strategy involves give and take,* with some hope that there will be amends. The idea is that cooperation will fulfill self-interests.

A *third strategy involves give and take with emphasis on problem solving.* In this case, balance is the *real* goal. As an example, The Benack Corporation is not *negotiating* a labor contract this year. For many years the union and the supervisors have been nose to nose.

They changed the negotiation climate completely. They—meaning labor and management—are *sharing* many discussions of their mutual needs; they meet at luncheons, and they plan to design, rather than hammer out, the next contract. Why the change? They have new union and management leaders with new ideas. It happens that both new leaderships are oriented toward *problem solving* rather than rewards. It is a lucky moment. The *resolution model* is the best one, because it focuses on problem solving.

In sum, if you have the opportunity to direct the energies of a group with internal conflicts, or if you can help conflicting groups, focus on problem solving. Above all, avoid a focus on rewards, and respect differences. *Do not argue; resolve!*

A very practical discussion of strategies for group conflict resolutions has found its way on to the best-sellers lists in recent years. *Getting to YES: Negotiating Agreement without Giving In* is as fine a set of procedures as you are likely to find for groups that are intent on problem solving and not on playing "hardball." This brief text by Roger Fisher, William Ury, and Bruce Patton (New York: Penguin Books, 1991) sets out to resolve conflict very much within the framework of task and social maintenance activities. The authors would agree with our observations about compromise, democratic or participatory workgroups, and a number of the "supportive" methods of approaching agreement. Let's examine their approach to *negotiation* for additional unique applications.

Essentially, when groups shift their focus from win-lose to win-win, they abandon the "me-first" attitude of so much traditional haggling that can harm committees or conflicting groups. The old style of negotiating compromise basically argues for "positions," which are statements that constitute a platform. **The traditional cause of the conflict was a *focus on rewards* rather than *problem solving*.** A more sensible tactic is to replace the win-lose, me-first positioning. This can be done without "selling out" by changing the focus of the meeting activities to *shared outcomes* based on a genuine understanding of *each other's interests* (*not* stated positions). Of course, this is easier said than done.

In a sense the Fisher and Ury strategy is an effort to humanize the confrontational nature of bargaining. The focus is on *people* (social maintenance) from the outset. The perception is really quite simple, if difficult to achieve. The idea is that *adversaries must cease to be in opposition*. Instead, groups should work in concert. The shift is the move from arguing nose to nose to working side by side—with the long-range objective of seeing eye to eye!

As a first strategy, we should respect the other group. Understanding people as people is the point of departure in replacing confrontation with compromise. We want to try to understand what the "other" group *feels* as well as thinks. Importantly, "openness" creates open dialog and a willingness to state or express feelings. Both groups must be able to understand the other group's feelings and hear their own feelings articulated as well. We speak of being "supportive" throughout this text; in the case of group conflict we must also learn to respect the feelings of others as *legitimate,* as equal to our own.

At the same time that we are making this effort to understand the opposition, **we should cultivate the relationship as a working relationship.** There is no room for the adversarial approach. Unique to the Getting to YES method for establishing the right environment is the very important concept of "interests." Beneath all the ballyhoo of political positioning are the *real* motives behind the defensive positions we try to protect. We observed that the real issues of real concern are *power, territory,* or *esteem.* These issues, and the motives behind them—and our feelings—are often veiled in the structure of arguing over "positions."

We need to look at the motives that create positions. One reason is obvious enough: the sincerity of our real interest is often hidden in the politics of our positioning. The reasons for subterfuge are many: decorum, embarrassment, company style, shrewdness, maneuvering. Whatever the reason, we must break the spell. The first and obvious reason to move our attention to interests is to get at the *real* reasons for a position. This gives us a better understanding of an adversarial position.

There is more, however. If groups discuss their interests, the dialog will lead to options that they will never see in defending their positions. Once a position is typed up it is locked in. One sheet of paper defies another sheet of paper from across the table. In the realm of interests we are not tied down. Far more options are available because the logic has not been channeled. It is strictly a dialog process, partly because no one would probably want their real interests to appear in writing. Besides, and this moves us to the next stage, the challenge of seeing *two* discrete sets of interests is critical if groups are to share the task of building solutions. **Then everyone on both sides can start to focus on options.** The real cards are on the table. The process of developing possible solutions to conflict begins to function as a *mutual interest.* The focus on mutual interest is simply the shift to *mutual gain* instead of individual gain. The situation becomes a win-win environment.

The pattern we are looking at in this strategy for conflict resolution is unique in still another way. Respecting and understanding others, even significant adversaries, should not be new to you at this point. Even the attention concerning interests (or motives or needs) is similar to the idea of understanding a "perceived need" or a "perceived threat," which are terms we discussed in *Workplace Communications.* But as a process that leads to problem-solving activity, this strategy does have a unique focus on *using* the interests or motives to build outcomes instead of conflict. We also have a unique method of creating a maximum number of available options to resolve differences. The remaining question is one any wily negotiator will ask: how do you keep one group from dominating the other even if the gains of the conflict resolution are mutual?

An additional condition for this sort of open negotiation strategy is perhaps the most unusual feature of this negotiating technique. We need a measure by which to judge the final agreements, one that would be akin to having a facilitator or an arbitrator strike a balance for the sides that are in opposition. The device is a mutually agreed to set of criteria that are used to judge the decisions. **The groups must determine what principles are to be used as guidelines to judge the outcomes.** At the very least, the conditions of settlement—the principles *all* members decide to accept for the judgment—should be independent of the two teams but derived from them. The teams must hammer out the criteria rather than argue over conflicts. If the teams build the criteria in writing, the mediating force is the written document. This is a unique tool for conflict resolution!

Essentially the document that states the objective criteria are to be used to evaluate proposals from both parties. The intent is to use the document to judge the outcomes rather than go through the typical "buying and selling" that usually goes on in this situation. Any of the following areas of concern could be included on a list of conditions of settlement:

The teams could mutually

- **agree to the efficiency of the best outcome,**
- **agree to the profit of the best outcome,**
- **agree to the cost of the best outcome,**
- **agree to the losses in the best outcome,**
- **agree to the management needs in the best outcome,**
- **agree to the employee needs in the best outcome,**
- **agree to the fairness in the best outcome,**
- **agree to the ethical standards in the best outcome,**
- **Agree to the legal implications in the best outcome. (There could be many others.)**

Using any of these criteria as the measure of the finished product should result in a just and respected outcome. Remember that the criteria should be agreed on long before outcomes are a possibility. No one gets to move the goalpost.

We sum up this very practical strategy in very practical terms:

- **Understand the people involved. Care.**

- **Share the interests, the concerns, and the motives of both groups.**

- **Build together; work-together. Construct all available options from the interests and not from platform positions.**

- **Agree to an objective method to measure the best possible outcomes and establish the objectives early on.**

The parks department of a major metropolitan area was planning on building a new recreational facility. It was to be located on park land on the edge of a service area. The citizens had not been consulted and they reacted negatively and demanded that the facility sit squarely in the middle of the service area so that it would be within walking distance of the maximum number of users. The parks department owned no land in the center of the service area, but the local school district did. A mothballed school sat on two acres of ground serving no purpose. Neither agency wanted to be involved in the complex option of putting a parks facility on school property. The community took up the challenge.

> *Neither agency would budge. They are very powerful and serve special interests. We made matters worse by trying to get the District Housing Authority involved to remodel the school as senior housing. Then we had a third agency that wouldn't budge! We wanted to transfer the school building to the housing authority and the property to the other departments for a recreational facility. We got all of them to the negotiation table only because our state congressional representative also happens to be a neighbor within walking distance. He had no civic authority but he was a powerful "observer."*

In meetings, the parks department, the school system, and the housing authority agreed that they are public service agencies that exist to serve the citizens. Important people were watching to see if they could prove the point. The school superintendent admitted that the mothballed school was a burden and sold the school and the acreage to the other departments for $1. The housing authority admitted that they never though of retrofitting city property before. They agreed to the "experiment" rather than build a new building for seniors. Once the parks department owned the land they had no argument for the other location. The locals got the recreation facility and a senior center as well. Win-win-win-win.

Inner Conflicts: Well-Being and the Workplace Lifestyle

Conflicts are not always of the interpersonal varieties that we have been discussing. We must realize that conflicts often exist for us because *we* see the conflict; something rubs us the wrong way. We create the conflict at times because we are stressed—"stressed out" as the expression goes—by some situation or another as we live out our role in the workplace. Our attitude controls the very *existence* of many conflicts, whether they are conflicts between us and our coworkers or conflicts between us and "system" problems where no one in particular is causing us any abuse: a noisy shop floor, inadequate medical coverage, a dress code. Perhaps the cafeteria food is the problem. It is a typical brunt of workplace complaints for some people. Many such conflicts obviously exist in the eye of the beholder.

Stress

The workplace setting cannot help but be a significant influence on your moods and attitudes. The influences may be practical matters such as the pay scale or commute snarls (system problems), but as we have seen in other chapters, the simple fact of community—and community politics—creates its own pleasures and stresses (the people problems). People have to *work* at getting along. We have analyzed a few approaches to conflict resolution that basically applied to resolution *between* parties. There is another path you can take—by looking at *your own* needs. You can make considerable strides in behavioral changes or attitude modification that can make a workplace setting work for you. There are times when the problem is in your hands. **At times, we must change *ourselves* to meet our own best interests.** Studies show that an overwhelming percentage of calls to the doctor's office are stress related. The results are all *too* real. The migraine headache patient is usually a victim of himself. The coworker who can hardly stay awake at work is keeping herself awake at night with stress. Stress will easily make you a victim, too. You must realize that you serve your interests poorly if you are a victim of your moods.

If a supervisor is absolutely impossible, then behavioral change means a change of scenery for the employee, but the frequency of little difficulties is the more likely source of our stresses—a little here, a little there. Soon the headaches start, and then the insomnia.

Then we take the troubles home and start to ruin our evenings and weekends. The question is, Can we adapt? Can we stay with the job, and stay healthy? Yes, if we reset the stage.

It is obvious from our earlier discussions that perception will condition attitude and behavior. If we are not happy employees it is because we *experience* our discontents. **The discontents may be "real"—external and observable to our fellow employees who share in the perception—or they may be personal and represent our own particular foibles.** Personally perceived conflicts are certainly real enough, but because they are personal they present us with the interesting opportunity to do a little psychological remodeling. Particularly in conflicts of our own making, we have the obvious option to resolve the conflict by simply making it *go away!* Our attitudes *can* be modified and we *can* improve our well-being. We can assert control over our own feelings and change them.

Stress is, of course, a response to a stressor. Stress is a response to a need for adaptation, which is a constant state for all living organisms. Because the environment changes, all living organisms must, over a period of time, be able to adapt. If an organism does not adapt it is likely to decline. We respond to the demands of our world. If these demands are particularly stressful to you or to me they can be quite harmful, as we well know. The highly stressed worker is an endangered species. High blood pressure is known to be stress related—as are headaches, digestion problems, heart problems, and a host of other ailments. Then there are emotional problems such as anger, anxiety, depression, and a wide variety of unsettling and potentially harmful moods. Whether the mood creates the ailment is of little concern here. Both conditions are harmful.

The point is that the stress manifests itself through body and mind. This is all very scientific by the way. You are perfectly familiar with the effects of adrenaline. We have also known for a long time, for example, that adrenaline (epinephrine) is a defense-response compound in the body, which is why synthetic doses of it and related compounds had to be banned from sports. Under stress, these powerful agents are released at sustained elevated levels. Too much aggressiveness is not good for either athletes or sports—much less the nine-to-five employee. Then there are the more recent discoveries such as endogenous morphines, which produce the marathon runner's high. The body *is* chemistry and so is the mind. Research on migraine headaches has identified the chemical causes of this destructive malady. Stress is as real as the nervous ticks we get in the corners of our eyes.

The obvious problem with being aroused to our constant defense is that our bodies and minds are always in some heightened state between alert and panic, the "fight-or-flight" effect. All our systems are racing all the time if we live in a state of stress. We simply burn out. The physiological effects are apparent in increased metabolic rates, increased heart rate, rising blood pressure, and similar "alert" responses. Then we neither eat nor sleep well. We make poor coworkers. We make life pretty awful on the home front. We can ultimately kill ourselves with stress: heart attacks, strokes, suicides, and similar disasters.

If we realize that many stress points are fictions of our own invention we see two apparent facts: (1) we often create the demons that plague us, and (2) we should seriously *control* the stress for our own good. We tend not to realize that we know friends, acquaintances, and coworkers who have died from their problems: suicides, heart attack victims, DWIs who did not live after that last drive. On the other hand, most of us get by. There is a positive side.

Attitude control is a possible solution, if not an easy one. Since a stressor is often not a shared response, we know that the perception may be the key to the conflict. Your idea of good music may not be mine. If I turn up my favorite station you may take flight. If you turn up your favorite station I may be ready to fight. Stress. Hostility. Since we can alter our perception of stressors, let's look at a few very practical controls you might use for relief. Here are a few useful steps to maintain your well-being.

Avoidance

Do you actually work a forty-hour week? It may be that you work far more and should cut back. Or it may be that you fret about worker-related stress for the entire drive home. Good time management is not a matter of doing more and more in less and less time. Then you have one of the worst stressors on your hands: overload. Good time management consists of good survival habits, an "off" switch and not an "on" switch.

Voice mail. Pagers. Cellular telephones. Answering machines. Accessibility to work-related matters means accessibility to stressors. Instead of being on a "digital leash," stressed people may need to get off the grid. Many people have an unlisted phone number and none of the convenience devices offered by phone companies. Many phones also have an on-off switch. Perhaps people should use the "off" position more frequently, or buy an answering machine also.

Limit your accessibility. The convenience items of the communications industry make all manner of stressors available to you day and night. Everyone will let their fingers walk all over you simply because it is so easy, so convenient. When our company e-mail system was installed, my message inputs grew exponentially. I do not have any critical need for e-mail, so I decided to analyze the inputs. The increase was *entirely* the result of convenience messaging, particularly "broadcasting," which resulted in time-consuming inconvenience to me. I delete close to 2000 messages a year, none of which are personal notes to me. Serious callers will use alternative, traditional paths: they will drop by, they will call, or they will otherwise ask for my whereabouts.

By extension, this same argument—avoidance—holds for actual commitments and not just for electronic time. **Just as we must shut down the gadgets to shut down our worries, we must finally cut down on any overtime or excess time allowed for work-related stressors.** Are you attending power breakfasts? Are you doing that extra office work in the car? Are you enjoying a good meal at lunch or talking shop? Use the off switch as much as you can. It takes practice, courage, and craft.

The last point is the tricky one. You cannot shut out your work world. Indifference can cost you your job. However, you can make real and sustained efforts to limit intrusions on your life—and health. Avoidance can put your job on the line. On the other hand you can make real and sustained efforts to limit intrusions on your well-being. Be crafty. Little will your employers realize that your efficiency will *increase* as a result of better attitude, better health, fewer sick days, and more employee and client rapport. Employers think in terms of quantity. Quality is *equally* as productive but perhaps not as accountable. Quality will improve if employees are not experiencing stress.

Exposure Limits

Cabin fever is a company dilemma also. Some people have strict tolerance limits for friends and coworkers. Skip, for example, simply shuts down when your time is up. The limit varies with each individual he encounters; you may get minutes or you may get hours. Or weeks if the going is good. Randy is another case. His threshold of attention for *anyone* is the length of any office appointment—in or out of his office. He has a twenty-minute threshold for any and all conversations. In other words you can control electronic time *and* actual contact time in order to limit the extent of your being around to be stressed.

You can limit your exposure to incidental stress by limiting the seriousness with which you take certain people who say "Oh Sue, seeing you reminds me that I wanted to ask . . . ." This particular type of request for employee time comes from someone who did not need Sue's time badly enough to even remember to look her up. The more you are *seen* the more likely you are to be inconvenienced by passerby encounters and their requests. **Low exposure levels eliminate trivial stress problems.** Trivial stressors may not provoke adrenaline, but they are just one more provocation for high blood pressure.

On the much more serious side, we have all had supervisors or workloads that are *major* stressors. Exposure limits become a real problem. Tolerance can reach zero. The causes may be many. Some are real among coworkers. Some are personal. Assertive managers cause some of the difficulties. Excessive workloads are another case. One way to handle the dilemma is flight. Avoid the personal annoyance by avoiding the person or abbreviating the workload. Limit your exposure to bothersome supervisors if you have the flexibility to be elsewhere and still get the job done. Also limit your visibility when more work is being farmed out.

For several years Phil was overinvolved in labor-management problems. He was gone days. He was gone nights. There was a great deal of stress—particularly when there was a strike, or media attention, or federal mediation. After several years, Phil had to face the fact that he had experienced acute overexposure. He had to simply *quit*. You see, the contract negotiation tasks were a voluntary overload well beyond his regular workload.

> *The extra work overwhelmed my circuits. I had no home life, but I had plenty of worries. The solution was to be normal! I removed myself from the activities completely so that I too was a forty-hour employee like anyone else. The stress was gone. It wasn't eliminated from my work world of course. Labor-management problems go on daily. But I eliminated such problems from my personal agenda. If you can pencil stress in, you can pencil stress out.*

Worry

If you take your problems home with you there is a very different control dilemma you have to handle: portable stress. Using *avoidance* and *exposure limits* will cut down your access to stressors that can pester you by way of media intrusion, overloads on your workday, and too much contact time between you and coworkers. If, however, you take all your stress home, then the internalized anxieties will reach the point where your workplace is not the major problem. Instead, there is a greater, self-induced problem: you cannot shake off the stressors. If the stress is floating around in your head, then additional strategies are critical. You are perhaps a hard-driving, achievement-oriented employee. Remember, stress is perceived, and perception is strictly what you make it. Because the perception is changeable, you are adaptable and can deal with take-home stress.

The tactics for internalized stress relief are obvious enough when you pull up to your apartment after work. Your spouse comes to greet you. Along comes the dog. The neighbor waves. We lose our commitments to these bonds in our lives because we center ourselves around work. **We should learn to center ourselves around our *support systems: our ties.*** We must reexamine our overall motive for working. We do not work because we "have to support a family." In truth, the family supports *us* in our efforts to deal with our world. We understand this truth exactly backward. Support systems are the critical tools for stress control. For the totally stressed out and overworked personality, support systems will help *repress* stress. A sincere and long-term commitment to our traditional values will *replace* stress, and the time and attention we have given it.

Natural Ties

To some extent the employee who is a workaholic has little else to do in life. The basic realities of family, friends, and social activities somehow end up pushed aside. This condition indicates that **one strategy for employee well-being is a healthy preoccupation with quality time and not company time.** For the hyper, hot-wired employee, life is little more than a spectator sport. We need to center ourselves on life's joys and not on life's labors. The labors should serve our pleasures rather than diminish them. We have all had friends who used unemployment benefits to have a rest—a sabbatical—and who have enjoyed the meager allowance to have a very good time with family, friends, and pleasures. It is interesting to see how few critics there will be among their friends. Somehow we see the good in the quality time and wish we could do the same. We can. We need to focus on our mates, our children, our neighbors, and our game—whether it is golf or bowling, boating or fishing. We need time for undivided attention to family, friends, and leisure. We need to allow for relaxation. We must unwind if we hope to live the good life. Work can be part of the good life if it is kept in its place. No one, so far as I know, ever sat up on a deathbed to state the regret "I only wish I had spent more time at the office."

The Wellness Trend

It is no trend at all to want to be healthy. The idea of the "sound mind" and "sound body" is as old as ancient Greece and probably as instinctive as our fight-or-flight response. We can exercise avoidance and exposure control only if we have a desire to do so, and the best way to construct a desire to do so is by developing options.

Family is one. *Friends* are another. And *fitness* makes a fine third partner for a better well-being. Tennis, soccer, biking and all the more highly aerobic activities have a *positive* stress effect on the body and the mind. Stress, we should note, is a condition of the environment, and it can be *good* for us as well as bad. Organisms improve themselves by responding to stressors in ways that positively alter their states. A good workout—even walking, bowling, and golf—will positively benefit the body and the mind.

There are predictable benefits from a healthy exercise program:

- the anticipation of fun
- the goal of fitness
- relaxation for the mind
- positive outcomes for the ego

It is nice to feel good and look good. And it is science. **The same chemistry that creates the flight response pumps up our sporting response, but with fulfilling, positive results.** Besides, if you can power down on stress you are more fun to work with as a result of all that new enthusiasm.

Mitch works for one of the northwest lumber companies as an executive. The corporation has a campus nestled in hundreds of acres with parklike trails. The pace looks quiet from the outside, but it can be hectic.

> *I thought running was mindless. And when I started running I was even more convinced. It is stupid. But compared to my smoking, it was brilliant, so I kept at it to get off tobacco. That was a lot of years ago. Now I run and love it, of course. Mainly, I figured it out. Runners run for no reason, and that I never understood. I am running from something to something. I run from stress to health. That gives me a mission. I can blow away a headache in the first mile—no problem.*

Simplicity

One path to improving the workplace lifestyle is apparent in all the preceding suggestions: streamline your work, but also try simplify your leisure. If you have to work frantically to earn the thousands of dollars for those two weeks in Mexico, then you are probably missing the point. If you buy your new car by calculating the payments in over-time, the result is the same. **Our consumer drives have a deadly effect on our well-being by forcing us to work more and more.** Work stressors often increase *not* because of work but because we entrap ourselves in the service of a complex and expensive lifestyle. Streamline. Simplify. Both serve to reduce work time and work stressors. The well-known book *Your Money or Your Life* by Joe Dominguez and Vicki Robin (New York: Viking Penguin, 1993) might be appropriate reading if you are interested in this particular issue.

Perception

If you have half a glass of water, is the glass half full or is it half empty? Remember that a stress point of *yours* is likely to mean *nothing* to someone else. Other people will not necessarily hear or see or experience the problem. Although behavior changes that allow you to reverse your perceptions are not always easy to achieve, such changes can and do happen. Turning points, for example, have odd effects of this sort. Turning twenty-one. Turning fifty. Marriage. Becoming a parent. Or a granddad. We change. Our perceptions change. What was important yesterday may not be important tomorrow. You *can* take control. We can also work on these perceptions when work-related stress is halfway up the glass. We must see both sides of the picture. For example a contract employee has no medical or dental plan—but what a joy to not work regularly! A contractor has no retirement program but can make fifty dollars an hour on a big job! **What you see *depends* on what you *want* to see.** Think positive.

Work Habits

You are back to work on Monday after a two-week vacation—on the cheap—with family and friends, workouts and reading, fishing and gardening, and peace of mind. And here you go again. Back to work. Stress is under control, but maintenance is important. Stressors return to haunt us. Keep pressure down with a few simple guidelines you will frequently see in magazines, hear from your doctor, and even see posted in the company lounge. Here are a few tips:

- **Do not try to achieve too much**

Overachievers are a problem for themselves. Do not offer more than you can produce in the *typical* workday or workweek.

- **Do not seek perfection**

Quality is not perfection. Do not confuse the two. Perfection, as we noted much earlier, is a problem. Quality is an outcome if we are good at what we do.

- **Avoid overload**

Two easy methods of eliminating overload are to create conditions where you cannot be interrupted, and to indulge your calendar by overscheduling all activities. In other words pad the timelines so you are not always behind.

- **Protect private time**

Breakfast, coffee breaks, and lunch are *not* company time. You will be better for it as an employee if you have a book with you, or take a walk, or plan your weekend.

- **Do not be sedentary**

Avoid static routines for mind and body by moving around during the day.

- **Do not disconnect**

There is a balance. The real goal is to enjoy work—by making it what it is, and no more.

Wellness is a huge industry—a larger version of fitness. You must prompt yourself with whatever allows you to seek a balance so you can neutralize points of stress at work. From touch football to meditation to letting-go therapy, you can use whatever works. We have looked at only a few of the basic options. Variations are endless, and relief is always available if you are perceptive to your needs for health and happiness.

Above all, do not take stressors for granted. Try to recognize stress factors. The stressor is not always apparent, and adaptation is not always easy. For example, we now use our eyes to make a living (instead of our shoulders) and then we stay in bright artificial light until the eleven o'clock news is over because, as we reason, this is our relaxation time. Yes and no. You would have to ask your eyes. For a thousand generations our eyes were exposed to almost equal volumes of light and darkness. Now, in three generations we have tried to adapt to artificial lighting and extended eye stress. But life is not so simple, and our eyes and our minds are greatly stressed by this basic change in the human lifestyle—*if* we do not recognize and respond to the problem. And what, after all, did our ancestors do for 10,000 years while they waited for Edison's invention? They got plenty of sleep—while studies point out that one in three of your fellow employees regularly sleeps six hours or less per night. Light stress. Eye stress. Sleep deprivation. All this plus the evening news—and most people do not recognize the dilemma such habits construct for them.

Summary

- Conflicts are often created as a result of changes that affect the perceived territory, power, or self-esteem of an employee.

- The best resolutions to such conflicts are win-win solutions that resolve conflicts to everyone's satisfaction.

- Help employees who are experiencing conflict by being supportive. Also seek a rational solution to the problem.

- Encourage *adaptation* as a resolution to conflict because the workplace community is not very flexible. It is structurally designed to meet collective needs rather than individual needs.

- The conventional strategy of conflict resolution in a group environment is a time-honored problem-solution analysis:

  Identify the problem (the conflict).

  Analyze the problem (the conflict).

  Offer solutions.

  Choose the best solution.

- The success of conflict resolution depends on a social contract of openness.

- Problem solving involves respect for all the task group members if the conflict is internal.

- Conflicts of interest also emerge between groups. These conflicts must be analyzed and resolved.

- A working relationship is cultivated by respect for opposition.

- Motives must be stated clearly and understood clearly.

- Conflicting members or groups must agree to criteria for resolution of a conflict.

- A group that experiences no conflict must guard against the opposite problem: exceptional group unity. Specifically, a cohesive group must be sure that the resources for a decision were thorough and correct. (See pp. 172-173.)

- Internal conflicts can be managed by removing your exposure to stress.

- Workplace pressures can be kept in perspective if your private support system is in place: family, friends, pets, and pleasures.

Activities Chapter 14

Present a memo to your instructor that develops a discussion of a personal experience that relates to the chapter. The memo should be 500 to 1000 words, typed. In it you can explain how you have related the text or the lecture discussions to some event you have experienced. This exercise will give you a better understanding of the material because you will explain incidents in terms of your perceptions of workplace structure and communication.

Select one of the following suggestions and develop an analysis that involves your current employment or a former position.

- *Explain the history of a conflict between two employees. Discuss the process of resolution.*

- *Explain the history of a conflict between a group of employees and an employee. What values were involved? Was rebellion a problem? Were misunderstandings a problem?*

- *Identify a workplace incident that did not resolve. This event probably involved the departure of an employee who could not get along or was not accepted, or both. Explore the issue.*

- *Discuss any experiences you have had in teams that were involved with conflict resolution: grievance committees, contract negotiations, and similar matters. These settlements or agreements also can involve dealings with agencies and can involve everything from building code permits to health authorities to OSHA.*

- *Explain your most fulfilling work experience. Why was it rewarding, and what was your response to such a feeling of well-being?*

- *Explain your worst work experience. Was it threatening to your welfare or well-being? Discuss.*

Work In Progress

15. The Cultural Matrix

I later found out some information about Tempo that I will include here as a kind of post-script. The majority of the production workers at Tempo are immigrants from Mexico or Southeast Asia. Tim and Sharon are proud of their employees and their production standards. They support their workers through good wages, benefits, and a safe workplace. Notices posted throughout the company are always written in at least three languages, and the employees are encouraged to come forward when any problems arise.

I didn't know it at the time of my original bid, but the real reason the wiring closet area was vetoed in the first place was not because of any "plot" by Steve, the nephew, but because Tim and Sharon were confidentially concerned that cutting the lunchroom by 20% would prove a hardship to the production crews who didn't have desks or offices to escape to for breaks or lunches.

After I forwarded the second proposal, Tim decided he would gather the entire production crew together in the lunchroom and show them what having a wiring closet would mean to their space. He let them know that if they did not want to give that space up, then he would not force them to. Tim let them know he valued their hard work and did not want to do anything that would seem to slight the employees.

Much to his surprise, the production crew did not have a problem with losing a part of the lunchroom. Instead, they were curious about what would be put there and how the partition wall was to be built. They liked the idea that the network would extend into the production area, since some of the production workers needed to learn some basic design software. It was this seal of approval that put the NetWorks NorthWest bid over the top! The three other bids had accepted Steve's demands and never proposed a wiring closet.

Here is a final note: Since I came to network engineering from the construction trades, I am used to being the one female on any given job site. High-tech businesses are more gender balanced, but only slightly so. I am really careful when I am having those first few phone calls and face-to-face meetings with potential clients. I weed out any who might give me (and NWNW) a hard time because of my gender. I still feel that perhaps I do have to prove myself a little more than my male colleagues do, but I don't waste much time or energy being upset by it. Like any other contractor, I am motivated to keep achieving my goals—and to win the next contract.

J. Q. C.

Under-standing a Misunder-standing

Of the main problem areas for faulty communications, one stands tall above the others for creating much of the difficulty, and that is the head on our shoulders. Our life histories have a profound conditioning effect on our way of seeing and interpreting the world—including, of course, all-communications with everyone in it. Two particular realities are at cross-purposes in every conversation event we happen into: language and the interpretation of language. Because the language we speak is a common ground, it holds us together as coworkers; but this unifying force can be very deceptive. Apart from the language and the shared company setting, each employee is distinct from fellow employees.

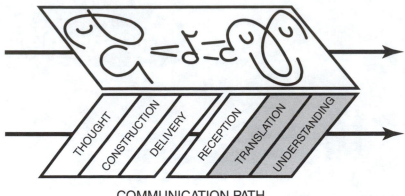

COMMUNICATION PATH

First, the *language* we use to bridge the two minds involved in a conversation is used with the assumption that both parties agree to the practical meaning of all the symbols or words we use. We assume, or our language assumes, that anyone we speak to is one of us, or like us. On the other hand, each of us is different. We *are* the same—sort of. We are enough the same to be speaking English let's say (260 million of us) or to live in Idaho (one million of us) or to live in New Orleans (300,000 of us) or maybe to work at the local pizza shop up the street from the college (20 of us). We could take this further, since life experience is, in fact, a solo journey and the final sum of people who will take your journey is exactly one.

The effect of conversation is therefore deceptive. Our language assumes—*depends*—on common experience. We have enough truly common experience to succeed—usually—in understanding each other. Yet the individual backgrounds of our life experiences are the likely causes of misunderstandings that can skew communications that would *usually* work, so we should look at this area closely.

Consider the two popular uses of the word *misunderstanding*. There are two types of misunderstandings. If you miss the point, or miss the understanding (viewpoint) of another person, you have one type of misunderstanding. If you argue about it, you have the other type of misunderstanding. The first type is more common and more subtle. In many respects, the *common* misunderstanding is more serious than the *argument* type of misunderstanding because it is a frequent condition and because people often do not realize that they have been misunderstood—until it is too late, so to speak.

Misunderstandings can cost a business enormous sums of money. One huge misunderstanding is the least likely cause of such a financial loss. Far more serious is the draining effect of endless lesser misunderstandings. Each small loss can chip away at the financial base of a business. "So we ordered fifty staplers too many for the office. It's not a big deal. I thought that was what Susan told me to order." Maybe it isn't a big deal—by itself. "Besides,

I'll return them." Okay, but Andy ordered them on company time; he returned them on company time; Sheryl typed the order for him; and where did he get the postage? These "costs" seem to be very minor, and nobody sees them. It is not a big deal, unless it is one of possibly hundreds or thousands of errors.

If this occurs with any frequency at work, the loss in time and money can be critical. An argument "misunderstanding" may be caused by much the same confusion but no one is likely to act until they agree or resolve the dilemma. The argument may be unpleasant, but at least it signals conflict. The endless little misunderstandings we confront often fail to send any signals.

Corporations are often frustrated in their efforts to stop communication errors, particularly spoken errors, because the causes are not on paper. We *have* the order for the staplers; the question is *how* the misunderstanding developed. But observe, there is probably no question about *where* it developed. Misunderstandings usually develop in *conversation*. It is quite difficult to improve efficiency in oral communications. Any workplace is burdened with various misunderstandings.

Understanding Cultural Identities

It would probably be difficult to find a distinct social or regional group in the United States that is not criticized for poor driving habits by other Americans. This pervasive stereotype clearly reflects a conflict in values and backgrounds. On a recent National Public Radio program, New Jersey natives were criticizing the driving of immigrants from India. In Fort Lauderdale, the fussing concerns the retired New Yorkers who—as the stereotype goes—drive at ten miles an hour. In Seattle, California drivers take the flak as bad drivers from the land of gridlock. This all-American type of criticism about "outsiders" goes on from coast to coast, because we are all different, and because we misunderstand each other. These misunderstandings are costly for corporations in the same way that numerical misunderstandings are costly. Morale and teamwork are critical to profit. Conflicts cost money.

American culture is a unique blend of many cultures that created the larger culture we know today. No one notices that the cheerful Tex-Mex music of the Southwest consists of Eastern European polkas sung in Spanish by Chicano singers who play Italian accordions because there once existed an equally unique American: the German cowboy, which incidentally, also explains how yodeling ended up in Texas. We seem to blend together—

particularly at work—and still go our separate ways. Our neighborhoods often reflect any differentiation that survives to make us distinct. We are similar, but our backgrounds are often a subtle cause of not quite understanding coworkers.

Neighborhoods reflect income, "class," cultural background, color, and even religion. Though our families may have long lost any trace of immigrant cultures and values, we remain unique. We are *not* all the same. Religion separates us. Income separates us. Color separates us. Values, attitudes, life experiences, family recipes. The list goes on. We always remain somewhat different. We lose sight of these distinctions at 8:00 a.m. sharp when all of us show up for work. The American workplace is a medley of Americans and the subgroups they reflect, not to mention immigrants, who enter the country at the rate of one million a year.

The problems that can emerge from grouping us together at work deserve your attention. Of course, any corporation or business is going to be a society of sorts. We know that already. As a result, the workplace setting will already be divided into many units that reflect group distinctions or classes. There will be executives, managers, secretaries, accountants, engineers, assemblers, clerks, janitors, and so on. All these ranks must interrelate, and that means one thing: good communication.

The divisions may be as simple as "white collar" and "blue collar." Behind the superficial divisions are the serious differences such as education or ethnic background. Because of these distinctions we need to be sensitive to the way people feel about issues, the way they understand us, the way we understand them, the way we communicate, and what we mean to say or do. We appear to be similar, and we are. We will all go home and watch the evening news. But we are also different, and the neighborhood you go home to may not be home to anyone else in your office building, which is a common carpool difficulty.

Our personal history becomes *the* background of our thinking. **Our history will color and condition the way we think, and the way we interpret what other people think.** Obviously there is a great potential for misunderstanding other people. It is just as likely that we may misunderstand our friends at work as it is that our coworkers may misunderstand us. Our personal history is a common source of such conflicts.

Our concern here is not the experiences we *share*. The issues are the experiences we *do not* share. For example, if I speak to a woman, there may be differences in the way we interpret what we say to each other. Yes, we are Americans, we live in the state of Ohio, we are urban, we are married, we have bills to pay, we have retired parents, we share a *hundred* experiences. But one of us is a male and one of us is a female. Does it matter? Suppose the conversation is not with a woman but with an older man—perhaps thirty years older than me. Will that matter? Of course it will, since he will reflect the values of a different generation. A very common problem in communication is often the interference that our own values create in our efforts to understand the people around us. Just how different are we, one from another? Just enough to cause occasional confusion—usually because we *do not* notice the differences, until we see a conflict.

Tony was at an office party and was looking at a magazine advertisement of an older Hollywood star who was in a full-page ad for fur coats.

Well, I got a little cynical I guess and called her "old" or something. I don't recall. Kent came rushing at me and almost dropped a tray of sandwiches in my lap. He was livid. I mean he has a temper anyway, but I didn't think about his age. Wow, did that set him off.

Understanding Historical Makeup

There are obvious elements in our historical luggage that greatly condition the way we communicate both in terms of speaking to others and listening to others. The historical elements in personal histories shape everyone's perceptions when they communicate. The conditioning effects are sometimes glaringly apparent to ourselves and to those around us. They have identifiable origins and can be analyzed. We need to have an awareness of these "preconditioning" elements. They will often explain where people are coming *from* in a conversation, as well as where people stand.

There are many historical influences, and we are well aware of them:

- **cultural background**
- **environment**
- **class**
- **income**
- **religion**
- **color**
- **gender**

Americans seem to have a cultural tendency to pretend that these matters are subtleties. Perhaps we should measure the importance of such topics by how *little* we allow ourselves to mention them. Then there are additional forces at work:

- **age**
- **geography**
- **life experiences**
- **interests**
- **attitudes**

We must be aware of these many forces. They influence *everyone* in the office. We do not rule our lives by them, but they remain meaningful in our understanding of our world and in our understanding of the people around us.

Cultural Background

The culture of our upbringing is a vast conditioner. It can determine family values, work values, social behavior, courtship ideas, and many other larger forces at work in a society. Consider the hour of the evening meal. Americans enjoy a fairly early dinner. South Americans will often eat the largest daily meal at 1:00 in the afternoon. Spaniards are well known for eating at 10:00 in the evening or perhaps 11:00 or maybe midnight.

Do such cultural practices shape a personality? They shape entire *cultures*. It is not a co-incidence that the early evening news is on television at a very specific hour in the United States. Even the eight-hour day that helped us get home in time for an early din-ner is essentially an American idea, as is the five-day workweek. The two-day weekend is, therefore, a social expectation, and it defines such cultural elements as the way sport-ing events are scheduled. By the millions, we line up in unison at the stadiums, and yet subgroups abound.

Cultural conditioning is also obvious among the ethnic groups that have stayed together in neighborhoods throughout the nation. It may also be a subtle force for older ethnic groups that have lost their old languages and their old neighborhoods, but whose grand-parents or great grandparents are still shocked when the pride of the family comes home with a fiancée from some other cultural background. The *Wall Street Journal* interviewed Chinese American executives about their apparent inability to reach the vice-president levels of major American corporations. It turned out that the positions were being turned down. The executive life is all-consuming, and "making it" can mean risking your mar-riage and your family. The young Chinese American executives explained that family life is too important for them to sacrifice for the hard corporate demands on their family time. The old values endure.

In the United States, there are hundreds of local magazines and newspapers printed in a mix of English and the native languages of many groups. There are Asian newspapers in the Northwest; Hispanic newspapers in the Southwest, California, and Florida; and European newspapers in the Northeast. My next-door neighbor fondly remembers making wine with his immigrant Yugoslavian father. That was sixty years ago. To keep the ties, he has subscribed to the Croatian newpaper from Pittsburgh—*Zajednicar Fraternalist*—for the last twenty years, although he has never been east of the Mississippi and has never seen a tamburitza ensemble and does not speak a word of his father's language from the "old country."

In my town you can go to a Chinese New Year parade to see the huge dragon, listen to an elegant mariachi band in May, watch a Scottish pipe band in July, and then go to the university to watch the university halftime show at an autumn football game, complete with cheerleaders and pom-poms. The United States is indeed something of a phenomenon.

Environment

All sorts of historical influences govern our perception, our behavior, and the way we communicate with those around us. Many influences—such as income level, social background (class), religion, color, and even geography—are what we might think of as conditions of environment. These forces have a persuasive effect on our lives, since we all live with them.

Class and Income

Chances are good that you can talk about football with your boss. You will communicate on equal footing. How about your bowling league? Or his golf? All of a sudden the conversation may fall flat. He is not too likely to bowl, and you are not too likely to golf, at least not at his country club. Let's turn to your other supervisor. Let's talk cars, since you both have been complaining about bills. You tell her that you are not happy with your $500 repair on your truck. She just laughs and says, "Try $3,000," which she just sank into her Range Rover. Her figures will not make much sense to you. Your figures will not make much sense to her. Class reflects shared values and will often be reflected by where we put our money—or *would* put it if we had it.

The two distinctions that seem to be commonly used to define "class" in the United States are money and education. Millionaires may be few, the luck of the draw. Wealth is usually the identifiable characteristic of the upper crust. A.A.'s, B.A.'s, M.A.'s, A.A.S.'s, B.S.'s, M.S.'s—the list goes on—are an ever growing group that is usually linked with whatever we mean by the term *middle class*. Our culture has a tremendous respect for education in general and looks to college education as the key to success for the young. Most families have an overwhelming wish: that their children go to college. They equate education with success, and a sort of prestige or status.

Class does not necessarily hold to what might be called demographic distinctions. Money and education are not entirely predictable indicators in the United States (though perhaps family background in these areas might be). You will see a great many contradictions if you visit an estate such as Graceland. "Class" and "economics" can be very oddly matched in a nation of opportunities such as ours. Our sports heroes become millionaires overnight, as do our actors and actresses in the film industry. Rarely do they "come from money." The same is true of rock stars, entertainers, and tens of thousands of successful merchants and entrepreneurs of endless sorts. Sketchy though class may be, there is a lot of labeling, or outright name-calling—"yuppie," "redneck"—associated with it that is regrettable. These terms reveal serious barriers in understanding and serious social barriers in our culture.

Religion

Faiths will influence our values and our communication. You can see this in the example of swearing. People who swear a lot usually overlook the fact that the people around them may swear little or not at all. Of course, this difference in word preferences is a complex issue and could be the result of a mixture of family values, class values, religious values, and other forces.

The faiths we uphold are persuasive forces in our lives, as are other peoples' images of those faiths. Consider the notion that Catholics oppose abortion rights. It is risky to make such assumptions. Indeed, the American Catholic leaders have for some years been at odds with the Vatican over all sorts of issues. Protestants on the other hand are thought by many to be pro-choice. Recent administrations of Republicans, whose ranks include large numbers of Protestants, are often pro-life. Activists who engage in protests against abortion are usually not Catholic and frequently reflect Protestant values of a very conservative nature. Religion is often stereotyped when in fact there may considerable and obvious evidence that the stereotype is false.

Color

Color usually reflects ethnic background, but the specific one that is visible. Needless to say, vast numbers of ethnic or cultural groups are not particularly identifiable, especially after a few generations in the United States. A new Russian or Iranian immigrant may not be particularly identifiable. An African American citizen, however, is part of a visible ethnic group. You would think that after generations of living in the United States, African Americans would not have an ethnic identity. In reality, only groups that are *accepted* by a host culture proceed to "melt" into the mainstream. People of color are often denied acceptance because of a particular bias against skin color.

Our melting pot accepted the huge European immigration waves of the 1880s and 1920s long before accepting the African Americans who were here all along. The result, within black culture, is obvious enough. The common struggle they experienced *created* an ethnic culture. Some authorities have even suggested that black English shows all the signs of a separate language structure. Black handshakes, or high fives, communicate and speak of their kinships.

These few observations overlook the reality of colors within color. Duke Ellington wrote a famous suite called *Black, Brown, and Beige.* You need look no further than your newspaper to see these distinctions among many cultures. Where we think we see the Yugoslavs, the Yugoslavs see Serbians, Croatians, Dalmations, and so on. The results can be very significant—even tragic.

The communication difficulties between a person of color and "whites" (a remarkably vague stereotype) is obvious to all of us. **Color barriers can influence serious issues such as employee relations.** Ethnic friends may pal around together, which should not be a problem—from their viewpoint. Other employees may see this behavior in negative terms. And possibly there are matters among ethnic employees that are complex, such as an ethnic coworker who does *not* associate with other ethnic employees. Consider the issue of a person of color misunderstanding a white superior. Is this going to be looked at as just any old misunderstanding? Or perhaps there is, let's say, an African American foreman who has conflicts with his white employees from time to time. People will talk. Color counts, by which I do *not* mean that it *should*. Our conditioning clearly *includes* color, since color sensitivity is an old problem in America.

Gender

Another workplace concept that can clash with the cultural backgrounds of individuals involves cultural attitudes toward the opposite sex. Is the boss who likes football always a guy, which I appeared to suggest earlier? Was that a stereotype or a reality? The overwhelming presence of women in the workforce today has resulted in many conflicts. Certainly, most women have a story or two to relate about encountering prejudice—which means, of course, "misunderstanding."

At our college we have long encouraged women to enter the engineering and vocational areas, which are sometimes called "nontraditional" employment sectors for women. Some years ago, the women from our electrical power program had problems with hazing, teasing, and other obstructive behavior when they first entered public utility jobs, particularly if the work involved manual labor. There were court cases as a result. Highway road crews obviously had to experience the same growing pains nationwide, but now women are common on the city, county, and state road crews. An army drill sergeant assured me that the armed services had to learn by the same process—in court. Military scandals in recent years clearly reveal an unresolved gender conflict. We know from the military court cases concerning harassment that the problem is not resolved.

Part of this cultural conflict is evident in the word *nontraditional*. *Tradition* means what is normal, what is expected, what we did before. When we make waves we upset the normal expected paths of daily lives. Court cases involving linemen or firefighters are conspicuous enough to not dwell on, but equally common in the workplace in today's world is the female supervisor. Women in command will find the that old values can be very subtle in white-collar sectors. The boss does not take no for an answer, but if men have not dealt with a female supervisor there can be conflict—because she is still the boss.

Remember also that because our country is a melting pot, **the gender attitudes of "Americans" will also vary with a great deal of possible ethnic variations concerning male perceptions of what the role of a woman should be and female perceptions of what the role of a man should be.** A male may say yes to his female boss, with a large internal no. This reaction will lead to possible conflict because of his cultural background and because his idea of what a woman's role should be will conflict with what it is.

Geography

The network television stations may create the image of a great massive America, but the size of our nation means precisely that regionalism matters. Regional distinctions and regional values are pervasive. Partly the issue is a simple one; the United States and Canada cover an entire continent, and regional character is the inevitable result of the size of the landmass. The regional character of any particular location—say, Texas—is shaped by its history and by its economics. A cattle ranching economy will abide by a lifestyle that evolved in the region. The values will match.

Joe was from Tennessee. Meredeth was from North Carolina. Both had university degrees. Both were working in the Northwest, where they lived together. What struck me as odd was that Meredeth had no accent, and Joe drawled on like a southern farmer. After dinner he was giving me a guided CD tour of Jimmy Rodgers and other early country-western singers. Finally, I could not resist the temptation, and I said,

> *Joe, how come you speak with an accent and Meredeth doesn't?*

He thought for a second and beamed. Then he said in a mighty drawl,

> *Cause I'm proud.*

Meredeth was making coffee and heard us and stuck her head in the door and said, with an equally powerful drawl,

> *Hush your face, Joe Anderson!*

Meredeth went back to the coffee and Joe whispered to me.

> *There, see, just get her angry and you'll hear plenty from the Carolinas.*

In the corporate world, these subtleties are no less formidable. As corporations have grown they have encompassed all regions of the country. The employees in these companies are supposed to respond to each other with professionalism, but that does not mean they respond uniformly. Regional differences can be important.

Mark was an executive for Sears Corporation. He was promoted again and again. Finally the big day came: he was promoted to New York. He and his wife were both natives of the Northwest. Within two months the great bubble of success broke and Mark's wife, Linda, went back to Seattle. She hated New York, and the people of The Big Apple were a close second.

I tried. I couldn't seem to get to know anybody. Everything was so fast and noisy. Well, that isn't right. Everything was so different! I can't explain it. I didn't even like the accent. I mean, it wasn't their fault. I just wanted to go home. My phone bills were huge. I called home everyday.

For anyone who has lived on both coasts, the problem was easy to see: culture shock. If Mark wanted to join her he knew what to do. He struggled with the issue of his career or his family. Finally he walked the plank and told the uppermost reaches of his company that he would have to go. His choice and his courage paid off handsomely when the company refused his resignation but sent him back to the West Coast, where he and his wife could reach a compromise. They live in Southern California near a major Sears headquarters.

Experiences

Notice that we constantly must recognize how hazy the definitions are because generalizations always admit of varying numbers of exceptions, sometimes vast numbers. To refer to a category as "experiences" is equally confusing, since all the categories are "experienced." Beyond those categories, however, are life's basic experiences—marriage, divorce, the way we are raised by our parents, our own child-rearing efforts, disabilities, and so on. Let's look briefly at two simple examples.

Consider "phobic learning." One of Anna's parents always feared stinging insects. She now fears stinging insects. You might think that it is a logical fear. True. Anna's family also lived in a state with a species of worm that could enter your foot if you walked barefoot. For generations, all children in the area were drilled on wearing shoes and sandals. Anna also does not like to walk barefoot on grass. Often the lifelong perception we have of, let's say insects, is one conditioned by family experience, personal experience, or even regional experience.

On the darker side, we all know that life experience can be traumatic. Recently divorced people get into all manner of trouble because they are preoccupied or insecure. They are not focused on what they are doing, such as driving a car. Divorced people are high-risk drivers. Certainly, the risk curve falls off after a year, but actuarial insurance data easily demonstrate such problems. Such troubled minds trip up steps. They walk into door frames. They leave a bag of groceries in the store—maybe the only bag. Anyone going through a divorce (or other crisis such as a family death, loss of a job, and so on) is a difficult person to *communicate* with. When our friends are in such a crisis they are not attentive. They are enormously preoccupied. In the workplace the effect can at times be costly, even dangerous.

Interests

In this category we might combine our interest in work, hobbies, sports, and possibly other areas such as community service or similar volunteer work. Many people take their sports interests right to work every morning. In fact they cannot wait to get there—not to work but to talk about the game. The game the day before is *the* topic of conversation. Loren is a baseball fan of the serious, serious kind. He goes to the preseason games and spends his two-week vacation enjoying the southern sun and taking in as much baseball as he can. He calls his new boss a "rookie" and contracts his engineering activities so that they will not interfere with his softball team commitments. The softball is more meaningful.

Work is also a major life interest for many men and women. Work defines older men so much so that a tragic number of men die within two years of retirement—and not because of old age. Tony was a bus driver. He could not wait to retire. Three years later he was almost dead from alcohol. Paul found time on his hands also, and drank himself silly. He was a surgeon, retired. Men often lose their meaning in life, their reason for getting up in the morning, their way of being the proud supporter of a family, their main focus of conversation—if they retire. Values are clearly reflected in the behavior of a person who is a serious career-oriented worker. At the very least, career people will admire the hard workers around them, and they may also vote against all manner of welfare.

Attitudes and Values

If we put all the forces together that are at work shaping and defining us each day we will discover that the outcomes include at least two more areas of conditioning. *Attitude* means psychological disposition. On the one extreme our coworkers might be cynics or fatalists, skeptics or whiners. They might be peppy, jolly, gung-ho. They might be blasé or indifferent. Optimistic or pessimistic. Patient or impatient. Let's include here the team player, the individualist, the overachiever, the underachiever and so on. Attitudes color conversations very quickly. The world view of anybody who has a distinct attitude is plain to see and not always pleasant to be around. I see Steve at work every day, but I avoid him most of the time. He snipes a lot, knows he is right about most matters, and always complains about the stupidity around him. He probably avoids me too, because I am often optimistic and tolerant, and he knows that I know that I am right. Values and attitudes are fused together in the way each of us labors each day.

Memory is luggage. We are burdened by it in many respects. We may remember very little of our past, but our total history is powerfully persuasive. The person you are here and now is standing on top of all that history. Each of us is the sum of our experiences. Differences among people are probably as sure as death and taxes, but differences are manageable. The message of the nineties was that differences are enjoyable, and that is probably a healthy attitude to take if we want to understand one another.

Recall that the mind—and all its historical influences—is only one of the six dimensions of a communication loop. No matter how simple the utterance, the six forces are at work. The first condition, the mind, tends to confuse or misunderstand communication from time to time. We must correct the errors by handling communication with skill. All the elements of the communication path can contribute to communication clarity. If one part of the loop creates a misunderstanding, we still have five other paths to clarity and truth.

Summary

- Many communication barriers are *shared*. You and the person you are speaking with will usually know if there is a communication problem. You know if vocabulary is not shared.

- Misunderstandings also occur because of problems you do not see or hear and may not understand.

- Personal history is a major force in the way people interpret their world. This historical background is made up of many social dimensions:

cultural background	age
class	geography
income	life experiences
religion	interests
color	attitudes
gender	

- Cultural background is often thought of in terms of ethnic groups, but the groups can include any identifiable community from military retirees to Nebraska farmers.

- Work is a key element in the cultural groupings of Americans.

- Class and income are important determinants in defining perceptions. In America, class has become associated with education, which means that spoken communication is socially sensitive and can create or inhibit communication between people with different levels of language skills.

- Religion persuades many people to lead the lives they lead, and religion asserts a subtle influence over the even larger group of people who have the historical values of faiths while practicing none.

- Color is culture in America. Many cultural groupings exist among people of color. Color makes people identifiable, whereas a cultural grouping may not. For historical reasons, this aspect of cultural heritage is a persuasive force in our perceptions.

- Gender would appear to be a simpler matter than cultural groups, but gender problems are among the most prevalent of the interpersonal conflicts that are found in the workplace.

- Life experiences shape every human being's perceptions. These experiences are part of the development of each person's attitudes, which shape his or her behavior.

- Where we live is a determinant of how we see our world and that includes the "part of town," the size of the town, the state, and the region.

- The complexity of the factors involved in our personal history makes shared meaning difficult at times. Remember that the loop of communication is fast and complicated but it can be slowed down, and each level can be put to work to avoid personal misunderstandings. For both speaker and listener—

  ✓ The thought process can be adaptive.

  ✓ The message-making process can be adaptive.

  ✓ The media are flexible.

Then we learn to understand.

Activities Chapter 15

Present a memo to your instructor that develops a discussion of a personal experience that relates to the chapter. The memo should be 500 to 1000 words, typed. In it you can explain how you have related the text or the lecture discussions to some event you have experienced. This exercise will give you a better understanding of the material because you will explain incidents in terms of our perceptions of workplace structure and communication.

Select one of the following suggestions and develop an analysis that involves your current employment or a former position. The following selection consists of communication conflicts that result from the historical influences upon people from different backgrounds.

- *Explain a miscommunication that clearly involved a distinction between a male employee and a female employee.*

- *Identify the conflicts that have emerged where you work that somewhat reflect the conflicting values of different age groups. These issues could involve computers, unions, promotions, and other matters.*

- *Have you seen any miscommunication that could have involved economics? These could be one-to-one misunderstandings or group conflicts such as you will see between employees at different levels of a company.*

- *Cultural background and color are the most publicly discussed issues in the employee profile. Do you have any experience with misunderstandings that could have had anything to do with heritage or color?*

Evaluation Forms

Performance Critiques

There are seven forms on the following pages. The forms are intended to be used in conjunction with three important activities that will involve your contributions, both as a participant and as an observer.

TASK GROUPS

FORM 1: Task-Group Observations

FORM 2: Task-Group Analysis

FORM 3: Group-Member Analysis

PRESENTATIONS

FORM 4: Committee Presentation Observations

FORM 5: Committee Presentation Analysis

INTERVIEWS

FORM 6: Interview Observations

FORM 7: Interview Analysis

During the performance of these activities you will watch other task committees, speakers, and interviews. There are observation forms for you to complete as you observe these activities (Forms 1, 4, and 6).

You will be asked to participate in each of the three performance activities. You will be a task-group member, a speaker, and an interviewee (candidate) on different occasions. There is a self-analysis form for you to complete after each of these activities (Forms 3, 5, and 7). In the case of the task-group committee activity, there is an additional form to complete. That analysis concerns your perceptions of the group performance of your team (Form 2).

All the forms are designed to draw your attention to important features of these commonplace speaking situations. *Only* your instructor will see the self-analysis forms and the task-group analysis form. The task-group observation (Form 1) form will be collected and given to the respective committees to review. The committee presentation observations (Form 4) will be given to the speakers to whom you listen. The interview observations (Form 6) will be given to the students you observe.

FORM 1

| TASK-GROUP OBSERVATIONS |

Name _____

This is a performance analysis of the task-group meeting that you observe. Be sure you have the correct name of each speaker. Fill out the form as you watch. Add comments on a third page as you see points of interest. Complete the third-page commentary before returning the observation form to the instructor. This form, along with those of the other observers, will be given to the task committee for review. The names of the observers will be removed.

A. Identify the order of the opening remarks by speaker.

1st to speak _____   3rd to speak _____   5th to speak _____
2nd to speak _____   4th to speak _____   6th to speak _____

B. Identify (with a ✔) the *types of contributions* you see in a ten- to fifteen-minutes segment from the meeting.

Identify the segment: early in the meeting ___, middle ___, late in the meeting ___.

Member Name	Provides Ideas	Provides data	Asks questions	Answers questions	Challenges	Leads	Passive
1 _____							
2 _____							
3 _____							
4 _____							
5 _____							
6 _____							

C. Identify which members contributed to important agenda tasks. Identify the most important contributor in the first column. the second most important in the second, and so on.

Most important ⟵⟶ *least important*

	Name	Name	Name	Name	Name	Name
Helped define problems	_____	_____	_____	_____	_____	_____
Helped identify solutions	_____	_____	_____	_____	_____	_____
Helped analyze solutions	_____	_____	_____	_____	_____	_____
Helped select best solution	_____	_____	_____	_____	_____	_____
Helped expedite a solution	_____	_____	_____	_____	_____	_____

D.

1. Identify the *leader* if there was one (or more):

2. Identify who appeared to be the *most frequent* speakers:

E. Draw a sociogram to illustrate the typical communication patterns you observed.

F. **Comments**

On a separate sheet of paper, develop several paragraphs to explain your response to the observations above. Comment on the nature of the leadership, the dominant channels in the sociogram, and the roles of the members.

TASK-GROUP ANALYSIS

Name _____

Respond to the following questions after you watch a playback of the videotape of your task group's final meeting. Answer the questions by considering *all* your meetings and not just the taped session.

---

1. At what point did members begin to assume roles? Who fulfilled what roles: leader, contributor, analyst, resource, prober, doubter, harmonizer, humorist, grump, monopolizer, arguer, and others?

2. Which of these members were task oriented. Why?

3. Which members were social maintenance oriented? Why?

4. What leadership pattern developed? Was the leader appointed or did he or she emerge? How did this come about?

5. Why was the leader a leader?

6. Was the hour used systematically for discussing problems, solutions, and best solutions?

7. Were all the options made available? Did the group rush the decision or deliberate the options?

8. Was the outcome decisive? Was it supported by all?

9. If there was an indecision, what barriers emerged?

10. Explain other perceptions if desired.

FORM 3

| GROUP MEMBER ANALYSIS |

Name _____

Respond to the following questions after you watch the playback of the videotape of your task group's final meeting. Answer the questions by considering *all* your meetings and not just the taped session. This is a self-analysis.

1. What was your contribution to the selection of the topic?

2. What were your initial reactions to each of the team members during the preliminary meetings?

3. Did you contribute to the development of the agenda? How so?

4. Do you see yourself as a task-oriented member or a social maintenance-oriented member? Explain.

5. Did you support the leader or leaders? In what way?

6. Did you repress sentiments? Did you hold back any important ideas? Did you agree with the outcome of the meeting? Explain.

7. Did your role or roles change during the various meetings? Did your relations with other members change? Why?

8. Did the committee style fit your preferences? (democratic and organized, authority oriented, casual, noisy, calm, aggressive, passive, hasty, cautious, and so on)

COMMITTEE PRESENTATION OBSERVATIONS

Name _____

Speaker: _____

Date: _____

Subject: _____

Running time: _____

This is a performance analysis of a presentation that you observe. Fill out the form as you watch. Add comments on a separate page as you see points of interest. Complete the commentary before returning the observation form to the instructor. This form, along with those of the other observers, will be given to the speaker for review. The names of the observers will be removed.

A. Delivery	Very Good	Good	Fair
Voice characteristics	▭	▭	▭
Body movements	▭	▭	▭
Eye contact	▭	▭	▭
Responsiveness	▭	▭	▭
Use of visuals	▭	▭	▭

B. Preparation	Very Good	Good	Fair
Understanding of subject	▭	▭	▭
Understanding of audience	▭	▭	▭
Quality of visual aids	▭	▭	▭
Organization of ideas	▭	▭	▭
Control of time	▭	▭	▭

C. Presentation

Organization	Very Good	Good	Fair
Structured opening	▆	▆	▆
Clearly defined key ideas	▆	▆	▆
Obvious transitions	▆	▆	▆
Vivid support of key ideas	▆	▆	▆
Appropriate summary	▆	▆	▆

D. Behavior (Circle two)

friendly	talkative	reflective	confused
enthusiastic	colorful	determined	embarrassed
dynamic	calm	timid	unconvincing
confident	restrained	quiet	other:

E. Problem Areas (Circle any of importance.)

knowledge	monotone	nervousness
preparation	image	acting
confidence	pronunciation	fear of audience
technical vocabulary	verbal skills	inability to take it seriously

F. Comments

On a separate sheet of paper, develop several paragraphs to explain your response to the preceding analysis. Comment on the speaker's strength and areas where you think there could be improvements. Discuss the use and management of the visual aids.

FORM 5

| COMMITTEE PRESENTATION |
| ANALYSIS |

Name _____

Respond to the following comments and questions after you watch a playback of the video-tape of your committee presentation.

1. Describe your preparation for the presentation.

2. Did you find the outline or notecards useful during the delivery? How so?

3. What organizational method did you use? Why? (topical, chronological, spatial, logical, other)

4. Did you clearly state your intentions and then follow the key ideas through to the end?

5. Did you make an effort to establish eye contact with the listener?

6. Did any of the listeners provide nonverbal cues? Explain.

7. What was your choice of visual aids? Why?

8. Do you think the aids did their intended job? How so?

9. Did you handle the visual materials skillfully?

10. What do you think of your voicing and your image as a speaker?

FORM 6

| INTERVIEW OBSERVATIONS |

Name _____

Candidate: _____

Interviewer: _____

Technical specialty: _____

Type of position sought: _____

This is a performance analysis of an interview that you observe. Fill out the form as you watch. Add comments on a separate page as you see points of interest. Complete the commentary before returning the observation form to the instructor. This form, along with those of the other observers, will be given to the candidate for review. The names of the observers will be removed.

A. Image	Very Good	Good	Fair
Voice pattern	▇	▇	▇
Body movements	▇	▇	▇
Pace	▇	▇	▇
Poise	▇	▇	▇
Dress	▇	▇	▇

B. Content	Very Good	Good	Fair
Precise technical answers	▇	▇	▇
Vivid general answers	▇	▇	▇
Used names of companies, locations, job sites, colleges, courses, etc.	▇	▇	▇

C. Content (continued)	Very	Good	Fair
Made specific time references: years of employment, college, etc.	▩	▩	▩
Limited yes/no answers.	▩	▩	▩
Was thoughtful about provocative questions: Why work for us? etc.	▩	▩	▩
Presented portfolio: references, transcripts, projects, awards.	▩	▩	▩

D. Behavior   (Circle two)

friendly	talkative	reflective	confused
enthusiastic	colorful	determined	embarrassed
dynamic	calm	timid	unconvincing
confident	restrained	quiet	other:

E. Problem Areas (Circle any of importance)

confidence	monotone	nervousness
listening	image	acting
understanding	pronunciation	fear of audience/camera
	verbal skills	

F. Comments

On a separate sheet of paper, develop several paragraphs to explain your responses to the preceding analysis. Comment on strengths and areas where you think there could be improvement.

FORM 7 | INTERVIEW ANALYSIS | Name _____

Respond to the following comments and questions after you watch a playback of the video of _your_ interview.

_____

1. How did you prepare for the interview?

2. Do you think the questions you were asked were appropriate?

3. Could you easily distinguish open questions (what, how, where, when, why) from closed questions (yes or no)?

4. Did you respond to each question fully and properly? Do you think you should change certain responses?

5. Were your responses to technical questions convincing? Did your response project an image of you as knowledgeable?

6. Were your responses to general questions appropriate? Did the answers contain enough explanation?

7. Were you specific in terms of names of employers, names of positions held, descriptions of skills, and so on?

8. Did you provide a time frame for schooling and employment?

9. Did you use your resume or portfolio items? What items were helpful?

10. What do you think of your image as an applicant? Did you look promising?

Acknowledgments

There are four texts and three instructor's supplements in the *Wordworks*™ series. In order to properly acknowledge the many people and organizations that have contributed to this project, I have chosen to first extend my thanks to those who assisted the endeavor in a larger context. Many more contributors helped shape the separate volumes.

The Wordworks™ Series

The *Wordworks*™ project would never have come about without the patience and generosity of my colleague and assistant Patricia Britz. There were 4000 pages of manuscript, endless keyboarding tasks, and elaborate page spreads. The project would have been impossible without Pat.

My colleague, Dr. Rita Smilkstein (published by Harcourt Brace), deserves very special thanks. Dr. Smilkstein has read every manuscript page of the project and contributed endless ideas and support.

To Stephen Helba, editor-in-chief at Prentice Hall, Pearson Education, I owe a special thanks. Some years ago I came home one day and found twenty-five pages of contracts spread all over the floor under an exhausted FAX machine. From the beginning Stephen took a personal interest in the *Wordworks*™ project and has been a source of encouragement for the many years it has taken to develop.

To Dr. David Mitchell, President of South Seattle Community College, I owe thanks for a special favor. When he was the former dean of my campus, I sought his help in finding a space where I could create a dedicated classroom for my program. His response was immediate. We surveyed a prospect and agreed to the experiment. He budgeted a fully provisioned room of tables and chairs and blackboards and other features I requested. That dedicated teaching and learning environment played a major role in developing the concepts embodied in the *Wordworks*™ series.

I would like to thank Marc Vassallo, Lee Anne White, Jeff Kolle, and other editors of the Taunton Press in Newtown, Connecticut. It was their faith in my ability to produce cover stories, feature articles, and shorter pieces for the nationally known Taunton magazines that give me the gumption to try to write *Wordworks*™. It was a great boost to realize that each publication had a circulation of several hundred thousand readers and that, all told,

I had reached several million readers thanks to the Taunton staff. I also discovered the excitement of writing copy that included photographic compositions and line art concepts.

It helps to have the support of a close friend when facing the misgivings involved in an enormous project. Rob Vinnedge, one of the Northwest's finest professional photographers, assisted me as I developed articles for the Taunton Press even when he was trying to meet his own book and magazine deadlines that were taking him as far away as Hong Kong. I regret that there was no time in our busy lives to share any shoots for *Wordworks,* but the deadlines were too tight. What matters the most is that Rob said, again, again, and again, "You can do it."

Workplace Communications for Engineering Technicians and Technologists

Special thanks to Jane Chateaubriand for taking the time out of her busy schedule to develop the NetWorks NorthWest case study. Jane also contributed the proposal that is an important feature of Chapter 11.

I would also like to thank the following publishing companies for permission to quote material or use illustrations from their publications.

Houghton Mifflin Company

Prentice Hall Inc., Pearson Education

Scott Foresman Publishing Co.

Simon and Schuster Publishing Co.

Warner Books Inc.

In addition, one illustration was used with permission of the Workforce Training and Education Coordination Board of the State of Washington.

Several sample documents that are used as illustrations are adapted from documents that I located at North Seattle Community College. I would like to thank Vince Offenback, former Associate Dean, for permission to use the documents.

Many of the diagrams in the text were first constructed with CAD by Stan Nelson, who generously donated his time to the project.

Prentice Hall/Pearson Education

Special thanks to the production staff at Prentice Hall in Columbus, Ohio, and the Clarinda Company of St. Paul, Minnesota. Two production teams saw to it that the *Wordworks Series* would go to press. At Prentice Hall, I would like to acknowledge the following staff:

> **Editor in Chief: Stephen Helba**
>
> **Executive Editor: Debbie Yarnell**
>
> **Associate Editor: Michelle Churma**
>
> **Production Editor: Louise N. Sette**
>
> **Design Coordinator: Robin G. Chukes**
>
> **Cover Designer: Ceri Fitzgerald**
>
> **Production Manager: Brian Fox**
>
> **Marketing Manager: Jimmy Stephens**

At Clarinda I would like to thank additional personnel for seeing the *Wordworks* through the production stages:

> **Manager of Publication Services: Cindy Miller**
>
> **Account Manager: Jennifer Graham**

For her patience and skills, a special thanks to Barbara Liguori for copyediting the *Wordworks*™ series.

I would also like to acknowledge the reviewers of this text:

> **Pamela S. Ecker**, Cincinnati State Technical and Community College (OH)
> **Dr. Harold P. Erickson**, Lake Superior College (MN)
> **Patricia Evenson**, Northcentral Technical College (WI)
> **Anne Gervasi**, North Lake College (TX)
> **Mary Francis Gibbons**, Richland College (TX)
> **Charles F. Kemnitz**, Penn College of Technology (PA)
> **Diane Minger**, Cedar Valley College (TX)
> **Gerald Nix**, San Juan College (NM)
> **M. Craig Sanders**, Bellevue Community College (WA)
> **Laurie Shapiro**, Miami-Dade Community College (FL)
> **Richard L. Steil**, Southwest School of Electronics (TX)
> **David K. Vaughan**, Air Force Institute of Technology (OH)

David W. Rigby

Index